影印版说明

本书介绍了近年来激光表面改性技术在腐蚀防护、腐蚀损伤零部件修复等方面的应用成果，讨论了钢、镍合金、钛合金等合金提高耐蚀性能的多种技术方法，分析了应用激光表面改性技术防止金属遭受液体冲蚀等不同腐蚀机制的腐蚀以及激光再制造损伤零件等的研究成果。

本书共分两部分内容，第一部分为提高耐腐蚀抗开裂性能，第二部分为改善耐磨和耐蚀性能。

本书主要适合从事表面改性处理的科研人员、技术人员使用，也可供高等院校相关专业的师生参考。

Chi Tat Kwok　中国澳门大学机电工程系副教授，在腐蚀和表面工程研究领域成果突出。

材料科学与工程图书工作室

联系电话　0451-86412421
　　　　　0451-86414559

邮　　箱　yh_bj@aliyun.com
　　　　　xuyaying81823@gmail.com
　　　　　zhxh6414559@aliyun.com

WOODHEAD PUBLISHING IN MATERIALS

影印版

激光表面改性处理合金的耐蚀性能

Laser surface modification of alloys for corrosion and erosion resistance

Edited by Chi Tat Kwok

哈爾濱工業大學出版社
HARBIN INSTITUTE OF TECHNOLOGY PRESS

黑版贸审字08-2017-080号

Laser surface modification of alloys for corrosion and erosion resistance
Chi Tat Kwok
ISBN: 978-0-85709-015-7

Elsevier (Singapore) Pte Ltd.
3 Killiney Road
#08-01 Winsland House I
Singapore 239519
Tel: (65) 6349-0200
Fax: (65) 6733-1817

First Published 2017

图书在版编目（CIP）数据

激光表面改性处理合金的耐蚀性能=Laser surface modification of alloys for corrosion and erosion resistance：英文/（美）迟·泰特·考克（Chi Tat Kwok）主编.—影印本.—哈尔滨：哈尔滨工业大学出版社，2017.10

ISBN 978-7-5603-6409-4

Ⅰ.①激… Ⅱ.①迟… Ⅲ.①合金-表面改性-耐蚀性-研究-英文 Ⅳ.①TG139

中国版本图书馆CIP数据核字（2017）第001596号

责任编辑 张秀华　许雅莹　杨　桦
出版发行 哈尔滨工业大学出版社
社　　址 哈尔滨市南岗区复华四道街10号 邮编 150006
传　　真 0451-86414749
网　　址 http://hitpress.hit.edu.cn
印　　刷 哈尔滨市石桥印务有限公司
开　　本 660mm×980mm 1/16 印张 25
版　　次 2017年10月第1版 2017年10月第1次印刷
书　　号 ISBN 978-7-5603-6409-4
定　　价 260.00元

Laser surface modification of alloys for corrosion and erosion resistance

---

Edited by

Chi Tat Kwok

Oxford Cambridge Philadelphia New Delhi

Contents

Contributor contact details

(* = main contact)

Editor

C. T. Kwok
Department of Electromechanical Engineering
University of Macau
Macau
China

E-mail: fstctk@umac.mo

Chapter 1

R. Vilar
Department of Chemical Engineering
Instituto Superior Técnico
Technical University of Lisbon
Av. RoviscoPais
1049-001 Lisbon
Portugal

E-mail: rui.vilar@ist.utl.pt

Chapter 2

K. Shinozaki*
Department of Mechanical Science and Engineering
Hiroshima University
1-4-1 Kagamiyama
Higashi-Hiroshima
Hiroshima
Japan

E-mail: kshino@hiroshima-u.ac.jp

T. Tokairin
Babcock-Hitachi K.K.
5-3 Takaramachi
Kure
Hiroshima
Japan

E-mail: tsuyoshi.tokairin.bu@hitachi.com

Chapter 3

W. K. Chan, C. T. Kwok* and K. H. Lo
Department of Electromechanical Engineering
University of Macau
Macau
China

E-mail: fstctk@umac.mo

Chapter 4

N. Semmar* and C. Boulmer-Leborgne
GREMI-UMR 6606 CNRS
University of Orléans
14 rue d'Issoudun
BP 6744
45067 Orléans cedex 2
France

E-mail: nadjib.semmar@univ-orleans.fr

Chapter 5

K. W. Ng and H. C. Man*
Department of Industrial and Systems Engineering
Hong Kong Polytechnic University
Hong Kong
China

E-mail: mfhcman@inet.polyu.edu.hk

Chapter 6

M. Duraiselvam
Department of Production Engineering
National Institute of Technology
Tiruchirappalli – 620 015
India

E-mail: durai@nitt.edu

Chapter 7

R. C. Shivamurthy and M. Kamaraj*
Department of Metallurgical and Materials Engineering
Indian Institute of Technology Madras
Chennai-36
India

E-mail: kamaraj@iitm.ac.in; rcshy123@yahoo.co.in

R. Nagarajan
Department of Chemical Engineering
Indian Institute of Technology Madras
Chennai-36
India

E-mail: nag@iitm.ac.in

S. M. Shariff and G. Padmanabham
Centre for Laser Processing of Materials
International Advanced Research Centre for Powder Metallurgy and Newer Materials (ARCI)
Hyderabad-05
India

E-mail: saabi@rediffmail.com; gpb@arci.ernet.in

Chapter 8

P. K. Wong and C. T. Kwok*
Department of Electromechanical Engineering
University of Macau
Macau
China

E-mail: fstctk@umac.mo

H. C. Man and F. T. Cheng
Department of Industrial and Systems Engineering
The Hong Kong Polytechnic University
Hong Kong
China

Chapter 9

J. H. Yao, Q. L. Zhang and F. Z. Kong
Zhejiang University of Technology
No.6 Chaohui District
Hangzhou
310014
China

E-mail: laser@zjut.edu.cn

Chapter 10

Buta Singh Sidhu
Punjab Technical University
Jalandhar-Kapurthala Highway
Jalandhar (Punjab)
India-144601

E-mail: butasidhu@yahoo.com

Introduction

C. T. KWOK, Department of Electromechanical Engineering,
University of Macau, Macau, China

Corrosion and erosion are material degradation processes which cause considerable environmental nuisance because they lead to valuable materials deteriorating, and becoming damaged and wasted. They can lead to failures in tools, machines, vehicles and infrastructure which are usually costly in terms of maintenance, environmental damage and human safety. Corrosion and erosion occur due to certain external actions on the surface of a material. According to the definitions, corrosion means the deterioration of a material because of chemical or electrochemical reaction with its environment [1] whereas erosion means the progressive loss of original material from a solid surface due to mechanical interaction between that surface and a fluid, a multi-component fluid, or impinging liquid or solid particles [2]. Corrosion and erosion often occur synergistically and material loss can be notably higher than the sum of the effects of the processes acting separately [3]. There are various types of corrosion including uniform corrosion, galvanic corrosion, pitting corrosion, crevice corrosion, intergranular corrosion, stress corrosion cracking, hydrogen embrittlement and selective leaching. Erosion involves several different processes such as liquid impingement erosion, slurry erosion (solid particle erosion), erosion-corrosion and cavitation erosion. The problems caused by corrosion and erosion can be avoided by various methods. Conventionally, numerous surface engineering techniques, such as electroplating, galvanizing, anodizing, diffusion coating, carburizing, nitriding and flame, induction hardening and thermal spraying possess several limitations such as high consumption of time, energy and materials, environmental-unfriendliness, poor-precision and flexibility, lack in scope of automation and complicated heat-treating procedures for fabricating protective coatings on the relative weak substrate alloys.

A laser is a clean heat source that exhibits a unique set of properties, namely high monochromaticity, coherence and directionality. It allows a wide range of surface treatments from mere heating to melting of coating materials on the substrate surface through the absorption of the laser energy. One of the limitations of laser surfacing is small laser beam size. Nevertheless, it is highly suitable for surface treating localized regions and remanufacturing undersized, corroded, worn or cracked engineering components. The generic term 'laser surface modification' includes laser transformation hardening

(LTH), laser surface melting/remelting (LSM), laser surface alloying (LSA), laser cladding (LC) and laser shock peening (LSP). When no additional materials are added to a substrate material, the laser energy can be used to heat the substrate surface to transformation temperature (for carbon steels) and even melting temperature for achieving LTH and LSM respectively. Owing to rapid quenching, the steels can be transformed into hard martensite by LTH while the structure and grain size of some alloys can be changed to the refined grain and homogeneous structure leading to enhanced corrosion or erosion resistance. When the additional alloying materials are alloyed or clad on the substrate, the tailor-made coatings can be fabricated for specific properties such as hardness, electrical and thermal conductivities, and corrosion and erosion resistance, to name a few. The service life of the laser-modified surface or laser-fabricated coatings for engineering components working in corrosive and/or erosive environments can be considerably extended. The laser-modified surface or laser-fabricated coatings on the substrate alloys can be effectively protected against the corrosive environments through the formation of protective passive layers or against the erosive environments by hard and energy-absorption coatings. The avoidance of corrosive or erosive damage on engineering alloys plays a crucial role in maintaining their quality, reliability, safety and profitability.

In the past three decades, a number of publications on laser surface modification of metallic materials for corrosion and erosion resistance have been reported. During 1970 to 2011, a quick survey with SCOPUS using the keywords 'laser AND corrosion OR erosion' gave about 3180 published papers. This indicates that laser surface modification for corrosion and erosion resistance is still an emerging topic deserving further investigation. Numerous studies were performed regarding a diversity of laser surface modification for fixing the corrosion and erosion problems in automobile, aerospace, nuclear plants, naval, defense, marine, electronic and biomedical applications. *Laser surface modification of alloys for corrosion and erosion resistance* intends to address the lack of a comprehensive book dealing with this topic and provide an impetus to the research on laser surface modification of various engineering alloys for enhancing their corrosion and erosion resistance. Scholars and experts of laser surface engineering from different countries/cities including China, France, Hong Kong, India, Japan, Macau and Portugal have generously contributed to this book.

The book is divided into two parts: Part I is about improving corrosion and cracking resistance and Part II is about improving erosion resistance for the laser-surface modified engineering alloys. In Chapter 1, Professor Vilar begins with theoretical approach on discussing phase transformation of steels and cast irons and then provides an extensive review on corrosion issues of the laser-surface modified steels and cast irons and their applications. In Chapter 2, Professor Shinozaki and his coworker study LSM of the weldments of

Ni-based alloys for improving the resistance against intergranular corrosion and cracking and propose a model describing the Cr depletion profiles near the grain boundary during heat treatment and LSM treatment. In Chapter 3, the published work related to the intergranular corrosion behavior of the laser-surface melted austenitic and duplex stainless steels are reviewed by the Editor and his coworkers. In addition, studies on intergranular corrosion of aged stainless steels including nickel-free austenitic stainless steel, Nb-stabilised austenitic stainless steel, and super duplex stainless steel after LSM are reported. In Chapter 4, pulsed excimer laser surface treatment of multilayer Au/Ni/Cu coatings for electronic components for corrosion resistance is reported by Professor Semmar and his coworker. Moreover, a model has been developed for simulating the thermal behavior of thin films obtained by the pulsed excimer laser. In Chapter 5, Dr Ng and Professor Man report laser surface alloying of NiTi alloy with molybdenum for improving its biocompatibility and corrosion resistance. The corrosion resistance of the laser-surface alloyed layer is found to be improved and it is intended to be used in biomedical implants. In Chapters 6, 7 and 10, Professor Duraiselvam, Professor Shivamurthy and his coworkers, and Professor Sidhu report that the laser-fabricated metallic and composite coatings have been applied to various substrates for resisting liquid impingement erosion, slurry erosion and erosion-corrosion respectively. In Chapter 8, the Editor and his coworkers describe LSA of copper with various metals for electrical erosion and corrosion resistance. In Chapter 9, field testing and applications of laser remanufacturing were investigated by Professor Yao and coworkers who indicate that laser remanufacturing is an effective and useful way to enhance the corrosion and erosion resistance of the engineering components.

The Editor would like to express his gratitude to all authors for their contributions in writing the book chapters and generosity in sharing research results. Finally, the Editor would like to take this opportunity to thank all the staff of Woodhead Publishing: the Publications Coordinators, Sarah Lynch and Helen Bradley, the Project Editor, Nell Holden, the Commissioning Editor Rob Sitton, and the Editorial Director Francis Dodds, for their assistance and efficient work in preparing this book.

References

[1] M.G. Fontana, *Corrosion engineering*, 3rd edn, McGraw-Hill, New York 1986.
[2] ASTM G73-10 Standard Test Method for Liquid Impingement Erosion Using Rotating Apparatus, ASTM.
[3] ASTM G119-09 Standard Guide for Determining Synergism Between Wear and Corrosion, ASTM.

Part I

Improving corrosion and cracking resistance

1 Laser surface modification of steels and cast irons for corrosion resistance

R. VILAR, Instituto Superior Téchnico and ICEMS, Technical University of Lisbon, Portugal

Abstract: The application of laser surface treatment to improve the corrosion resistance of ferrous alloys is discussed. By dissolving precipitates of phases present in small proportions in the alloy's microstructure (carbides, nitrides, intermetallic compounds, etc.) or preventing their precipitation during cooling, laser surface treatment improves, in general, the localised corrosion resistance of ferrous alloys, in particular of stainless steels. However, in order to achieve this improvement, chemical and microstructural inhomogeneities resulting from the laser treatment process must be avoided.

Key words: corrosion, laser surface melting, laser alloying, laser cladding, ferrous alloys.

1.1 Introduction

The wear, corrosion and oxidation resistance required by tools, machine parts and other mechanical components is normally obtained by designing complex alloys employing large amounts of expensive and strategically important metals. This solution is not only costly but also often technically inadequate, because the desired surface characteristics can only be obtained by sacrificing important bulk properties such as ductility, strength and toughness. A sounder approach is to select the material on the basis of its bulk properties and to apply a surface treatment to it, aiming at tailoring its surface properties to the application requirements. By doing so the consumption of costly, scarce and strategically important metals is also minimised. Surface treatment is particularly important for the corrosion resistance of ferrous alloys, because corrosion resistance is achieved in these alloys by intensive use of Cr, Ni and Mo, three elements that present exactly those characteristics.

Surface treatment operates through controlled modification of a material's microstructure and/or chemical composition, or by coating it with a different material. Among presently available surface engineering techniques, laser-based methods have achieved prominence owing to their remarkable characteristics. In fact, thanks to their spatial and temporal coherence, laser beams can be focused on a spot with a diameter close to the diffraction limit, generating extremely high-power densities. As a result, materials can be heated rapidly and with extreme accuracy, allowing a wide variety of effects to be achieved by varying only a few parameters [1]. The use of high-power lasers opened up new and exciting possibilities in surface engineering and led to the development of a wide range of surface treatment methods with considerable technical and economic interest, many of them already extensively applied in industry. Laser surface treatment allows the refinement of microstructures, dissolution of inclusions and precipitates and formation of non-equilibrium supersaturated solid solutions, quasi-crystalline and amorphous materials [2], microstructural changes often leading to improved corrosion resistance. The purpose of this chapter is to present and discuss laser surface treatment methods applicable to enhancing the corrosion resistance of ferrous alloys.

1.2 Laser surface treatments enhancing the corrosion resistance of ferrous alloys

The two main parameters in laser materials processing are the power density (power per unit area) and the interaction time (time during which the radiation interacts with a certain area of the material surface) [1]. When laser radiation is applied for a certain time to the surface of a material, part of it is absorbed by the conduction electrons in a surface region a few nanometres thick and rapidly converted into heat through collisions between the electrons and the lattice ions. The heat generated in this shallow material surface layer is transferred to the bulk by conduction [2]. Depending on the maximum surface temperature attained (T_s), the following situations may occur:

- $T_s < T_m$ (where T_m is the melting temperature): the material remains in the solid state, but it may undergo solid-state phase transformations. This regime is applied in the laser hardening of steels and cast irons.
- $T_m < T_s < T_v$ (where T_v is the vaporisation temperature): the material is melted with no appreciable vaporisation. This is the regime corresponding to liquid phase laser treatments.
- $T_s > T_v$: the material is partially vaporised. This regime is used, for example, in laser drilling and laser welding.

Of particular interest to this chapter are the liquid phase surface treatment methods, namely laser surface melting, alloying and cladding and, to a lesser extent, solid-state laser surface hardening. Surface hardening is achieved by austenitising a surface layer of steel or cast iron, the austenite being transformed into martensite during cooling. Laser surface melting consists of creating a melt pool with the laser beam, which is left to solidify spontaneously. It modifies only the microstructure of the material, while in laser surface alloying the chemical composition and microstructure are modified simultaneously, by adding suitable proportions of alloying elements to the melt pool before solidification. These elements may be either injected into the melt pool in powder form, as a wire or predeposited as a coating by electroplating, evaporation, sputtering, thermal spraying and other similar methods. Alloying may also be carried out by melting the material in contact with a reactive gas, which diffuses into the liquid (for example in laser nitriding [3]). The chemical composition of the surface alloy depends on the relative amounts of alloying elements and substrate melted and combined per unit time so, in order to achieve a required composition, the dimensions of the melt pool must be accurately controlled and kept stable in time. In laser cladding the laser beam melts simultaneously the clad material, usually injected into the melt pool in powder or wire form, and a thin layer of the substrate in order to form a coating. The processing parameters are chosen so that the clad material is metallurgically bound to the substrate, but dilution is kept as low as possible. In all these processes the laser beam scans the substrate in a continuous motion, to form a track of approximately semi-circular cross-section. To achieve complete surface coverage consecutive tracks are partially overlapped. More detailed information on these processes may be obtained by consulting other books and review papers [2, 4–11].

1.3 Transformation and microstructure of laser-treated steels and cast irons

The corrosion resistance of a material with a given composition is determined by its microstructure. Consequently, it is of utmost importance to understand how laser surface treatments affect the microstructure of materials in order to devise treatments capable of optimising their properties. The effect of laser hardening on corrosion resistance being only moderate, laser treatment processes involving melting are the most appropriate to enhance the corrosion resistance of materials. They are usually performed using continuous wave CO_2, Nd:YAG, diode or fibre lasers, using power densities and interaction times in the ranges 10^3–10^9 W/mm^2 and 10^{-1}–10^{-8} s, respectively, leading to the cooling rates that vary between 5×10^3 to 5×10^6 K/s, depending on the processing parameters. At these high cooling rates solid-state diffusive transformations are usually suppressed and the microstructure and properties of the material are often those that result from solidification.

1.3.1 Solidification

The solidification mechanisms in laser materials processing are relatively well understood and the microstructure of laser-treated materials can be predicted with reasonable accuracy [12–23]. The corresponding theoretical basis was established, among others, by Kurz and co-workers [12–16, 24]. Since in laser surface treatment the liquid is in permanent contact with a crystalline substrate and heat is extracted to the bulk (positive temperature gradient at the solid–liquid interface), solidification usually starts by epitaxial growth on the substrate, without the need for a nucleation step. Moreover, growth is columnar, with the position of the S/L interface being defined approximately by the location of the alloy *liquidus* and *solidus* isotherms [12]. For a specific alloy composition the solidification microstructure depends essentially on the local solidification parameters (solidification rate, R, and temperature gradient at the solid–liquid interface, G), which, in turn, depend on the heat and mass transfer in the system. The solidification parameters cannot be measured easily, but they can be estimated by approximate analytical solutions to the heat conduction equation [25, 26] or, preferably, by numerical simulations based on finite element methods [27–29]. The melt pool created by a quasi-Gaussian laser beam is roughly semi-circular for low scanning speeds (~1–5 mm/s), becoming progressively more elongated as the scanning speed increases [13, 30]. If the process occurs in steady-state conditions, the melt pool shape remains constant. R depends on the laser scanning speed (V), melt pool shape and the crystal preferential growth directions, R and V being related by (Eq. 1.1) [31]:

$$R = V \frac{\cos\theta}{\cos\phi} \qquad [1.1]$$

where θ and ϕ are the angles between the R and V vectors and between the normal to the interface (the temperature gradient direction) and the nearest preferential growth direction, if any (for example the <100> directions in the dendritic growth of cubic crystals), respectively. If no preferential growth direction exists $\phi = 0$. Since $\cos\theta = 0$ at the bottom of the melt pool, $R = 0$. R increases rapidly as solidification proceeds, up to a value lower than the scanning speed ($\cos\theta < 1$) near the surface. G is highest at the bottom of the melt pool and decreases as the solidification front approaches the surface. If R and G are known, two other important solidification parameters can be calculated: the G/R ratio, which controls the solid–liquid interface morphology at solidification rates lower than the limit of absolute stability [32], and the cooling rate, given by (Eq. 1.2):

$$\dot{T} = G * R \qquad [1.2]$$

G/R presents an infinite value at the bottom of the melt pool and decreases as solidification proceeds, while the cooling rate reaches its maximum value at the surface. In laser processing [12], as in welding [30], solidification starts with a plane S/L interface at the bottom of the melt pool due to the high value of G/R in this region and, as solidification proceeds, the interface evolves to a cellular and, eventually, to a dendritic morphology due to the decrease of that ratio. The variation of the solidification parameters is fast near the bottom of the melt pool, but it slows down rapidly and the solidification parameters remain approximately constant during most of the solidification process. As a result, the plane front solidification and the cellular microstructure regions are usually very narrow and the microstructure of the re-solidified layer is predominantly dendritic or dendritic + eutectic, depending on the alloy composition and solidification conditions and relatively uniform in morphology and characteristic dimensions. When the laser beam moves away from a particular point the temperature of the liquid decreases and may become sufficiently undercooled for equiaxed growth to occur, in particular if there are particles in suspension in the liquid that induce heterogeneous nucleation of the equiaxed crystals [24]. The impingement of cold powder particles on the melt pool in blown powder laser surface alloying and cladding may also have this effect [33]. The cooling rate controls the solidification time and, hence, the characteristic dimensions of the microstructure, in particular the primary and secondary dendrite arm spacing. The dependence of the primary dendrite arm spacing (λ_1) on the solidification parameters is complex, because λ_1 has different functional relationships with G and R [12], but the secondary dendrite arm spacing (λ_2) varies with the cooling rate according to (Eq. 1.3) [34, 35]:

$$\lambda_2 = c(G * R)^{-n} \qquad [1.3]$$

where c and n are constants. In general the exponent n is in the range 0.33 to 0.5. This equation allows the cooling rate to be estimated from secondary dendrite arm spacing measurements, because k and n are similar for a wide range of alloys and their values are known for many systems. By using special treatment methods (ultra-short pulse duration lasers or very high scanning speeds) very high cooling rates can be reached (10^5 to 5×10^6 K/s), allowing, for example, amorphous [36, 37] or quasi-crystalline [38, 39] alloys to be formed.

It is important, when the corrosion behaviour of laser-treated materials is being discussed, to recognise that, as result of solute partition during solidification, resolidified materials often are chemically inhomogeneous. In the absence of solid-state diffusion, the distribution of the dentrite arms of a solid-solution can be calculated by Scheil's equation (Eq. 1.4) [40]:

$$C_s = kC_0(1 - f_s)^{k-1} \qquad [1.4]$$

where C_s is the concentration of solute in the solid, k the partition coefficient, C_0 the initial concentration of solute in the liquid and f_s the fraction of solid. Alternatively, a dendritic solidification model such as the KGT (Kurz–Giovanola–Trivedi) model [12] may be used [41]. The effective partition coefficient, which characterises the redistribution of solute between the solid and the liquid, depends on the solidification parameters according to (Eq. 1.5):

$$k = \frac{k_0}{k_0 + (1 - k_0)\exp\left(^{-R\delta}/_{D}\right)} \qquad [1.5]$$

where k_0 is the equilibrium partition coefficient, R the solidification rate, D the solute diffusion coefficient in the liquid and δ the thickness of the stationary boundary layer. The effective partition coefficient varies within the range k_0 to 1 and is approximately equal to 1 when $\delta > 2D/R$. This situation occurs when liquid convection and the solidification rate are low (for example in the mushy zone that forms in the dendritic solidification of large solidification range alloys) and is particularly favourable to corrosion resistance because the chemical composition within the dendrite arms will be approximately uniform, reducing galvanic effects. A more detailed analysis of segregation during the solidification of alloys can be found, for example, in the book by Grøng [30].

By using a suitable microstructure selection criterion [42] and the response functions of all phases and interface morphologies liable to form in the system, Kurz and co-workers [14–16, 42] calculated microstructure selection maps that allow the solidification microstructure to be predicted if the solidification parameters R and G are known. Examples will be given later in this chapter.

1.3.2 Solid-state transformations

At the high cooling rates prevailing in laser surface treatment, solid-state diffusive transformations are usually suppressed, but in steels and cast irons austenite may transform into martensite by a martensitic (diffusionless) transformation, which is sufficiently fast to occur even in these conditions. In a typical martensitic transformation the proportion of martensite formed (y_m) does not depend noticeably on the cooling rate but on the undercooling below the martensite start temperature. It can be given, for example, by the Koistinen and Marburger equation (Eq. 1.6) [43]:

$$y_m = 1 - \exp\,[-\alpha(M_s - T)] \qquad [1.6]$$

where T is temperature, M_s the martensite start temperature and α a constant that depends on the material and is typically of the order of 0.011.

When laser-treated tracks are overlapped to cover extensive areas, considerable reheating of the previously deposited material occurs and diffusive transformation may be induced, in particular martensite tempering. Knowing the thermal cycles at each point, the phase transformations that may occur can be identified and their extent calculated using a finite element heat transfer model [23]. The expectable transformations in a typical martensitic stainless steel (AISI 420 – Fe-0.33%C–13.5%Cr) are represented in Fig. 1.1. The microstructure after solidification consists essentially of primary austenite dendrites [17]. A proportion of this austenite given by the Koistinen and Marburger equation transforms into martensite during cooling to room temperature. The transformations occurring during cooling in this first

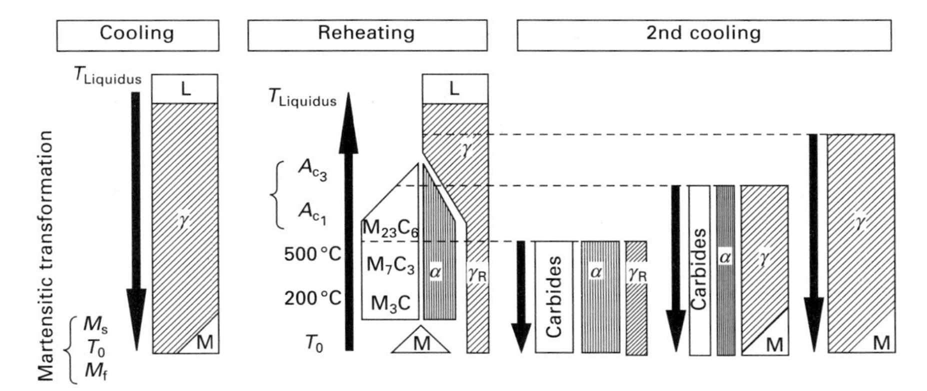

1.1 Diagram indicating the phase transformation occurring during the repeated thermal cycles generated due to single layer overlapping in an AISI 420 martensitic stainless steel. M – martensite; α – ferrite; γ – austenite; L – liquid phase [23].

thermal cycle are indicated as *Cooling* in Fig. 1.1. During overlapping, the previously deposited material undergoes a new thermal cycle, which may originate martensite tempering, a complex sequence of precipitation reactions that, eventually, leads to the decomposition of martensite into ferrite and carbides [18]. If the temperature exceeds Ac_1 temperature, austenitisation occurs. These transformations are indicated as *Reheating* in Fig. 1.1. The transformations observed during the corresponding cooling period depend on the transformations that occurred during heating and are indicated as *2nd cooling* in the diagram. If no austenite is formed during heating, no significant phase transformations occur, but if the material was austenitised ($Ac_1 < T < Ac_3$), austenite will partially or totally transform into martensite. Reheating may also cause destabilisation of retained austenite and its transformation into martensite [18].

1.3.3 Microstructure of martensitic stainless steels

Colaço and Vilar [17] developed a model that can predict the solidification microstructure of laser-melted steels with compositions in the peritectic region of the equilibrium phase diagram, in particular martensitic stainless steels. In equilibrium the solidification of these materials starts by the precipitation of δ-ferrite, followed by the transformation of δ-ferrite into austenite (γ) by peritectic and/or allotropic transformations. However, the experimental results contradict this prediction: for example, in DIN X42Cr13 steel (equivalent to UNS S42000, with composition Fe–0.43C–13Cr) γ is the primary solidification phase [19], while a transition from δ to γ primary solidification was observed in AISI 440C steel (Fe–1C–18Cr–0.75Mo) for a critical value of the scanning speed [8], showing that the solidification mode depends on the solidification rate. This behaviour was explained on the basis of an analysis of δ and γ dendritic growth kinetics. In this type of steel the liquid is in contact with a substrate containing both δ-ferrite and austenite. As a result, any of these phases may grow epitaxially, provided that the required undercooling exists [44]. Colaço and Vilar [17] calculated the variation of γ and δ dendrite tip temperatures as a function of the solidification rate, using the Kurz, Giovanola and Trivedi dendritic growth model [12, 45], and showed that, for both steels, the dendrite tip temperature of δ is higher than the dendrite tip temperature of γ below a critical solidification rate, $R_{\gamma/\delta}$, while the opposite occurs above this rate. One may then conclude that the first phase to solidify must shift from the equilibrium phase, δ-ferrite, to non-equilibrium austenite as the solidification rate increases [42], in agreement with experimental results.

The critical solidification rates for $\delta \rightarrow \gamma$ transition are $\sim 10^{-2}$ and 10 mm/s for DIN X42Cr13 and AISI 440C steels, respectively. In practice this means that the equilibrium primary solidification mode (δ) will never be observed in laser-treated DIN X42Cr13, but it will appear in AISI 440C steel, when treated at low scanning speeds. These predictions were corroborated by experimental observations. In DIN X42Cr13 steel solidification starts by the precipitation of austenite for scanning speeds between 1 and 200 mm/s [19] and the microstructure consists of austenite dendrites with some interdendritic carbides. In AISI 440C steel treated with very low scanning speed the dendrites present an inner core of δ-ferrite and an outer layer of austenite, while for higher scanning speeds the core of δ-ferrite is observed only near the fusion line, where the solidification rate is lower. The authors showed that the volume fraction of retained austenite in laser surface melted of DIN X42Cr13 martensitic stainless steel increases with decreasing power density and with increasing scanning speed and may reach about 100% in some tool steels. They also proposed processing maps that allow the solidification microstructure in laser surface melted martensitic stainless steels to be predicted [19]. These results can be generalised to any steel with representative points in the peritectic region of the corresponding phase diagram and composition near the peritectic *liquidus*, because this behaviour is due to the fact that, in this region of the phase diagram, the γ-phase partition coefficient of most alloying elements is closer to unity than the δ-phase partition coefficient. As a consequence, the solute build-up ahead of the δ-dendrite tips is larger than for γ dendrites, leading to a larger constitutional undercooling and lower dendrite tip temperature.

The prevalence of γ dendritic solidification over δ dendritic solidification at high solidification rates helps explain why laser-melted steels often present a much higher proportion of retained austenite than conventionally treated materials [19, 46–48], but it is not the only reason, because for the retained austenite content to be so large the martensitic transformation must also be avoided. On the basis of the existing experimental evidence, Colaço and Vilar [19] concluded that the refinement of austenite dendrites due to rapid solidification (see Eq. 1.3) is the most probable explanation for the observed variation of the retained austenite proportion with laser scanning speed and power density, but other factors may play a role, too. Important factors that affect the M_s temperature are the concentration of alloying elements [49], residual stresses [50] and the density of crystallographic defects such as dislocations and grain boundaries [51].

Laser surface melted martensitic stainless steels and tool steels may undergo secondary hardening due to tempering [7]. The secondary hardening temperature of laser melted alloy steels is usually higher than after conventional quenching [8, 18, 52–54], while the maximum secondary hardness may be higher or lower. For example, the secondary hardening temperature of laser-treated DIN X42Cr13 martensitic stainless steel is 600 °C and the maximum hardness 620 HV, as compared to 500 °C and 570 HV for the same steel in the quenched condition [18]. The microstructure after laser surface melting, consisting of about 40% lath-type martensite, 59% retained austenite and 1% $M_{23}C_6$ carbide, does not change noticeably during 1 hour isochronous tempering treatments up to 500 °C but, at higher temperatures, martensite decomposes and M_7C_3 and $M_{23}C_6$ carbides precipitate within the martensite laths and at the lath boundaries. Above 575 °C M_7C_3 carbide precipitates within austenite as well, leading to destabilisation of this phase. Destabilised austenite decomposes into a mixture of ferrite + carbides if the material remains at high temperature or transforms into martensite if the material is cooled below M_s temperature. Recrystallisation occurs at temperatures higher than 600 °C. Similar behaviour is observed for AISI 440C steel (Fe–1Cr–18Cr) [7]. The precipitation of Cr carbides due to tempering probably leads to a Cr-depletion of the matrix and lowers the corrosion resistance of the material, as experimentally observed by several authors in the overlap regions of laser-treated materials [47, 55, 56]. These transformations can be predicted using suitable tempering models [23].

1.3.4 Microstructure of austenitic and duplex stainless steels

Austenitic and dual phase stainless steels present compositions near the eutectic valley in the ternary Fe–Cr–Ni *liquidus* [57] and their equilibrium solidification path depends on the position of their representative points in relation to that valley. When the alloying element's content is lower than that limit, equilibrium solidification starts by the precipitation of δ-ferrite. This δ-ferrite transforms partially during cooling into austenite by a peritectic reaction and/or an allotropic transformation, leading to a $\delta + \gamma$ microstructure.

For higher alloying elements contents, γ is the first phase to solidify. Solidification may proceed by the precipitation of a small proportion of interdendritic δ-ferrite (vermicular δ-ferrite), favoured by the segregation of ferrite-stabilising elements to the interdendritic region. In complex stainless steels it is usual to express the overall chemical composition in terms of Cr-equivalent (Cr_{eq}) and Ni-equivalent (Ni_{eq}), the empirical Schaeffler–DeLong diagram [58, 59] often being used to predict the solidification mode.

Similarly to martensitic stainless steels, several authors observed in Fe–Cr–Ni alloys a transition from equilibrium δ solidification to metastable γ solidification when the solidification rate increases [20, 21, 60, 61], an effect that was also explained by the influence of the solidification rate on δ and γ dendrite tip temperatures [62]. Based on modelling and experimental research Vilpas [63] studied the solidification mechanism in a range of austenitic stainless steels with different Cr_{eq}/Ni_{eq} ratios and concluded that steels with $Cr_{eq}/Ni_{eq} > 1.74$ solidify in the equilibrium δ-ferrite solidification mode within a vast range of solidification rates, while steels with $Cr_{eq}/Ni_{eq} = 1.46$ undergo the equilibrium primary austenitic solidification at all practical solidification rates. In contrast, steels with $Cr_{eq}/Ni_{eq} = 1.62$ (such as AISI 316) solidify in the equilibrium primary ferritic solidification mode at low solidification rates, but shift to metastable austenitic solidification when the solidification rate increases. A theoretical explanation for this behaviour was proposed by Fukumoto and Kurz [22] on the basis of an analysis of the δ and γ dendritic solidification kinetics. Their conclusions can be summarised as follows (temperature gradient at the S/L interface = 700 K/mm, typical of laser cladding experiments):

- For solidification rates higher than 1 m/s (absolute stability limit) solidification occurs with plane solid–liquid interface. The phase formed is γ for $Cr_{eq}/Ni_{eq} < 1.64$ and δ for larger values of the Cr_{eq}/Ni_{eq} ratio.
- For solidification rates between ~ 0.1 and 1 m/s a banded structure parallel to the solidification front and oscillating between plane front solidification and cellular solidification appears. The microstructure consists predominantly of γ phase for $Cr_{eq}/Ni_{eq} < 1.66$ and δ phase for larger values of the Cr_{eq}/Ni_{eq} ratio.
- For low solidification rates ($R < 10^{-4}$–10^{-3} m/s) a lamellar $\delta + \gamma$ eutectic forms, independently of the Cr_{eq}/Ni_{eq} ratio. For some alloys, the lamellar eutectic is replaced by a rod-like eutectic at high solidification rates.
- Within the precedent solidification rate range, alloys with $Cr_{eq}/Ni_{eq} <$ 1.54, solidify in the γ solidification mode and alloys with $Cr_{eq}/Ni_{eq} >$ 1.62, solidify in the δ-ferrite solidification mode, independently of the solidification rate.
- Alloys with $1.54 < Cr_{eq}/Ni_{eq} < 1.62$ undergo a transition from stable δ-ferrite

solidification to metastable γ-solidification at a critical solidification rate that increases with increasing Cr_{eq}/Ni_{eq} ratio. This transition is controlled by δ and γ dendritic growth kinetics.

The authors presented their predictions in microstructure selection maps for alloys with 18 wt% Cr and $1.3 < Cr_{eq}/Ni_{eq} < 1.8$. Their analysis shows that, despite their apparent simplicity, stainless steels present a wide variety of solidification microstructures, containing variable proportions of δ and γ. Plane front solidification structures stabilised by the absolute stability criterion and banded structures and eutectic structures will not form in steels treated with continuous wave lasers in normal processing conditions, because the solidification rates required for their formation lie outside the solidification rate range achievable in these conditions, but all other microstructures may form. The solidification mechanism and the proportion of δ and γ have a direct influence on the distribution of alloying elements in the microstructure and, hence, on the corrosion resistance of stainless steels [64]. Vilpas [63] observed that when solidification occurred in the primary austenite mode, the dendrite cores show the lowest pitting corrosion resistance, due to the segregation of Cr and Mo (ferrite stabilising elements) to the interdendritic region, while in steels with primary ferrite solidification mode, the lowest pitting corrosion resistance occurs in the austenite formed at the end of solidification or by solid-state reaction, near the δ/γ interface, where a sharp solute concentration gradient exists.

1.4 Applications: steel

1.4.1 Laser surface alloying and cladding of carbon and low alloy steels

The chromium content of carbon and low alloy steels is not sufficient to ensure passive behaviour, independently of the material's microstructure, so their corrosion resistance can only be improved by laser surface alloying or laser cladding. The literature on laser surface alloying and cladding of carbon and low-alloy steels is too extensive to be reviewed here, so only a few examples will be presented. Of particular relevance to the scope of this chapter is the production of low-cost analogues of high corrosion resistance stainless steels by laser surface modification of carbon steels and low-alloy structural steels. Carbon and low-alloy steels are ubiquitous materials in all branches of industry. They are cheap, widely available, strong, easily shaped, weldable and very versatile, but they present an important limitation: their low oxidation and corrosion resistance. Laser surface alloying and laser cladding allow the production, at relatively low cost, of surface layers with compositions similar to almost any grade of stainless steel on carbon and low alloy steels in order to create a composite material with a good compromise between corrosion resistance and mechanical strength. Surface layers with specific compositions may be produced by laser surface alloying, adding alloying elements such as Cr, Ni and Mo to a laser-generated melt pool. However, in order to achieve high corrosion resistance, galvanic effects must be avoided, so the surface must present uniform chemical composition and be free of defects. This requires that the laser beam/material interaction time to be sufficiently long for the alloying element particles to dissolve completely and the melt to homogenise. Surface reflectivity must also be uniform in order to avoid variations of the proportion of the incident radiation that is absorbed by the material, and, hence, of the melt pool size. Laser cladding is advantageous under this point of view, because uniform surface composition is easily ensured by using pre-alloyed powders. In any case, contamination of the surface layer by carbon must be avoided to reduce the risk of localised corrosion and sensitisation due to chromium carbide precipitation.

Moore and McCafferty [65, 66] were among the first to demonstrate that surface alloys with compositions similar to stainless steels could be produced by laser surface melting. The authors used a 0.18% C carbon steel substrate previously coated with chromium by sputtering or electrodeposition. Unlike subsequent authors, they used very high power densities and scanning speeds, in order to achieve cooling rates as high as 10^6–10^7 K/s. Using this method, shallow layers of Fe–Cr alloys were produced, with chemical compositions that were controlled by varying the coating thickness and melt depth. Typically, the surface alloys contained about 20%Cr. As expected from this composition, they presented passive behaviour in anodic polarisation tests carried out in a deaerated 0.1 M Na_2SO_4 solution at 25 °C. The critical passivation current density and the current density in the passive region decreased with increasing chromium content, but the values achieved were always inferior to those observed in bulk ferritic steels of similar composition, due to chemical inhomogeneity and micro-cracks in the surface alloys. A similar approach was adopted by Renauld *et al.* [67] to prepare Fe–Cr–Ni surface alloys containing up to 24%Cr and 24%Ni. The laser was used to melt coatings of Ni and Cr of suitable thickness deposited electrochemically on a mild steel substrate. The best surface alloys showed a corrosion rate of 1 μm/year and a passivity zone of more than 800 mV in diluted NaCl solutions, better than 304L stainless steel. The pitting corrosion resistance was better, too.

The limitations of laser surface alloying for creating surface analogues to stainless steels were clearly demonstrated by Anjos *et al.* [68]. The authors prepared surface alloys by the blown powder laser surface alloying method on 0.2% carbon steel substrates, using two premixed mixtures of Ni, Cr and Mo powders containing relative proportions of these elements similar to those observed in popular Mo-containing stainless steel grades, namely 54 wt% Cr, 38 wt% Ni and 8 wt% Mo and 45 wt% Cr, 42 wt% Ni and 13 wt% Mo. The scanning speed and the power density were varied in the ranges 2.5–20 mm/s and 0.3–3 kW/mm^2 in order to find the parameters that optimise the microstructure and properties of the material. The tracks produced at higher scanning speeds (15–20 mm/s) and lower power densities (0.3 kW/mm^2) presented non-uniform composition, undissolved Mo particles and some cracks. In the experimental conditions used, surface layers with uniform microstructure and composition were achieved for a power density of 1.5 kW/mm^2 and scanning speeds of 5 and 10 mm/s, these relatively low scanning speeds being required to dissolve completely the Mo particles. The surface alloys obtained presented compositions in the ranges $2.7 < \text{Cr\%} < 35.4$, $2.6 < \text{Ni\%} < 29.4$, $0.65 < \text{Mo\%} < 6.23$, $0.10 < \text{C\%} < 0.18$, often differing considerably from the target composition. This wide variation of the chemical composition is explained by changes of the melt pool size, mass of alloying elements incorporated in the melt pool per unit time, powder catchment efficiency and alloying elements losses due to evaporation or oxidation. Moreover the surface alloys presented a relatively large carbon content, due to contamination by the substrate. Even so, surface alloys with composition similar to AISI 316 stainless steels showed better pitting corrosion resistance than the corresponding bulk material. This work provides evidence of the main advantages and limitations of laser surface alloying as a method for the production of high corrosion resistance surface alloys: good corrosion resistance, but difficulty in achieving a targeted chemical composition, low reproducibility of results, non-uniform composition and contamination of the surface alloy by the substrate.

These problems are avoided in laser cladding. The preparation of pore- and crack-free coatings of AISI 316L austenitic stainless steel on carbon steel substrates by laser cladding was reported in an early paper by Weerasinghe *et al.* [69]. The microstructure of the coatings consisted of austenite dendrites with a small fraction of interdendritic δ-ferrite, formed due to Cr and Mo segregation to the interdendritic regions. The stress and the pitting corrosion resistance of the clad material were better than those of a bulk alloy with similar composition, but its resistance to generalised corrosion in boiling acid solutions was lower. This property was recovered by a solution heat treatment at 1060 °C, which eliminated δ-ferrite. Anjos *et al.* [70] and Li *et al.* [71] studied the microstructure and corrosion resistance of coatings of UNS S31254 austenitic stainless steel prepared by laser cladding, using pre-alloyed powders. By properly selecting the processing parameters, dilution of the coating material by the substrate and alloying elements evaporation were avoided, resulting in a coating with a chemical composition identical to the composition of the powder used in the process. Dilution was limited to a region thinner than 20 μm, situated at the clad/substrate interface. The distribution of alloying elements along the depth of a clad layer was homogeneous. No significant loss of nitrogen occurred, independently of using nitrogen or argon as shielding gases. The clad material presents a highly refined microstructure consisting of austenite dendrites. Due to solute partitioning during solidification, Cr and Mo segregated from the dendrite's core to the interdendritic region, while Ni showed the opposite trend. Some precipitates of Fe–Cr–Mo χ-phase and Mn_5Si_3 were observed, as well. The anodic polarisation behaviour of the clad material and of a bulk alloy with the same composition in 4 M NaCl solutions was similar. Both materials appeared immune to pitting corrosion in the electrochemical tests and presented excellent passive behaviour, with similar transpassivation potential [70]. Slight pitting corrosion was observed in both materials upon immersion for 3 months in NaCl [70] and $FeCl_3$ [71] solutions, but the pits were considerably smaller in the coatings than in the bulk alloy. This is because the pits, which start at the dendrite core due to Cr and Mo depletion, do not propagate across the interdendritic region, where the Cr and Mo concentrations are higher, so their growth is restrained [71].

Coatings can also be synthesised *in situ* by melting mixtures of elemental powders combined in suitable proportions. This approach allows preparing coatings with compositions different from existing commercial alloys or unavailable in powder form, for technical or commercial reasons. It also allows new coating materials, unavailable in the marketplace, to be prepared for research or industrial purposes [72]. Li *et al.* [73] used this approach to prepare coatings with composition similar to superferritic stainless steel UNS S44700 on 0.2%C carbon steel substrates. The coatings were prepared by melting a mixture of Ni, Cr and Mo powders in nominal proportions and using processing parameters chosen to avoid dilution. However, owing to the large difference between the melting temperatures of Mo on one hand and Fe, Cr and Ni on the other hand, Mo particles tended to remain undissolved and an additional laser surface melting processing step was necessary in order to homogenise the coating material. After laser surface melting the coating presented a homogeneous structure and was free of defects. Its chemical composition was slightly different from the nominal composition, showing that some adjustments of the powder composition are required to reach a specified nominal composition. The microstructure consisted of fine dendrites of δ-ferrite, with some degree of chemical segregation, due to solute partition during solidification. The coating and a reference sample of commercial UNS S44700 wrought steel were studied by potentiodynamic anodic polarisation tests, using 0.5 M NaCl, 1 M $FeCl_3$ and 0.6 M HCl solutions, as well as by long-term immersion in 1 M $FeCl_3$ solutions. The corrosion behaviours of the coating and bulk material were similar, with excellent passivity and pitting resistance in the anodic polarisation tests. Some pitting occurred in the immersion tests that, in the surface alloy, tended to concentrate in the dendrite's core, where the concentration of Mo was lower.

The achieved results show that, when alloying or cladding to target specific surface compositions with outstanding corrosion resistance, it is imperative to control the processing parameters in order to avoid deviations from the required composition, since these may substantially reduce the corrosion resistance of the material. Defects such as porosity and interdendritic cracking have also been reported in laser-clad austenitic stainless steel coatings [69] and must be avoided as well. The presence of undissolved alloying element particles [73] and the heat-affected region created when consecutive tracks are overlapped are also critical, because they lead to microstructural inhomogeneity and, consequently, to reductions in corrosion resistance [47, 56]. As to the chemical composition, no significant losses of alloying elements were observed when using pre-alloyed powders, provided that the shielding gas protected the surface from oxidation efficiently and the melt pool temperature remained sufficiently low to avoid vaporisation of alloying elements. Nitrogen loss is a special concern when depositing high-nitrogen stainless steels, but existing results [68, 74] suggest that the nitrogen content of N-containing stainless steels deposited by laser cladding is independent of the shielding gas used, probably because the melt pool lifetime is too short for significant nitrogen loss to occur. The formation of chromium carbides is also deleterious for the corrosion resistance of stainless steels, because the Cr-depleted regions surrounding chromium carbide particles are preferential sites for the initiation of corrosive attack [75]. Therefore, when depositing stainless steels on carbon-steels dilution must be avoided as much as possible. By properly selecting the processing parameters, dilution can be minimised, so that the amount of carbon from the substrate incorporated in the clad tracks is negligible. Laser surface alloying and laser cladding being solidification processes, chemical segregation due to solute partition cannot be entirely avoided. This segregation, which appears often as non-uniform etching of the dendrites [70], can be minimised by using high scanning speeds [41].

The corrosion resistance of coatings deposited by laser cladding is, in general, better than the corrosion resistance of coatings of the same material deposited by thermal spraying methods, due to their better homogeneity, finer microstructure and absence of defects, but the deposition rates achieved by laser cladding are usually lower. The advantages of laser cladding and thermal deposition methods can be combined by predepositing the coating material by thermal spraying and laser surface melting the coating in order to eliminate porosity and oxide inclusions and improve its cohesion and adherence [76]. Obviously a major concern here is to select laser-processing parameters that allow the coating to be melted throughout its complete thickness while avoiding excessive substrate melting. The melt depth in the laser remelting step must be only slightly larger than the coating thickness, and this melt depth must be kept constant during processing, a requirement that can be difficult to satisfy, owing to variations of the surface reflectivity and the progressive increase of the system temperature.

In conclusion, relatively low-cost materials with exceptional resistance to corrosion can be produced by laser cladding carbon or low alloy steel substrates with a surface layer of a high corrosion-resistant stainless steel. Cladding with high-grade stainless steels may also considerably improve the corrosion behaviour of components built from cheaper, lower-grade materials. The economic advantages of these techniques depend, obviously, on the material's cost as compared with the investment and running costs of the laser processing equipment. The increasing scarcity and cost of elements such as Cr, Ni and Mo may justify a wider application of these technologies in the near future.

1.4.2 Laser surface melting and alloying of martensitic stainless and tool steels

Laser surface treatment is particularly well adapted to tool manufacturing, since it frequently allows remarkable improvements in hardness, wear and corrosion resistance of tool steels, without degradation of their bulk properties [8]. Moreover, the high temperature resistance of the material may also be improved, due to the high secondary hardening temperature of laser-melted steels as compared to quenched and tempered steels. Laser surface melting is particularly beneficial for sintered steels, because residual porosity is eliminated [77], leading to greatly improved corrosion and fatigue behaviour. The microstructure formation mechanisms in martensitic stainless steels and their relation with the materials' properties were previously discussed in this chapter and are the object of several review papers [7, 8]. Laser surface melting causes the dissolution of the large carbide particles existing in steels containing large proportions of carbon and carbide-forming elements, homogenises the material and leads to the formation of a very fine dendritic microstructure, with consequent improvements of the mechanical [78], wear [79], erosion [80] and corrosion resistance [81, 82]. Laser processed tool steels present a dendritic structure, consisting of austenite, martensite and, in some cases, extremely fine carbides [7, 8]. These carbides may appear as interdendritic precipitates or thin films or as an interdendritic eutectic [7, 8]. In some conditions a small proportion of δ-ferrite may appear too, as previously discussed.

By dissolving large Cr-rich carbide particles and non-metallic inclusions responsible for pitting initiation, laser surface melting improves the pitting corrosion resistance of the material, but, owing to the inhomogeneous temperature distribution in the material during the laser treatment, the resulting microstructure is non-uniform, with a potentially negative effect on the corrosion resistance. In an early paper on laser surface melting of UNS

S42000 steel martensitic stainless steel (equivalent to AISI 420 and DIN X40Cr13), Damborenea *et al.* [81] showed that pitting corrosion resistance was generally improved by laser surface melting, but later studies revealed a more complex behaviour [47, 55, 56, 62]. Escudero and Belló [55, 56] compared the corrosion resistance in a 0.05 M aqueous NaCl solution with pH = 6.8 ± 0.1, at room temperature, of a martensitic stainless steel containing 0.36%C and 13.1% Cr (AISI 420) submitted, on one hand, to laser surface melting and, on the other hand, to quenching and tempering at different temperatures. They found that the best overall corrosion resistance was achieved when the material was submitted to a conventional heat treatment involving complete austenitisation followed by tempering at low temperature (150 °C), because chromium carbide particles are dissolved by this heat treatment and a homogeneous structure with a high Cr content in solid solution is created. The corrosion resistance of the laser-melted material was only slightly inferior to the corrosion resistance after heat treatment and much better than the corrosion resistance after quenching and high temperature (650 °C) tempering. In the laser-melted region the steel presented a passive behaviour, but the heat-affected zone (HAZ) in the substrate and in track overlap regions, where the material was reheated, presented active behaviour and fast corrosion rates. In what concerns pitting corrosion resistance, the rupture and passivation potentials were similar for the laser-melted and the low tempering temperature samples but, again, the HAZ and the track overlap regions presented higher pitting susceptibility. In the laser-treated samples pitting consistently occurred in regions where the material had been reheated, a result that can be explained by carbide coarsening due to Ostwald ripening and formation of martensite due to austenite destabilisation by reheating [18], i.e. pitting was essentially associated to microstructural inhomogeneities.

These conclusions were confirmed by Ion *et al.* [47], who studied the pitting corrosion behaviour in artificial seawater of the same steel after laser hardening and laser surface melting with scanning speeds in the range 6–40 mm/s. In the laser-hardened material severe pitting was observed in the HAZ, in areas adjacent to the limits of tracks, while the track's central region exhibited only slight general corrosion. The corrosion pits were initiated at the interface between undissolved carbide particles and the matrix, as well as in regions where precipitation due to reheating had occurred. The poor corrosion resistance in these areas was explained by Cr depletion of the matrix due to carbide precipitation or low Cr concentration due to incomplete dissolution of the primary carbides. The microstructural inhomogeneities resulting from the non-uniform temperature field led to galvanic effects between the centre and the periphery of the tracks, with the track centre being effectively protected from corrosion by the more anodic peripheral regions. In laser-melted samples, where carbides were completely dissolved and the material homogenised, the corrosion behaviour depended on the laser beam scanning speed. For high scanning speeds an outstanding pitting corrosion resistance was observed and the pits were small and uniformly distributed. In contrast, for lower scanning speeds light uniform corrosion occurred at the centre of the tracks accompanied by intense pitting corrosion at their periphery. The authors explained this behaviour by the evaporation of Cr at low scanning speeds, but the smaller size of the HAZ, on one hand, and the shorter duration and lower maximum temperature of the reheating thermal cycle created when tracks are overlapped at higher scanning speeds, on the other hand, may also significantly contribute to the observed dependence. In fact, lower scanning speeds will lead to a denser precipitation of chromium carbides and larger carbide particles size in the HAZ, hence leading to a lower Cr content in solid solution. Moreover, retained austenite is destabilised by reheating at temperatures higher than approximately 580 °C [19], so the amount of newly formed martensite in the HAZ will be larger for lower scanning speeds. Since both factors negatively affect the corrosion resistance and favour pit nucleation, the corrosion resistance must decrease with decreasing laser beam scanning speed.

The influence of retained austenite on the corrosion resistance of UNS S42000 martensitic stainless steel was studied by Kwok *et al.* [80]. The authors tested samples submitted to different heat treatments, including laser surface melting with scanning speeds between 5 and 25 mm/s. The corrosion tests were carried out in 3.5% deaerated NaCl solution. The samples that were air-cooled after austenitisation at 1020 °C showed the lowest pitting potential, due to the presence of large carbide particles favouring pit initiation in the presence of Cl^- ions. Laser surface melting shifts the pitting potential in the noble direction, indicating an improvement in the pitting corrosion resistance. The beneficial effect of austenite was demonstrated by the existence of a linear relationship between the pitting potential and the volume fraction of retained austenite. The pits were nucleated at the interdendritic regions, where carbides precipitated during solidification, and propagated into the austenite dendrites. The results confirm that the beneficial effect of laser surface melting on the pitting corrosion resistance is due to the elimination and refinement of carbides.

Since Cr-rich precipitates play such an important role on the corrosion resistance of martensitic stainless steels, the influence of the laser treatment on corrosion resistance must depend on the carbon and nitrogen contents of the material. Pyzalla *et al.* [83] compared the corrosion resistance of a high-nitrogen DIN X30CrMoN15 1 martensitic stainless steel in the laser hardened condition with the corrosion resistance of a similar low-nitrogen steel, DIN X39CrMo1 7, in identical treatment conditions. Corrosion was evaluated by potentiodynamic tests using diluted sulphuric acid (uniform corrosion) and NaCl (pitting corrosion) solutions as electrolytes. The laser treatment had only a negligible influence on the uniform corrosion resistance of the low-nitrogen

X39CrMo1 7 steel, revealed by a slight decrease of the passivation current. The effect was much more important for the high-nitrogen steel, and it depended on the maximum temperature attained during the laser treatment. Similarly, the pitting corrosion resistance of the X39CrMo1 7 steel was independent of the austenitisation temperature, while in DIN X30CrMoN15 1 this effect was temperature dependent. The corroded surfaces presented shallow pits, mainly at the former austenite grain boundaries and at the interface between the precipitates and the matrix. The influence of the laser treatment parameters on the corrosion resistance was explained by their effect on the maximum temperature and cooling rates achieved in the laser treatment, which control the concentration of alloying elements in solid solution. When the maximum laser hardening temperature is lower than Ac_1, the laser treatment leads to an increase of the amount of precipitates and their particle size. These effects are more noticeable in the high-nitrogen steel than in X39CrMo1 7, because precipitation is more important in the former material.

Corrosion tends to concentrate on former austenite grain boundaries, where precipitates appear preferentially. When the maximum temperature reached in the laser treatment is higher than Ac_1 the corrosion resistance increases because the precipitates dissolve in the matrix, increasing its alloying elements content. This effect is more noticeable in the high-nitrogen steel. The influence of the cooling rate was explained by similar arguments: lower cooling rates favour precipitation, reducing the alloying elements concentration in the matrix. The atmosphere used in laser treatment can be expected to affect considerably the corrosion resistance of laser-treated steels. An excellent inert gas protection during laser treatment is of the utmost importance to ensure good corrosion resistance in laser-treated stainless steels, because excessive oxidation reduces their chromium content and, consequently, their corrosion resistance [84]. Cr depletion due to surface oxidation is also responsible for poor corrosion resistance of medical tools made of DIN X30Cr13 martensitic stainless steel laser marked in the oxidative regime (power density 4.55×10^4 W/cm^2, interaction time 210 µs) under normal atmosphere, but corrosion resistance deterioration can be avoided by operating in the ablative regime, using much shorter interaction time and higher power density (interaction time 10 µs, power density 1.9×10^7 W/cm^2) [85]

Another martensitic stainless steel whose corrosion behaviour was studied in detail is UNS S44044 (or its AISI standard equivalent 440C). This steel undergoes primary δ-solidification at solidification rates lower than about 10 mm/s, shifting to primary γ-solidification at higher solidification rates [17]. If the solidification path follows the equilibrium diagram, solidification starts by the precipitation of primary δ ferrite dendrites, followed by the formation of austenite by an incomplete peritectic reaction, and the precipitation of carbides at the interdendritic region. Austenite will then transform partially into martensite (M) during cooling down to room temperature, leading to a microstructure comprising $\delta + \gamma + M +$ carbides. Conversely, if solidification start by the formation of γ phase the final microstructure will contain only $\gamma + M +$ carbides. Kwok and co-workers [86, 87] compared the corrosion resistance of UNS S44044 martensitic stainless steel submitted to different heat treatments, including laser surface melting and laser hardening, in a 3.5% NaCl solution, and concluded that both laser treatments improve the corrosion resistance of this steel, most noticeably the laser surface melting treatment. The corrosion resistance after laser surface melting was better than after conventional heat treatment, as evidenced by a shift from active corrosion to passivity. The laser surface melted material showed a large passive range and low passive current density. Its corrosion behaviour was dictated by the proportions of retained austenite and undissolved carbides in the microstructure, the pitting corrosion resistance increasing with decreasing proportion of carbides and increasing proportion of retained austenite. The influence of the processing parameters on the corrosion resistance was explained in terms of their influence on the microstructure. The corrosion resistance was approximately independent on the power density, but depended considerably on the scanning speed, which is the main factor controlling, on one hand, the interaction time and, hence, the extension of carbide dissolution in the melt and, on the other hand, the solidification rate, which controls the proportion of austenite in the microstructure.

Corrosion plays an important role in the deterioration of tools such as plastic moulds and rock drills used in sea or brackish water, for example. On the other hand, since laser powder deposition is becoming increasingly important as a tool manufacturing and repair method [88], the corrosion resistance of laser-melted tool steels is becoming increasingly important for the tooling industry. Unfortunately, studies on the corrosion resistance of laser-treated tool steels are scarce. The corrosion resistance of laser-melted UNS S44044 and UNS T51621 (a low-alloy steel containing 0.2%C–4.1%Ni–0.25%Cr and 0.2%V) plastic mould steels in 3.5% NaCl and 1 M sulphuric acid solutions was studied by electrochemical tests by Kwok *et al.* [89]. Independently of the heat treatment, UNS T51621 undergoes uniform corrosion and pitting in both media, while UNS S44044, due to its high Cr content, exhibits passive behaviour. It suffers, however, from pitting corrosion, because of its high carbide content. Laser surface melting improves the corrosion resistance of both materials considerably, an effect that was explained by the dissolution and refinement of carbide particles.

1.4.3 Laser surface melting of austenitic and duplex stainless steels

Stainless steels are ferrous alloys specially developed to resist oxidation and corrosion. The initial design of stainless steels was based on the recognition that adding a certain proportion of Cr to iron protects it from atmospheric oxidation, thanks to the formation of a self-healing passive film of chromium oxide. However, the excellent corrosion resistance of modern stainless steels in extremely aggressive environments is achieved by designing complex alloys, adapted to specific environmental conditions, and the wide variety of presently available stainless steels reflects the diversity of applications of these materials. Despite their high corrosion resistance, stainless steels remain susceptible to local breakdown of the passive film, leading to localised corrosive attack, namely pitting, crevice corrosion, intergranular corrosion and stress corrosion cracking [75], as well as general corrosion due to the dissolution of the passive film. The influence of chemical composition and microstructure on the corrosion behaviour of stainless steels was analysed by Sedriks [75, 90]. In order to protect iron from atmospheric corrosion more than 12% of chromium in solid solution is required. The resistance of Fe–Cr alloys to uniform corrosion and pitting in chloride ions containing solutions increases with increasing Cr content. Passivity is also considerably improved by the addition of several other alloying elements, in particular Ni, Mo, Si and nitrogen. Molybdenum and nitrogen have a beneficial impact on the pitting corrosion resistance as well, shifting the pitting potential in the noble direction and extending the passive potential range. Other elements that have a beneficial influence on the pitting corrosion resistance are nickel, copper, vanadium, silicon and tungsten. All these elements may be considered for laser surface alloying of stainless steels.

Contrary to uniform corrosion, localised corrosion is associated with the presence of minor constituents in the microstructure, in particular particles of intermetallic or interstitial compounds containing large concentrations of passivating elements. This is because the precipitation of these phases (Cr_7C_3 and $Cr_{23}C_6$ carbides, σ and χ phases, etc.) creates a zone depleted in alloying elements around their particles, which is more susceptible to passive film breakdown than the bulk matrix, leading to localised corrosion. MnS inclusions and δ-ferrite also affect the corrosion resistance, MnS because it dissolves in some corrosive media nucleating pits, and δ-ferrite because the ferrite stabilising elements Cr and Mo segregate from austenite to ferrite during solidification, depleting austenite and creating steep concentration gradients at the γ/δ interface.

An early demonstration of the beneficial influence of laser surface treatment on the corrosion resistance of austenitic stainless steels is due to McCafferty and Moore [66]. These authors carried out laser surface melting experiments on AISI 304 stainless steel, using a power density of 10^7 W/cm^2 and scanning speeds up to 200 cm/s. Because of these drastic conditions, alloying elements partially evaporated and debris were ejected and redeposited, leading to a non-uniform surface. Anodic polarisation tests in 0.1 M NaCl solution at 25 °C showed that the laser treatment increased the pitting potential substantially, but the improvement was much more noticeable when the surface was polished to eliminate any artefacts originating from the laser treatment. A detailed analysis of the polished surface showed that its Cr and Ni contents were unaffected by the laser treatment, allowing compositional changes to be ruled out as the explanation for the observed improvement in the corrosion resistance. However, the metallographic analysis of the laser-melted layer revealed that large MnS inclusions had been eliminated by the laser treatment and the sulphur concentration had decreased. On the basis of these results, the authors explained the observed pitting corrosion resistance improvement by the elimination of manganese sulphide inclusions, which are preferential sites for pit initiation in stainless steels [91], due to sulphur evaporation. A reappraisal of these, as well as later, results suggests that the observed improvement in pitting corrosion resistance may be due, on one hand, to a refinement of MnS precipitates and, on the other hand, to a reduction of the total amount of MnS in the microstructure. In fact, for pitting to be initiated on MnS inclusions in 304 stainless steel in contact with diluted Cl^--containing solutions, the diameter of these inclusions must exceed about 0.5 μm [92, 93]. Since the melting point of MnS (1610 °C) is only slightly higher than the melting point of steel, this compound melts during the laser treatment and its constitutive elements are dissolved in the melt. MnS will reprecipitate during solidification, but, owing to the high solidification rate, the precipitates will be much smaller than in materials produced by conventional methods.

The resistance to localised corrosion of laser-melted stainless steels is also affected by the redistribution of alloying elements during solidification. By using the Kurz–Giovanola–Trivedi dendritic solidification model [12], Nakao and Nishimoto [41] estimated the redistribution of Cr and Mo during the primary austenite solidification of JIS 904L austenitic stainless steel (Fe–21Cr–25.5Ni–2Mn–1Si–1.5Cu) and concluded that Cr and Mo are segregated to the interdendritic regions ($k < 1$, Eq. 1.4), while Ni concentrates in the dendrite' core ($k > 1$). The segregation decreases with increasing solidification rate and, hence, with increasing scanning speed, the distribution of alloying elements becoming almost uniform above a critical solidification rate. This effect was explained by the tendency of the effective partition coefficient to approach 1 when the solidification rate increased. The authors evaluated the pitting corrosion susceptibility of samples treated with scanning speeds varying over a wide range, by measuring the pitting potential in 5% H_2SO_4 aqueous solutions at 303 K according to JISG0579 standard and by immersion tests in 10% $FeCl_3$ aqueous solution and confirmed that the onset of pitting was retarded by the laser treatment. The pitting corrosion resistance increased with the laser beam scanning speed, while the pit density decreased, in agreement with the theoretical predictions. Pitting was initiated at the austenite dendrite's core where Cr and Mo concentrations were lower and propagated within the dendrite's arms.

The influence of the solidification mode and of the δ-ferrite content of laser-treated austenitic stainless steels on their pitting and passivation behaviour was investigated by Akgün *et al.* [94]. Samples of AISI 304L steel were laser-melted with a scanning speed of 7 mm/s, to produce a $\delta + \gamma$ microstructure. Some of the laser-melted samples were further annealed at 1050 °C to eliminate δ-ferrite. The corrosion behaviour of the material was studied by potentiodynamic tests in deaerated 1N and 3.5 wt% NaCl aqueous solutions. The polarisation curves of the laser surface melted samples presented in the cathodic zone, two Tafel regions with slopes –127 and –86 mV/decade, while the wrought samples and the laser-melted and heat-treated samples showed only a single region. The Tafel slopes of the wrought material and of the laser surface melted + heat treated material were –110 and –113 mV/decade, respectively. The active regions of the polarisation curves were different as well. The corrosion potential was shifted to more positive values by laser surface melting, an effect that was not modified by the annealing treatment. On the other hand, the corrosion current density of the laser surface melted and of the laser-melted and heat-treated samples was similar and lower than the corrosion current density of the wrought alloy. Transpassivity started at 906 mV for all the microstructural conditions used. In what concerns the pitting behaviour in 3.5 wt% NaCl solutions, laser surface melting increased the pitting potential of the alloy by 145 mV and annealing increased it further by 35 mV. The pit morphology was changed as well: after the laser treatment the pits became wider and shallower than in the wrought condition. A semi-continuous network of interdendritic δ-ferrite was visible inside the pits in laser surface melted alloys, which was not observed in the wrought and laser surface melted + heat treated samples. The authors related the existence of the two Tafel regions in the anodic polarisation curves of laser surface melted samples to the presence of δ-ferrite in the material microstructure (between 5 and 10%), the difference observed in the Tafel slopes being explained by the difference in the chemical composition of those phases. However, since annealing does not change the active-passive behaviour of these alloys, the authors concluded that the improvement in the pitting corrosion resistance was not explained by the presence of δ-ferrite but by the reduction of MnS inclusions content due to the formation of an amorphous Mn–Si phase, which traps Mn and reduces the amount of this element available to form MnS. Anjos *et al.* [70] detected Mn_5Si_3 precipitates in a coating of ASTM S31254 stainless steel produced by laser cladding, but did not relate the presence of this compound with the corrosion resistance of the material. In a later paper [95], Akgun and Inal analysed in more detail the role of the solidification mode and of the percentage of δ-ferrite in austenitic stainless steels, using AISI 304L as the study material.

For the processing parameters used AISI 304L steel solidifies in the δ-ferrite primary solidification mode, corresponding to the solidification path $L \rightarrow L + \delta \rightarrow \gamma + \delta$. Since the solidification rate decreases with increasing depth in the laser-melted layer (Eq. 1.1) and the volume fraction of δ-ferrite decreases with increasing solidification rate [20, 21], the proportion of δ-ferrite varies from 10% near the bottom of the melt pool to 5% at its surface. This variation leads to a decrease of the critical passivation current density and an increase of the corrosion potential with depth, showing that the corrosion resistance improvement is roughly proportional to the δ-ferrite content of the material. The authors explained the beneficial influence of δ-ferrite on the pitting corrosion resistance by its influence on sulphur solid/liquid partition during solidification [95, 96]. Since the partition coefficient of sulphur for ferrite is lower than for austenite, for the same initial sulphur content, the concentration of sulphur in the melt will be lower for primary δ-ferrite solidification mode than for primary γ solidification mode, and will decrease with an increasing proportion of δ-ferrite. According to Battle and Pehelke [97], the partition coefficient of sulphur for austenite varies between 0.001 and 0.02 in the corresponding solidification temperature range, while, for austenite, it varies between 0.04 and 0.06. As a result, the amount of sulphur segregated to the liquid during the solidification of δ-ferrite is lower than during the solidification of austenite and the amount of MnS precipitated from the liquid will be lower, as well. However, some experimental results contradict this explanation. For example, Kwok *et al.* [82] observed similar improvements in the pitting corrosion resistance of UNS S31603 and S30400 austenitic stainless steels by laser surface melting, despite the fact that the first steel solidifies in the primary austenite mode, while the second solidifies in the primary δ-ferrite mode. The fact that in S31603 the pits start at the core of the dendrites and are smaller and shallower than in the wrought material, while in S30400 steel the pit density is lower in laser surface melted samples than in the wrought material but the pit morphology and depth are similar, suggests that other factors such as Cr, Ni and Mo segregation during solidification and the microstructure morphology and size play an important role in determining the nucleation density and the development of pits.

The proportion of δ-ferrite in laser-treated austenitic stainless steels will depend, on one hand, on the amount of δ-ferrite formed in the first laser path (in general the proportion of δ-ferrite is inversely proportional to the scanning speed [22]) and, on the other, on the amount of ferrite transformed during the consecutive thermal cycles created by track overlapping. Since the material corrosion resistance increases with increasing proportion of δ-ferrite [95, 96], this transformation will have a deleterious influence on the corrosion resistance [98]. Track overlapping can be reduced by shaping the laser beam in order to create tracks with a rectangular cross-section. Chong *et al.* [99] used a line laser source created by a segmented parabolic mirror to generate a shallow melt pool with an approximately rectangular transverse cross-section. This particular melt pool shape allows treating large areas with contiguous tracks, thus minimising overlapping. The system was applied to a 304L stainless steel, creating a $\delta + \gamma$ microstructure consisting of a proportion of δ-ferrite considerably larger than those found by previous authors who used quasi-Gaussian laser beams [95]. The amount of δ-ferrite that increased with depth and decreased with increasing scanning speed, reaching 27.3 and 12.9% for 500 and 1000 mm/min, respectively, as compared to 5 and 10% found by Akgun and Inal [95] for the same steel. The electrochemical behaviour of the material was characterised using a 3.5% deaerated NaCl solution. The pitting potential in the as-received condition was 285 mV, but after laser surface melting and slight polishing this value increased to 530–600 mV. The pitting potential was inversely proportional to the scanning speed, while the passive current density followed the opposite trend. The pits were initiated at the δ/γ interface, but δ-ferrite corroded at a slower rate than austenite, limiting pit propagation. The corrosion resistance of the surfaces treated with the laser line source was better and more uniform and reproducible than the results reported in the literature, a difference that the authors explained by a higher proportion of δ-ferrite and a more uniform distribution of this phase. Taking into consideration that the proportion of δ-ferrite in many austenitic stainless steels decreases with increasing solidification rate [95] due to a shift from primary δ to primary γ solidification [22], the larger proportion of δ-ferrite observed by Chong *et al.* [99] is probably explained by a lower solidification rate in their experiments as compared to materials treated with a quasi-Gaussian beam, due to a difference in the melt pool shape for the same scanning speed, leading to values of the angle Θ in equation 1.1 nearer to zero. Another possible explanation is that, the overlap being smaller when a line source is used, a larger proportion of retained δ-ferrite is preserved [98].

The localised corrosion resistance of duplex ($\delta + \gamma$) stainless steels was discussed in a recent paper by Kwok *et al.* [100]. The authors studied UNS S31803 and S32950 steels previously aged at 800 °C for 40 hours for sensitisation. This heat treatment causes the precipitation of a relatively large proportion of σ phase at the grain boundaries, due to the eutectoid decomposition of δ phase. Carbides and traces of other intermetallic compounds are also present. Laser melting dissolves these phases and restores the $\delta + \gamma$ microstructure As a result, the intergranular corrosion resistance of the aged steel is greatly improved.

A different approach to the laser surface treatment of austenitic stainless steels was tried by Peyre *et al.* [101]. The laser treatment was carried out on AISI 316L grade steel using a high-brilliance Nd : glass laser capable of delivering several GW/cm^2 during a few nanoseconds, generating shock pressures in the gigapascal range (laser shock hardening). This laser treatment leads to the formation of a dense network of slip bands and deformation twins and large compressive residual stresses of ~470 MPa. It noticeably improved the pitting corrosion resistance of the material in a dilute NaCl solution.

1.4.4 Laser surface alloying of austenitic and duplex stainless steels

Taking into consideration the remarkable influence of alloying elements such as Cr, Mo, Ni and N on the corrosion resistance of stainless steels, laser surface alloying is an attractive surface treatment method to enhance their corrosion resistance. Even if the cost of alloying elements such as Mo excludes their use in high concentrations in bulk alloys, laser surface alloying makes this application much more affordable. On the other hand, the rapid cooling rates found in laser surface treatment allow avoiding the precipitation of intermetallic and interstitial compounds, which cause brittleness and reduce the corrosion resistance of stainless steels [95].

McCafferty and Moore [65, 66] were among the first to demonstrate the potential of laser surface alloying in improving the corrosion resistance of stainless steels. They prepared surface alloys with nominal compositions Fe–18Cr–10Ni–3Mo (similar to AISI 316 stainless steel) and Fe–19Cr–12Ni–9Mo on a 304 stainless steel substrate by alloying with elements predeposited on the substrate as thin films. The corrosion resistance of the Fe–18Cr–10Ni–3Mo alloy was only slightly better than the corrosion resistance of a bulk alloy with similar composition, but the alloy containing 9% Mo could resist pitting up to the oxygen evolution potential, confirming the effectiveness of molybdenum in enhancing the pitting corrosion resistance of Fe–Cr–Ni alloys and the possibility of adding large concentrations of this element to stainless steels by laser surface alloying.

Similar experiments were carried out by Akgun and Inal [95, 96] on 304L austenitic stainless steel, using molybdenum and tantalum in concentrations up to 4.3 and 6.8%, respectively, as alloying elements. The addition of these ferrite-stabilising elements shifted the solidification mode of 304L stainless steel from primary γ solidification to primary δ-ferrite solidification, leading to an increase in the proportion of retained δ-ferrite from 10 to 80%. The addition of Mo improved the pitting corrosion resistance and reduced the weight loss by immersion in $FeCl_3 \cdot 6H_2O$ solution during 96 h at room temperature to zero, as compared to 23.4 g/m/day for wrought 304L steel, but Ta failed to provide a comparable enhancement. The addition of this element shifted the corrosion potential in the noble direction and reduced the critical current density for passivation in 1N H_2SO_4 solution, but the immersion weight loss was similar to the weight loss of the substrate and its pitting corrosion resistance in a 3.5 wt% NaCl solution was only slightly better.

A comparative study of the influence of C, Co, Cr, Mn, Mo, Ni and Si on the microstructure and corrosion resistance of laser surface alloyed UNS S31603 stainless steel was carried out by Kwok *et al.* [102]. The alloying elements were added to the melt in various proportions, in elemental form or as compounds and alloys. The largest improvement in the corrosion resistance, evaluated according to ASTM standard G61-86, using a 3.5% NaCl solution as electrolyte, was achieved by alloying with Si. The addition of Si or Si_3N_4 to the bulk material shifted the pitting potential in the noble direction by 170 and 221 mV, respectively, as compared to the substrate, an effect that was explained by the enrichment of the passive film in Si and N. In low concentrations, Mo remained in solid solution and had a beneficial influence on the pitting corrosion resistance, but in high concentrations it led to the formation of intermetallic compounds, which degraded the pitting corrosion resistance of the material. Cr and Co did not noticeably affect the corrosion resistance, while Mn, C and AlSiFe caused considerable degradation of the corrosion resistance due to the formation of Cr-rich phases during solidification, with the consequent decrease of the concentration of this element in solid solution.

Other alloying elements that are known to improve the corrosion resistance of stainless steels are Cu and Cr and Ni, in particular if employed synergistically. Kamachi Mudali *et al.* [103] studied the corrosion properties of coatings containing Cr and Ni concentrations in the ranges ($25 \leq Cr \leq 30\%$; $8 \leq Ni \leq 10\%$) and ($25 \leq Cr \leq 30\%$; $25 \leq Ni \leq 30\%$) deposited by laser cladding on AISI 304L austenitic stainless steel substrates. The corrosion resistance was tested by potentiodynamic polarisation tests, while the dielectric properties of the passive films were investigated by electrochemical impedance spectroscopy. The electrolytes were 0.5 M NaCl and 1N H_2SO_4 solutions. Typical coating compositions were Fe–24.3Cr–9Ni and Fe–24.4Cr–21.7Ni. The Fe–24.3Cr–9N alloy presented a vermicular ferrite $\delta + \alpha$ microstructure at the bottom of the melt pool, owing to dilution of the coating material by the substrate, and an acicular ferrite microstructure near the surface, where the coating composition was closer to the nominal composition. The Fe–24.4Cr–21.7Ni alloy consisted essentially of austenite. The polarisation tests in the Cl^- solution revealed that alloying with Cr increased the pitting potential from 310 to 720 mV, while the Fe–24.4Cr–21.7Ni coating presented a pitting potential of 980 mV. The electrochemical impedance measurements indicated an increase of the passive film stability. Moreover, the coating/substrate system capacitance was lower than for the bulk alloy, suggesting that the passive film was thicker and, consequently, more stable in the long term than the substrate.

Nitrogen has also a beneficial effect on the corrosion resistance of austenitic stainless steels [104] and is an abundant and inexpensive element, which can easily be alloyed into a range of metallic materials by laser surface melting in a nitrogen atmosphere (laser nitriding [3]). Conde *et al.* [74] showed that about 0.5 wt% nitrogen can be incorporated in AISI 304 steel by laser surface melting in a flowing nitrogen atmosphere, resulting in a considerable improvement of the corrosion resistance. The concentration of nitrogen and the alloy's corrosion resistance increased with increasing nitrogen gas flow and, for the maximum gas flow used, shifts of the corrosion and pitting potentials of + 300 and +200 mV, respectively, were observed. Simultaneously, the passivation current density decreased by nearly two orders of magnitude with respect to untreated AISI 304 stainless steel. The beneficial influence of laser nitriding on the corrosion resistance of stainless steels was confirmed by Mudali *et al.* [105].

1.5 Applications: cast iron and other materials

1.5.1 Laser surface treatment of cast irons

Cast irons are relatively inexpensive and widely available materials, which can easily be formed into complex shapes by casting, a near-net shape process that allows the manufacture of large series of relatively low cost parts used in many industrial sectors, such as the automotive industry, pipe and tube fittings, machinery, etc. The low added-value of many cast iron parts could discourage the application of laser surface treatment to these materials, but its application may be justified in special situations, for example in the localised treatment of expensive large cast iron parts for the chemical or energy-production industries and in the refurbishment of used components. Solidification of laser-melted grey [106] and nodular [107] hypoeutectic cast irons is characterised by a shift from the Fe–C equilibrium phase diagram to the metastable Fe–Fe_3C phase diagram, leading to microstructures formed of proeutectic austenite dendrites and interdendritic austenite–cementite eutectic (usually called ledeburite). If the processing parameters are properly selected, the graphite existing in the material prior to the laser treatment dissolves completely in the melt, an essential requirement to minimise cracking and surface defects caused by the combustion of graphite. In unalloyed cast irons austenite transforms, at least partially, into martensite during cooling to room temperature, but in alloyed cast irons large proportions of austenite may persist. Retained austenite may have a beneficial effect on the mechanical properties of the surface layer because, due to the brittleness of cementite and the large residual stresses that develop during the laser treatment, cracks often form in the laser-treated layer and retained austenite may reduce the propensity to cracking, due to its ductility.

One of the first studies on the corrosion resistance of laser-treated grey cast iron was by Damborenea *et al.* [108]. The corrosion resistance was evaluated by electrochemical methods using concentrated sulphuric acid, 3 and 10% NaCl, and 10% NaOH aqueous solutions as electrolytes. The authors observed a shift of the open circuit potential towards more positive values, indicating an improvement of the corrosion resistance, which was explained by the refinement and homogenisation of the microstructure. However, the results achieved depended on the surface finish, much better corrosion resistance being achieved when the surface defects resulting from the laser treatment (pores, oxide particles, etc.) were eliminated by polishing. The importance of the surface finish was confirmed by later work [109]. Gadag *et al.* [110] also observed a marked improvement of the corrosion resistance of a pearlitic ductile cast iron in diluted sulphuric acid by laser surface melting, provided that the surface layer was crack-free and structurally homogeneous. Laser surface melting reduced the corrosion rate in acid solutions by nearly 40%, but the improvement in synthetic seawater was less noticeable. Laser surface alloying with copper was also tested as a potential method to improve the corrosion resistance of Ni-alloyed cast irons, with positive results [111].

1.5.2 Other applications of laser surface treatment

Sensitisation of austenitic stainless steels occurs when carbon-containing alloys are exposed to temperatures in the 450–900 °C range. The possibility of recovering sensitised steels by laser surface treatment was considered by several authors. Anthony and Clyne [112] were the first to report the recovery of the intergranular corrosion resistance of sensitised type AISI 304 stainless steel by laser surface melting. Akgun and Inal [113] carried out laser surface melting experiments on samples of AISI 304 stainless steel sensitised by ageing for 24 h at 650 °C. The intergranular corrosion resistance was evaluated according to ASTM 262 standard and by exposure to boiling copper sulphate–16% H_2SO_4 for 48 h followed by tensile testing. Metallographic examination of the laser processed samples showed that the chromium carbides responsible for sensitisation were completely eliminated in the laser-melted region and partially eliminated in the HAZ adjacent to the melt pool. Owing to this microstructural alteration the surface corrosion resistance was completely recovered. Sensitisation is particularly dangerous in the process and energy production industries, where stainless steels are submitted to temperatures within the precipitation temperature range during long periods of time.

Mudali *et al.* [114] investigated the possibility of recovering sensitised parts of high-nitrogen 316L stainless steel by *in situ* laser surface melting with a Q-switched ruby laser with 30 ns pulse duration and 6 J/pulse pulse energy. Two laser pulses were applied to each treated spot. The laser treatment led to a substantial improvement of the critical pitting potential in 0.5 M NaCl and

H_2SO_4 solutions at room temperature and, in some conditions, the properties of the steel were entirely recovered. The laser treatment changed the pitting morphology significantly as well: while in the sensitised material pitting concentrated at grain boundaries, in particular at triple points, where Cr-depletion due to the precipitation of chromium carbides was more noticeable, in the laser-treated regions pits were uniformly distributed, smaller, and shallower. In the laser-treated material, pits initiated not around chromium carbide particles but at inclusions of Ti oxides and sulphides.

Prolonged ageing of some stainless steels also leads to the precipitation of intermetallic compounds such as σ, χ, μ and η Laves phases [115]. As a result of the formation of these compounds, the alloying element concentration in the matrix decreases, lowering the material's corrosion resistance. These intermetallic compounds can be eliminated by laser surface melting, potentially leading to considerable improvement of the material's localised corrosion resistance [115].

An important application of AISI 316L austenitic stainless steel is in the medical field. Because of the risk of long-term tissue contamination by metallic ions and the allergenic reaction of many patients to nickel, stainless steels are mainly used for temporary orthopaedic devices, such as fixation plates, pins and screws, titanium alloys being preferred for long-term and permanent implantable devices. The body environment is extremely aggressive and several types of corrosion occur in implanted materials, which may lead to implant fracture, tissue necrosis and implant rejection. To overcome these limitations stainless steels can be surface treated to improve their corrosion resistance. Laser surface treatment allows alleviating localised corrosion, which is one of the main metallic implant failure mechanisms in the body, and it has the advantage over other surface engineering techniques of ensuring excellent adhesion of the treated layer to the bulk, avoiding coating detachment and contamination of the implant surrounding tissues by metallic particles.

Surprisingly, there are only very few studies on the corrosion resistance of laser-treated biomaterials. Singh *et al.* [116] studied the effect of laser processing parameters on the corrosion resistance and metal-ion release of laser surface melted AISI 316L austenitic stainless steel in a simulated physiological medium but, unlike general practice, the authors chose not to use a protective atmosphere, because they assumed that the thick oxide layer formed by surface oxidation in the liquid phase would constitute a stable passive film. Unfortunately, this oxide film did not ensure passivation, because it was a poor adherent and contained a large proportion of Fe_2O_3. The corrosion resistance was evaluated by electrochemical polarisation tests carried out in a normally aerated Ringer's fluid (a solution containing NaCI, $CaCl_2$, KCl and $NaHCO_3$) at a temperature of 37 °C to simulate body conditions. Despite the extensive surface oxidation of the laser-treated samples, both the untreated and the laser-treated samples were expected to present a passive behaviour in typical in-body conditions, but the results of anodic polarisation tests revealed the limitations of the laser treatment method used by the authors. The pitting potential of the laser-treated specimens with unpolished surfaces was consistently lower than the pitting potential of untreated 316L, while their passivation current was consistently higher, showing that the laser treatment decreased the pitting corrosion resistance. However, if the surface oxide layer was eliminated by slight polishing, the pitting corrosion resistance of the laser-treated material became better than the corrosion resistance of the bulk. For example, polished surfaces treated with a laser beam power higher than 1200 W show pitting potential in the range +369 to +688 mV, while the pitting potential of the unpolished surfaces varied between +172 and +198 mV, as compared to +367 mV for the untreated material. More importantly, the pitting potential of the unpolished surface is of the same order of magnitude or lower than the *in vivo* potential (+200 to +350 mV), indicating that these materials may suffer pitting corrosion in the body. On the contrary, after polishing, the pitting potential is much higher than the *in vivo* potential. These results demonstrate the importance of performing the laser treatment within an efficient inert gas protection system in order to avoid oxidation and degradation of the corrosion resistance of the material.

1.5.3 Cladding with metallic glasses

Metallic glasses present outstanding corrosion resistance in a variety of hostile environments [117]. Consequently, cladding low corrosion resistance bulk materials, such as carbon or low-alloy steels, with metallic glasses presents obvious advantages. However, several obstacles to the success of such treatment exist:

- The cooling rates typically used in laser cladding are considerably lower than those found in common amorphous alloys preparation methods, such as melt spinning and splat quenching, and may be insufficient to avoid crystallisation.
- In laser surface processing solidification starts by epitaxial growth on a (crystalline) substrate, a process that requires a low undercooling. As a result, solidification of the crystalline phases may occur before reaching the liquid–glass transition temperature.
- Since the T_g temperature depends considerably on the alloy composition, dilution, alloying element losses by evaporation and oxidation, and contamination must be avoided in order to achieve an amorphous material.

As a result of these difficulties, the coatings obtained are often totally or partially crystalline. For example, coatings of a Fe–13Cr–9P–2C alloy, which is easily amorphised by the melt-spinning technique, prepared by laser cladding on a low-carbon steel substrate presented a partially crystalline structure even for extremely high scanning speeds, due to compositional changes caused by dilution [118]. When dilution was low, the coating was predominantly amorphous, but, even so, its corrosion resistance was inferior to the corrosion resistance of the same alloy produced by melt spinning. Laser remelting improved the coating properties to some extent, but even with this additional treatment the clad layer failed to present the outstanding corrosion resistance of the amorphous melt-spun alloy. Similar results were achieved by subsequent authors [37]. Probably, better results can be achieved by using bulk metallic glasses [119] and improved laser cladding methods.

1.6 Acknowledgements

The author is grateful to Dr Carmen Rangel, Head of the Fuel Cell and Hydrogen Unit of LNEG, Portugal and Prof. Amélia Almeida, Department of Chemical Engineering and ICEMS, Instituto Superior Técnico, Technical University of Lisbon, for their helpful comments on the manuscript.

1.7 References

[1] Ion, J.C., Shercliff, H.R., and Ashby, M.F., Diagrams for laser materials processing, *Acta Metallurgica et Materialia*, 1992, **40**, 1539–1551.

[2] Steen, W.M. and Mazumder, J., *Laser Material Processing*, 4th edition, 2010, Springer-Verlag Ltd., Berlin.

[3] Schaaf, P., Laser nitriding of metals, *Progress in Materials Science*, 2002, **47**, 1–161.

[4] Ion, J., *Laser Processing of Engineering Materials: Principles, Procedure and Industrial Application*, 2005, Elsevier Butterworth-Heinemann, Burlington, MA.

[5] Kannatey-Asibu Jr., E., *Principles of Laser Materials Processing*, Wiley Series on Processing of Engineering Materials, 2009, J. Wiley & Sons, New Jersey.

[6] Mazumder, J., Conde, O., Vilar, R., and Steen, W. (ed.), *Laser Processing: Surface Treatment and Film Deposition*, NATO Science Series E, (ed.), edition, Vol. 307, 1996, Kluwer Academic Publishers, Dordrecht.

[7] Vilar, R., Colaço, R., and Almeida, A., Laser surface treatment of tool steels, *Optical and Quantum Electronics*, 1995, **27**, 1273–1289.

[8] Vilar, R., Colaço, R., and Almeida, A., laser surface treatment of tool steels, in *Laser Processing: Surface Treatment and Film Deposition*, Mazumder, J., Conde, O., Vilar, R., and Steen, W. (ed.), NATO ASI E Series, Vol. 307, 1996, Kluwer Academic Publishers, Dordrecht, 453–478.

[9] Vilar, R., Laser cladding, *J. Laser Applications*, 1999, **11**, 64–79.

[10] Vilar, R., Laser alloying and laser cladding, *Materials Science Forum*, 1999, **301**, 229–252.

[11] Vilar, R., Laser cladding, *International Journal of Powder Metallurgy*, 2001, **37**, 31–48.

[12] Kurz, W., Giovanola, B., and Trivedi, R., Theory of microstructural development during rapid solidification, *Acta Metallurgica*, 1986, **34**, 823–830.

[13] Kurz, W. and Trivedi, R., Microstructure and Phase Selection in Laser Treatment of Materials, Transactions of ASME, 1992, **114**, 450–458.

[14] Kurz, W. and Trivedi, R., Rapid Solidification processing and microstructure formation, *Materials Science and Engineering A*, 1994, **179/180**, 46–51.

[15] Gilgien, P. and Kurz, W., Microstructure selection in binary and ternary alloys, *Materials Science and Engineering A*, 1994, **178**, 199–201.

[16] Gilgien, P. and Kurz, W., Microstructure and phase selection in rapid laser processing, in *Laser Processing: Surface Treatment and Thin Film Deposition*, Mazumder, J., Conde, O., Vilar, R., and Steen, W.M. (ed.), NATO ASI E Series, Vol. 1996, Kluwer Academic Publishers, Dordrecht, 77–92.

[17] Colaço, R. and Vilar, R., Phase selection during laser surface melting of martensitic stainless tool steels, *Scripta Materialia*, 1997, **36**, 199–205.

[18] Colaço, R. and Vilar, R., Effect of laser surface melting on the tempering behaviour of DIN X42Cr13 stainless tool steel, *Scripta Materialia*, 1997, **38**, 107–113.

[19] Colaço, R. and Vilar, R., Effect of the processing parameters on the proportion of retained austenite in laser surface melted tool steels, *J. Materials Science Letters*, 1998, **17**, 563–567.

[20] Fukumoto, S. and Kurz, W., The δ to γ transition in Fe–Cr–Ni alloys during laser treatment, *ISIJ International*, 1997, **37**, 677–684.

[21] Fukumoto, S. and Kurz, W., Prediction of the δ to γ transition in austenitic stainless steels during laser treatment, *ISIJ International*, 1998, **38**, 71–77.

[22] Fukumoto, S. and Kurz, W., Solidification phase and microstructure maps for Fe–Cr–Ni alloys, *ISIJ International*, 1999, **39**, 1270–1279.

[23] Costa, L., Vilar, R., Reti, T., and Deus, A.M., Rapid tooling by laser powder deposition: process simulation using finite element analysis, *Acta Materialia*, 2005, **53**, 3987–3999.

[24] Kurz, W., Bezençon, C., and Gäumann, M., Columnar to equiaxed transition in solidification processing, *Science and Technology of Advanced Materials*, 2001, **2**, 185–191.

[25] Ashby, M.F. and Easterling, K.E., The transformation of steel surfaces by laser beams – I. Hypo-eutectoid steels, *Acta Metallurgica*, 1984, **32**, 1935.

[26] Li, W.-B., Ashby, M.F., and Easterling, K.E., The transformation of steel surfaces by laser beams – II. Hypereutectoid steels, *Acta Metallurgica*, 1986, **34**, 1533.

[27] Hoadley, A.F.A. and Rappaz, M., A thermal model of laser cladding by powder injection, *Metallurgical Transactions B*, 1992, **23**, 631– 642.

[28] Picasso, M. and Hoadley, A.F.A., Finite element simulation of laser surface treatments including convection in the melt pool, *International Journal of Numerical Methods for Heat & Fluid Flow*, 1994, **4**, 61–83.

[29] Toyserkani, E., Khajepour, A., and Corbin, S.F., 3D finite element modeling of laser cladding by powder deposition: effects of powder feedrate and travel speed on the process, *J. Laser Applications*, 2003, **15**, 153–161.

[30] Grøng, O., *Metallurgical Modelling of Welding*, 1994, The Institute of Materials, London.

[31] Rappaz, M., David, S.A., Vitek, J.M., and Boatner, L.A., Development of microstructures in Fe−15Ni−15Cr single crystal electron beam welds, *Metallurgical and Materials Transactions A*, 1989, **20**, 1125–1138.

[32] Mullins, W.W. and Sekerka, R.F., Stability of a planar interface during solidification of a dilute binary alloy, *J. Applied Physics*, 1964, **35**, 444–451.

[33] Vilar, R., Santos, E., Ferreira, P., Franco, N., and Silva, R., Structure of NiCrAlY coatings deposited on single-crystal alloy turbine blade material by laser cladding, *Acta Materialia*, 2009, **57**, 5292–5302.
[34] Bower, T., Brody, H., and Flemings, M., Measurement of solute redistribution in dendritic solidification, *Trans. AIME*, 1966, **236**, 624.
[35] Flemings, M., *Solidification Processing, Materials Science and Engineering*, 1974, McGraw-Hill, New York.
[36] Audebert, F., Colaço, R., Vilar, R., and Sirkin, H., Production of glassy metallic layers by laser surface treatment, *Scripta Materialia*, 2003, **48**, 281–286.
[37] Carvalho, D., Cardoso, S., and Vilar, R., Amorphization of Zr60Al15Ni25 surface layers by laser processing for corrosion resistance, *Scripta Materialia*, 1997, **37**, 523–527.
[38] Audebert, F., Colaço, R., Vilar, R., and Sirkin, H., Laser cladding of quasicrystalline alloys, *Scripta Materialia*, 1999, **40**, 551–557.
[39] Gargarella, P., Almeida, A., Vilar, R., Afonso, C.R.M., Rios, C.T., Bolfarini, C., Botta, W.J., and Kiminami, C.S., Microstructural characterization of a laser remelted coating of Al19Fe4Cr3Ti2 quasicrystalline alloy, *Scripta Materialia*, 2009, **61**, 709–712.
[40] Scheil, E., Bemerkungen zur schichtkristallbildung, *Zeitschrift für Metallkunde*, 1942, **34**, 70–72.
[41] Nakao, Y. and Nishimoto, K., Effects of laser-surface melting on corrosion-resistance of stainless-steel and nickel-base alloy clad layers in cast bimetallic pipes, *ISIJ International*, 1993, **33**, 934–940.
[42] Kurz, W. and Gilgien, P., Selection of microstructures in rapid solidification processing, *Materials Science and Engineering A*, 1994, **178**, 171–178.
[43] Koistinen, D. and Marburger, R., A general equation prescribing the extent of the austenite–martensite transformation in pure iron–carbon alloys and plain carbon steels, *Acta Metallurgica*, 1959, **7**, 59–60.
[44] Löser, W. and Herlach, D.M., Theoretical treatment of the solidification of undercooled Fe–Cr–Ni melts, *Metallurgical Transactions A*, 1992, **23**, 1585–1591.
[45] Rappaz, M., David, S.A., Vitek, J.M., and Boatner, L.A., Analysis of solidification microstructures in Fe–Ni–Cr single-crystal welds, *Metallurgical and Materials Transactions A*, 1990, **21**, 1767–1782.
[46] Lamb, M., West, D.R., and Steen, W.M., Residual stresses in two laser surface melted stainless steels, *Materials Science and Technology*, 1986, **2**, 974–980.
[47] Ion, J.C., Moisio, T., Pedersen, T.F., Sorensen, B., and Hansson, C.M., Laser surface treatment and corrosion behaviour of martensitic stainless AISI 420 Steel, *J. Materials Science*, 1991, **26**, 43–48.
[48] Song, Q. and Shen, L., Microstructural feature of laser surface melting 1Cr17Ni2 stainless steel, *Scripta Materialia*, 1997, **36**, 531–534.
[49] Kishitake, K., Era, H., and Otsubo, F., Structures and tempering behaviour of rapidly solidified high-carbon iron alloys, *Metallurgical Trasactions A*, 1991, **22**, 775–782.
[50] Xie, Z., Sundqvist, B., Hanninen, H., and Pietkainen, J., Isothermal martensitic transformation under hydrostatic pressure in an Fe–Ni–C alloy at low temperatures, *Acta Metallurgica et Materialia*, 1993, **41**, 2283–2290.
[51] Hayzelden, C., Melt Spinning of transformable steels, Ph.D thesis, University of Sussex, 1983,
[52] Colaço, R. and Vilar, R., Laser surface melting of bearing steels, in *Laser Applications*

for Mechanical Industry, 1992, Erice, Italy, Kluwer Academic Publisher, Dordrecht, The Netherlands, 305–314.

[53] Peng, Q.F., Shi, Z., Bloyce, A., and Bell, T., Surface electron-beam melting and alloying of tool steels, *Materials Science and Technology*, 1990, **6**, 999–1004.

[54] Peng, Q.F., Shi, Z., Hancock, I.M., and Bloyce, A., Energy beam surface treatment of tool steels and their wear, *Key Engineering Materials*, 1990, **46 & 47**, 229–244.

[55] Escudero, M.L. and Belló, J.M., Corrosion behaviour of laser surface-treated martensitic steel X40Cr13, in Eleventh International Corrosion Congress, 1990, Florence, AIM, 5.83–5.90.

[56] Escudero, M.L. and Belló, J.M., Laser surface treatment and corrosion behaviour of martensitic stainless AISI 420 steel, *Materials Science and Engineering A*, 1992, **158**, 227–233.

[57] Folkhard, E., *Welding Metallurgy of Stainless Steels*, 1998, Springer Verlag, Wien.

[58] DeLong, W., Ferrite in austenitic stainless steel weld metal, *Welding Journal*, 1974, **53**, 273s–286s.

[59] Washko, S. and Aggen, G., Wrought stainless steels, in *Metals Handbook*, Vol. 1, 1990, ASM International, Metals Park, Ohio, 841–907.

[60] Suutala, N., Effect of solidification conditions on the solidification mode in austenitic stainless steels, *Metallurgical and Materials Transactions A*, 1983, **14**, 191–197.

[61] Elmer, J.W., Allen, S.M., and Eagar, T.W., Microstructural development during solidification of stainless steel alloys, *Metallurgical Transactions A*, 1989, **20A**, 2117–2131.

[62] Bobadilla, M., Lacaze, J., and Lesoult, G., Influence des conditions de solidification sur le deroulement de la solidification des aciers inoxidables austénitiques, *J. Crystal Growth*, 1988, **89**, 531–544.

[63] Vilpas, M., Prediction of microsegregation and pitting corrosion resistance of austenitic stainless steel welds by modelling, Ph.D thesis, Helsinki University of Technology, 1999, Espoo.

[64] Brooks, J.A., Williams, J.C., and Thompson, A.W., STEM analysis of primary austenite solidified stainless steel welds, *Metallurgical and Materials Transactions A*, 1983, **14**, 23–31.

[65] Moore, P. and McCafferty, E., Passivation of Fe/Cr alloys prepared by laser-surface alloying, *J. Electrochemical Society*, 1981, **128**, 1391.

[66] McCafferty, E. and Moore, P., Corrosion behavior of laser-surface melted and laser-surface alloyed steels, *J. Electrochemical Society*, 1986, **133**, 1090.

[67] Renaud, L., Fouquet, F., Millet, J.P., Mazille, H., and Crolet, J.L., Microstructural characterization and comparative electrochemical behavior of Fe–Ni–Cr and Fe–Ni–Cr–P laser surface alloys, *Materials Science and Engineering A*, 1991, **134**, 1049–1053.

[68] Anjos, M.A., Vilar, R., Li, R., Ferreira, M.G., Steen, W.M., and Watkins, K., Fe–Cr–Ni–Mo–C alloys produced by laser-surface alloying, *Surface and Coatings Technology*, 1995, **70**, 235–242.

[69] Weerasinghe, V.M., Steen, W.M., and West, D.R.F., Laser deposited austenitic stainless steel clad layers, *Surface Engineering*, 1987, **3**, 147–153.

[70] Anjos, M.A., Vilar, R., and Qiu, Y.Y., Laser cladding of ASTM S31254 stainless-steel on a plain carbon-steel substrate, *Surface and Coatings Technology*, 1997, **92**, 142–149.

[71] Li, R., Ferreira, M.G.S., Anjos, M.A., and Vilar, R., Localized corrosion of laser

surface cladded UNS S31254 superaustenitic stainless steel on mild steel, *Surface and Coatings Technology*, 1996, **88**, 90–95.

[72] Vilar, R. and Colaço, R., Laser-assisted combinatorial methods for rapid design of wear resistant iron alloys, *Surface and Coatings Technology*, 2009, **203**, 2878–2885.

[73] Li, R., Ferreira, M.G.S., Anjos, M.A., and Vilar, R., Localized corrosion performance of laser surface cladded UNS S44700 superferritic stainless steel on mild steel, *Surface and Coatings Technology*, 1996, **88**, 96–102.

[74] Conde, A., Garcia, I., and de Damborenea, J.J., Pitting corrosion of 304 stainless steel after laser surface melting in argon and nitrogen atmospheres, *Corrosion Science*, 2001, **43**, 817–828.

[75] Sedriks, A., Effects of alloy composition and microstructure on the passivity of stainless-steels, *Corrosion*, 1986, **42**, 376–389.

[76] Pujar, M., Dayal, R., and Sing Raman, R., Microstructural and aqueous corrosion aspects of laser surface-melted type 304 SS plasma-coated mild steel, *J. Materials Engineering and Performance*, 1994, **3**, 412–418.

[77] Vilar, R., Sabino, R., and Almeida, M.A., Laser surface melting of sintered AISI T15 high-speed steel, in ICALEO 91, 10th International Congress on Applications of Lasers and Electro-Optics, Laser Materials Processing Symposium, 1991, San Jose, CA, USA, Laser Institute of America.

[78] Åhman, L., Microstructure and its effect on toughness and wear resistance of laser surface melted and post heat treated high speed steel, *Metallurgical Transactions*, 1984, **15A**, 1829–1935.

[79] Colaço, R., Pina, C. and Vilar, R., Influence of the processing conditions on the abrasive wear behaviour of a laser surface melted tool steel, *Scripta Materialia*, 1999, **41**, 715–721.

[80] Kwok, C.T., Man, H.C., and Cheng, F.T., Cavitation erosion and pitting corrosion behaviour of laser surface-melted martensitic stainless steel UNSS42000, *Surface and Coatings Technology*, 2000, **126**, 238–255.

[81] Damborenea, J.d., Marsden, C., West, D., and Vazquez, A., Pitting resistance of 420 stainless steel after laser surface treatment, in 9th European Congress on Corrosion, 1989, Utrecht, Fu–172.

[82] Kwok, C.T., Man, H.C., and Cheng, F.T., Cavitation erosion and pitting corrosion of laser surface melted stainless steels, *Surface and Coatings Technology*, 1998, **99**, 295–304.

[83] Pyzalla, A., Bohne, C., Heitkemper, M., and Fischer, A., Influence of a laser rapid heat treatment on the corrosion resistance of the high nitrogen steel X30CrMoN15 1+0.3% N, *Materials and Corrosion – Werkstoffe Und Korrosion*, 2001, **52**, 99–105.

[84] Van Ingelgem, Y., Vandendael, I., Van den Broek, D., Hubin, A., and Vereecken, J., Influence of laser surface hardening on the corrosion resistance of martensitic stainless steel, *Electrochimica Acta*, 2007, **52**, 7796–7801.

[85] Steyer, P., Valette, S., Forest, B., Millet, J.P., Donnet, C., and Audouard, E., Surface modification of martensitic stainless steels by laser marking and its consequences regarding corrosion resistance, *Surface Engineering*, 2006, **22**, 167–172.

[86] Kwok, C.T., Lo, K.H., Cheng, F.T., and Man, H.C., Laser surface melting of martensitic stainless steel 440C for enhancing corrosion resistance, in First International Symposium on High-Power Laser Macroprocessing, 2003, 493–498.

[87] Lo, K.H., Kwok, C.T., Cheng, F.T., and Man, H.C., Corrosion resistance of

laser-fabricated metal–matrix composite layer on stainless steel 316L, *J. Laser Applications*, 2003, **15**, 107–114.

[88] Costa, L. and Vilar, R., Laser powder deposition, *Rapid Prototyping J.*, 2009, **15**, 264–279.

[89] Kwok, C.T., Leong, K., Cheng, F.T., and Man, H.C., Microstructural and corrosion characteristics of laser surface-melted plastics mold steels, *Materials Science and Engineering A*, 2003, **357**, 94–103.

[90] Sedriks, A., Corrosion resistance of austenitic Fe–Cr–Ni–Mo alloys in marine environments, *International Metals Reviews*, 1982, **27**, 321–353.

[91] Sedriks, A., Role of sulphide inclusions in pitting and crevice corrosion of stainless steels, *International Metals Reviews*, 1983, **28**, 295–307.

[92] Ke, R. and Alkire, R., Initiation of corrosion pits at inclusions on 304 stainless steel, *J. Electrochemical Society*, 1995, **142**, 4056–4062.

[93] Stewart, J. and Williams, D., The initiation of pitting corrosion on austenitic stainless steel: on the role and importance of sulphide inclusions, *Corrosion Science*, 1992, **33**, 457–463, 465–474.

[94] Akgün, A., Ürgen, M., and Çakir, A., The effect of heat treatment on corrosion behavior of laser surface melted 304L stainless steel, *Materials Science and Engineering A*, 1995, **203**, 324–331.

[95] Akgun, O. and Inal, O., Laser surface melting and alloying of type 304L stainless steel. Part I: Microstructural characterization, *J. Materials Science*, 1995, **30**, 6097–6104.

[96] Akgun, O. and Inal, O., Laser surface melting and alloying of type 304L stainless steel. Part II: Corrosion and wear resistance properties, *J. Materials Science*, 1995, **30**, 6105–6112.

[97] Battle, T. and Pehelke, R., Equilibrium partition coefficients in iron-based alloys, *Metallurgical Transactions B*, 1989, **20**, 149–160.

[98] Wang, X.Y., Liu, Z., and Chong, P.H., Effect of overlaps on phase composition and crystalline orientation of laser-melted surfaces of 321 austenitic stainless steel, *Thin Solid Films*, 2004, **453**, 72–75.

[99] Chong, P., Liu, Z., Wamg, X., and Skeldon, P., Pitting corrosion behaviour of large area laser surface treated 304L stainless-steel, *Thin Solid Films*, 2004, **453–454**, 388–393.

[100] Kwok, C., Lo, K., Chan, W., Cheng, F., and Man, H., Effect of laser surface melting on intergranular corrosion behaviour of aged austenitic and duplex stainless steels, *Corrosion Science*, 2011, **52**, 1581–1591.

[101] Peyre, P., Scherpereel, X., Berthe, L., Carboni, C., Fabbro, R., Beranger, G., and Lemaitre, C., Surface modifications induced in 316L steel by laser peening and shot-peening. Influence on pitting corrosion resistance, *Materials Science and Engineering A*, 2000, **280**, 294–302.

[102] Kwok, C.T., Cheng, F.T., and Man, H.C., Laser surface modification of UNS S31603 stainless steel. Part I: Microstructures and corrosion characteristics, *Materials Science and Engineering A*, 2000, **290**, 55–73.

[103] Mudali, U.K., Kaul, R., Ningshen, S., Ganesh, P., Nath, A., Khatak, H.S., and Raj, B., Influence of laser surface alloying with chromium and nickel on corrosion resistance of type 304L stainless steels, *Materials Science and Technology*, 2006, **22**, 1185–1192.

[104] Gavriljuk, V.G. and Berns, H., *High Nitrogen Steels. Structure, Properties, Manufacture, Applications, Engineering Materials*, 1999, Springer Verlag, Berlin.

[105] Mudali, U.K., Khatak, H.S., Raj, B., and Uhlemann, M., Surface alloying of nitrogen to improve corrosion resistance of steels and stainless steels, *Materials and Manufacturing Processes*, 2004, **19**, 61–73.

[106] Trafford, D., Bell, T., Megaw, J., and Bransden, A., Laser treatment of grey iron, *Heat Treating J.*, 1983, 198–206.

[107] Gadag, S.P., Srinivasan, M.N., and Mordike, L.B.L., Effect of laser processing parameters on the structure of ductile iron, *Materials Science and Engineering A*, 1995, **196**, 145–151.

[108] Damborenea, J., Gonzalez, J., and Vazquez, A., Corrosion behaviour of cast iron after laser surface treatment, *J. Materials Science Letters*, 1988, **7**, 1046–1047.

[109] Panagopoulos, C.N., Markaki, A.E., and Agathocleous, P.E., Excimer laser treatment of nickel-coated cast iron, *Materials Science and Engineering A*, 1998, **241**, 226–232.

[110] Gadag, S.P. and Srinivasan, M.N., Surface properties of laser processed ductile iron, *Applied Physics A*, 1996, **63**, 409–414.

[111] Zeng, D., Xie, C., Hu, Q., and Yung, K., Corrosion resistance enhancement of Ni-Resist ductile iron by laser surface alloying, *Scripta Materialia*, 2001, **44**, 651–657.

[112] Anthony, T.R. and Clyne, H.E., Surface normalization of sensitized stainless-steel by laser surface melting, *J. Applied Physics*, 1978, **49**, 1248–1255.

[113] Akgun, O. and Inal, O., Desensitization of sensitized 304 stainless steel by laser surface melting, *J. Materials Science*, 1992, **27**, 2147–2153.

[114] Mudali, U.K., Pujar, M.G., and Dayal, R.K., Effects of laser surface melting on the pitting resistance of sensitized nitrogen-bearing type 316L stainless steel, *J. Materials Engineering and Performance*, 1998, **7**, 214–220.

[115] Subba Rao, R., Parvatharvarthini, N., Pujar, M., Dayal, R., Khatak, H., Kaul, R., Ganesh, P., and Nath, A., Improved pitting corrosion resistance of cold worked and thermally aged AISI type 316L(N) SS by laser surface modification, *Surface Engineering*, 2007, **23**, 83–92.

[116] Singh, R., Martin, M., and Dahotre, N.B., Influence of laser surface modification on corrosion behavior of stainless steel 316L and Ti–6Al–4V in simulated biofluid, *Surface Engineering*, 2005, **21**, 297–306.

[117] Scully, J., Gebert, A., and Payer, J., Review: Corrosion and related mechanical properties of bulk metallic glasses, *J. Materials Research*, 2007, **22**, 302–313.

[118] Virtanen, S., Bohni, H., Busin, R., Marchione, T., Pierantoni, M., and Blank, E., The effect of laser surface modification on the corrosion behaviour of Fe and Al base alloys, *Corrosion Science*, 1994, **36**, 1625–1644.

[119] Suryanarayana, C. and Inoue, A., *Bulk Metallic Glasses*, 2010, CRC Press, Boca Raton, FL.

2
Laser surface melting (LSM) to repair stress corrosion cracking (SCC) in weld metal

K. SHINOZAKI, Hiroshima University, Japan and
T. TOKAIRIN, Babcock-Hitachi K.K., Japan

Abstract: This chapter discusses repairing stress corrosion cracking (SCC) in weld overlaid Inconel 182 weld metal at nuclear power plants, using laser surface melting (LSM). The effects of microstructure, chemical composition and residual stress on the corrosion resistance of the weld metal undergoing LSM treatment are described, and theoretical models not only of Cr depletion zone at grain boundaries but also of residual stress analysis are discussed to identify why LSM treatment of welded metals improves corrosion resistance.

Key words: corrosion resistance, laser surface melting, Ni-base superalloys, precipitation, residual stress.

2.1 Introduction

Intergranular cracking (IGC) and intergranular stress corrosion cracking (IGSCC) are common types of damage in nuclear power plants.[1–6] The cladding materials used in some parts of nuclear power plants built several decades ago are usually nickel-based superalloys such as Inconel 182, and they are now suffering from IGC or IGSCC after long-term operation. For the safe working of these plants, it is important to find an efficient method to repair these components to prolong their life and reduce the costs associated with component replacement after long-term use. A practical repair method is urgently needed.

The tungsten inert gas (TIG) welding process is one candidate method for this repair procedure. The alloy near the position where IGC or IGSCC has occurred is removed and then preparatory welding is carried out by TIG welding, as shown in Fig. 2.1(a). However, the length of time needed for this repair procedure, together with the high heat input, mean that the TIG welding technique has many disadvantages for this application.

In recent years, laser beam processing has brought many technological and economic advantages, including its high precision, reliability, efficiency and productivity, and consequently it has attracted much attention as a new method. Laser surface melting (LSM) treatment is considered to be one of the most powerful surface modification techniques available to improve the surface properties of materials in cases of corrosion or wear, for example, as

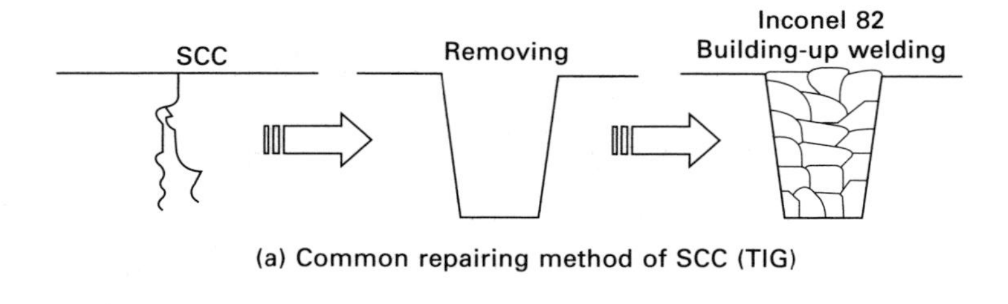

(a) Common repairing method of SCC (TIG)

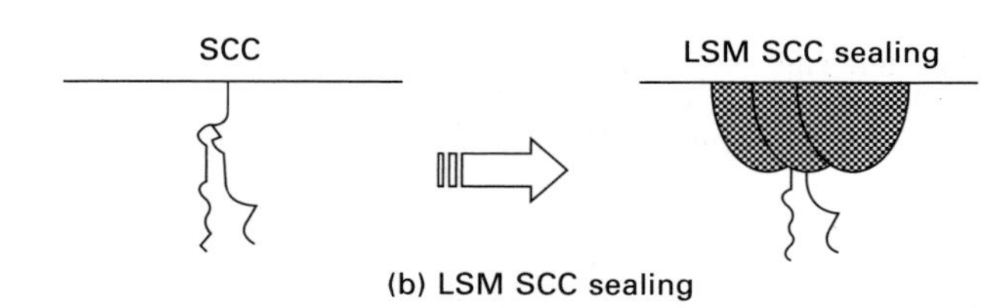

(b) LSM SCC sealing

2.1 Schematic illustration of repair method by TIG welding and LSM process.

a result of homogenization and refinement of microstructures. LSM treatment can be applied to repair tubes degraded by IGC/IGSCC during normal operation of nuclear power plants, as shown in Fig. 2.1(b), since a laser beam can easily be directed to the failed parts through a beam transmission system such as an optical fiber. Moreover, much research on improvement of stress corrosion cracking (SCC) resistance by LSM treatment has been reported[3–11] and results indicate that the IGC/IGSCC resistance of the base alloy Inconel 600 can be improved by LSM treatment; some research also refers to repairing procedures using LSM treatment, especially in the case of microstructure evolution and the relationship between the microstructure and the IGC/IGSCC susceptibility of Inconel 182 during the repairing procedure itself.

However, it is obvious that the composition of an alloy has a large influence on its corrosion resistance. Alloys with different chemical compositions can have different corrosion behaviors. Commonly, the function of niobium in the nickel-based superalloy is used to stabilize carbon by formation of NbC, so that the formation of Cr carbides is avoided and the susceptibility of the material to IGC/IGSCC is lowered. For Inconel 600, the formation of NbC seemed to correlate with increased protection from IGC, where the suitable addition of Nb/C has a positive influence.[12, 13] For Inconel 182, the addition of Nb gives better IGC/IGSCC resistance compared with Inconel 600 with the same thermal treatment.[13] However, the effect of the addition of Nb on the microstructure and IGC/IGSCC susceptibility of Inconel 182 after the LSM process has not yet been reported.

In addition to the effects of service environment and weld structure, the presence of tensile residual stresses in the weld zone has been identified as a significant factor leading to the occurrence of IGSCC, since IGSCC mainly occurs in regions where the residual stresses due to manufacturing are highest; however, only a few papers about the effect of the residual stress of LSM treatment on SCC susceptibility are available.[7]

In this chapter, application of the LSM treatment method as a repair procedure for nickel-base superalloys (Inconels 182 and 600) will be discussed. The factors which affect SCC susceptibility are studied from both a metallurgical and a mechanical viewpoint. Moreover, the effect of microstructure and residual stress on the SCC susceptibility of nickel-base superalloy are discussed separately.

2.2 Materials and experimental procedures

2.2.1 Materials used

Two kinds of Inconel 182, containing 1.92% Nb (1.92% Nb Inconel 182) and 1.10% Nb (1.10% Nb Inconel 182), were clad onto a Inconel 600 base plate by shielded metal arc welding (SMAW). The chemical composition of the alloys is shown in Table 2.1.

2.2.2 Specimen preparation

The thermal cycle flow of the repair procedure is shown in Fig. 2.2. SMAW was used to clad Inconel 182 and then stress relief (SR) treatment (898 K × 86.4 ks) was performed to reduce the residual stress. These two processes were used to simulate the actual multi-pass welding process used for repairing. Low temperature sensitization (LTS) treatment (773 K × 86.4 ks) was carried out to simulate the sensitization process after long-term operation. Next, LSM treatment was performed on the specimen surface before, finally, the LTS treatment was carried out again to verify the corrosion resistance of the material after the LSM treatment.

An yttrium aluminum garnet (YAG) laser with 2 kW maximum power was used for the LSM treatment. Ar was used as the shielding gas to prevent oxidation of the melted region. Various parameters were tested (including combinations of laser power of 1.5 kW and a speed of 4.2 to 33.3 mm/s, with a defocus length of 0 or 10 mm; see Table 2.2) to find the optimum manufacturing conditions for repairing using LSM treatment. The average laser spot diameter was 0.5 mmϕ, whilst it was 4 mmϕ at the focal point and 10 mm at the defocus point. The lapping rate of the LSM treatment bead changed from approximately 25 to 75%; for a definition of the LSM treatment lapping rate; see Fig. 2.3.

Table 2.1 Chemical compositions of materials used (mass %)

	C	Si	Mn	P	S	Cu	Ni	Cr	Fe	Nb+Ta	Ti
1.92% Nb Inconel 182	0.04	0.21	2.76	0.003	0.003	0.03	71.4	14.54	8.66	1.92	–
1.10% NB Inconel 182	0.06	0.18	2.7	0.007	0.003	0.01	72.4	14.68	8.68	1.1	–
Inconel 600	0.02	0.89	0.21	0.004	0.001	–	73.51	15.1	9.6	0.04	0.33

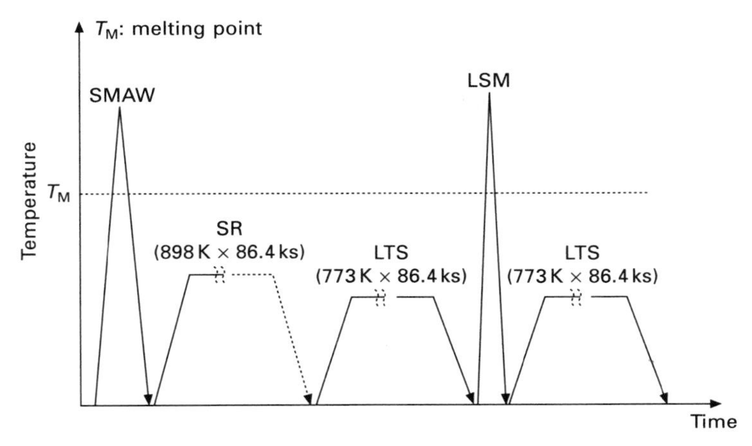

2.2 Thermal cycle flow of repairing procedure.

Table 2.2 Laser surface melting (LSM) conditions

Laser power (kW)		1.5		
Traveling speed (mm/s)	4.2	8.3	16.7	33.3
Lapping rate (overlapped ratio by laser beam diameter) (%)	25	50		75
Defocus length (distance off specimen surface) (mm)	0		10	
Shielding gas (10^{-3} m^3/s)	0.5 (axial), 0.3 (side)			

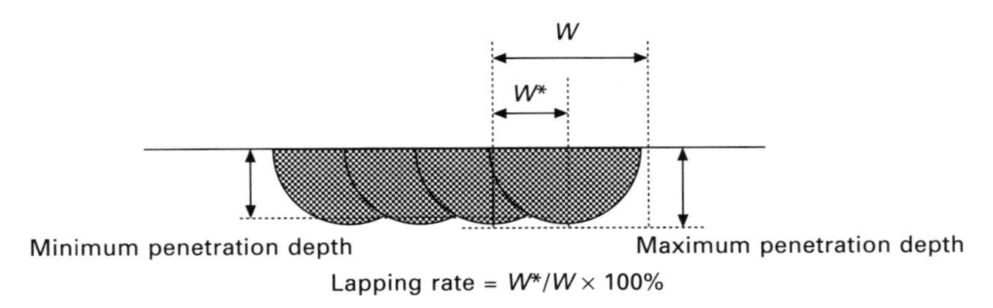

2.3 Schematic illustration of lapping rate of LSM process.

Figure 2.4 shows a schematic illustration of the preparation of a specimen with SCC. To prepare the specimen for SCC sealing, SMAW was used to deposit SR and LTS treatments, then a tensile strain of 1% was loaded on the specimen surface, which was immersed into a caustic solution ($K_2S_4O_6$ 10 g/l and NaCl 1 g/l). After a few days, cracking (of 2–3 mm) was obtained and used to simulate the occurrence of SCC after long-term operation. An LSM sealing treatment was then performed on the specimen surface in the direction parallel to the surface cracks, as shown in Fig. 2.5.

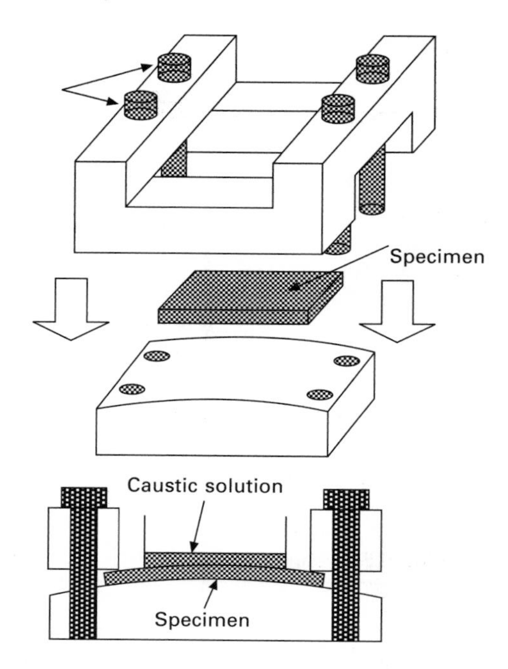

2.4 Schematic illustration of SCC specimen preparation.

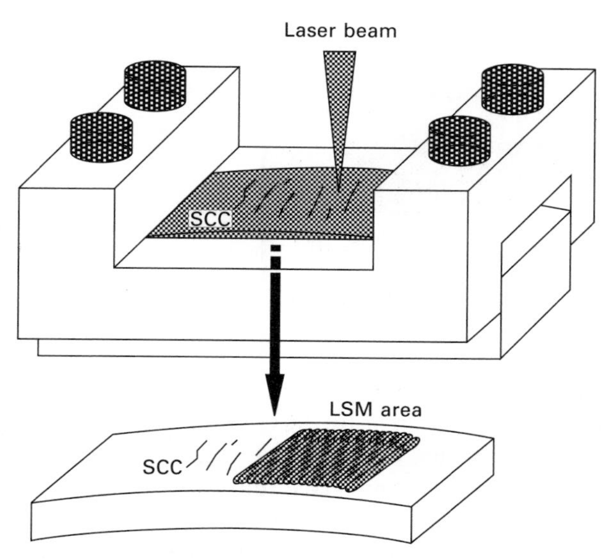

2.5 Sketch map of SCC sealing by LSM process.

2.2.3 Corrosion tests

Streicher test (modified ASTM G28)

Tests for IGC and IGSCC were performed in a solution of 400 ml H_2O, 233 ml H_2SO_4 and 50 g $Fe_2(SO_4)_3$ using a Streicher test. The specimens were cut into small samples of size 40 mm × 8 mm × 5 mm (length × width × height), and put into the boiling solution for 86.4 ks. The samples were then bent to reveal the IGC/IGSCC for observation. After the Streicher test, the maximum IGC depths in two cross-sections were measured by optical microscopy to evaluate the IGC/IGSCC susceptibility of the samples.

Open bend beam (OBB) test

In addition, an open bend beam (OBB) test was used to evaluate the IGSCC resistance of the specimens. The specimen was cut into small samples of size 40 mm × 10 mm × 3 mm, and a tensile strain of 1% was loaded on the surface of the sample. The LSM treated surface of the specimen was immersed in a solution of $K_2S_4O_6$ 10 g/l and NaCl 1 g/l, and the other parts covered by silicone rubber. The experiment was carried out at room temperature and the solution was replaced every 432 ks. The time duration before SCC occurrence was used to evaluate the SCC susceptibility of the material.

2.3 Laser surface melting (LSM) treatment conditions for repair procedures

During LSM treatment, many different variables (including laser power, welding speed, defocusing distance and lapping rate) have to be considered, any of which may have an significant effect on heat flow and fluid flow in the weld pool which, in turn, affects the penetration depth. The shape and final solidification structure of the fusion zone influence the properties of the weld bead considerably.

2.3.1 Weld bead shape

LSM treatments with different parameters (as shown in Table 2.2) were carried out to obtain the optimum conditions for repair. Two types of weld bead shape cross-section are observed: a deep penetration type and a heat conduction type, as shown in Fig. 2.6. When the defocus length is 0 mm, a deep penetration type is achieved and the weld bead becomes a nail-head shape, whilst the shape of the weld bead becomes a heat conduction type when the defocus length is 10 mm.

When the laser speed increases, the penetration depth of the weld bead becomes shallow and the width of the weld bead becomes narrow since the

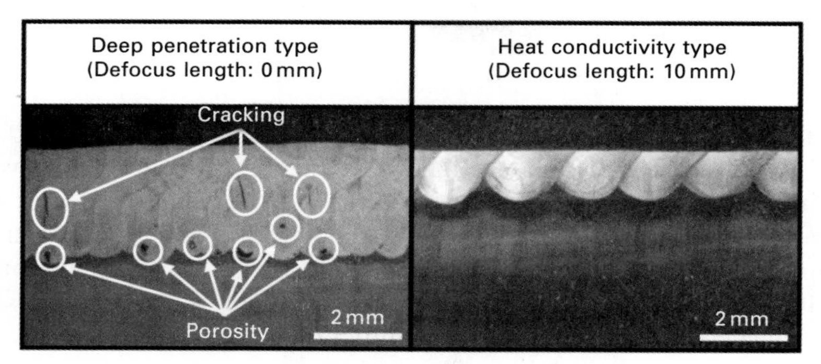

2.6 Welding pattern and defects of LSM process (1.92% Nb Inconel 182).

heat input decreases, as shown in Fig. 2.7. For the deep penetration type, the maximum penetration depth of the weld bead is about 3.3 mm, whilst for the heat conduction type, the maximum penetration depth of the weld bead is only about 1.5 mm. Considering the demands required for repair, the deeper penetration of the weld bead is preferred for sealing an existing SCC.

2.3.2 Welding defects

Weld defects such as cracking and porosity have a great effect on the performance of the weld bead. Samples taken after the LSM process, with different manufacturing parameters, were checked to verify the occurrence of porosity and cracking. Figure 2.6 shows a typical solidification crack and porosity in the weld bead. Figures 2.8 and 2.9 show the relationship between LSM parameters and porosity and cracking defects, respectively.

It has been found that the defects of porosity and cracking are markedly reduced when the welding pattern changes from a deep penetration type to a heat conduction type. The weld bead of the heat conduction type is almost free of defects whilst, for deep penetration welding, it is easy for bubbles to be trapped and for solidification cracks to be produced since the molten pool is much deeper than that of the heat conductivity type. Moreover, it can also be seen clearly that defects increase as the laser traveling speed decreases, because the slower traveling speed induces a greater penetration depth which in turn results in more defects.

2.3.3 Discussion

A deep weld bead with no defects, such as porosity and cracking, is preferred for repair procedures. Although deep penetration welds produce a deeper

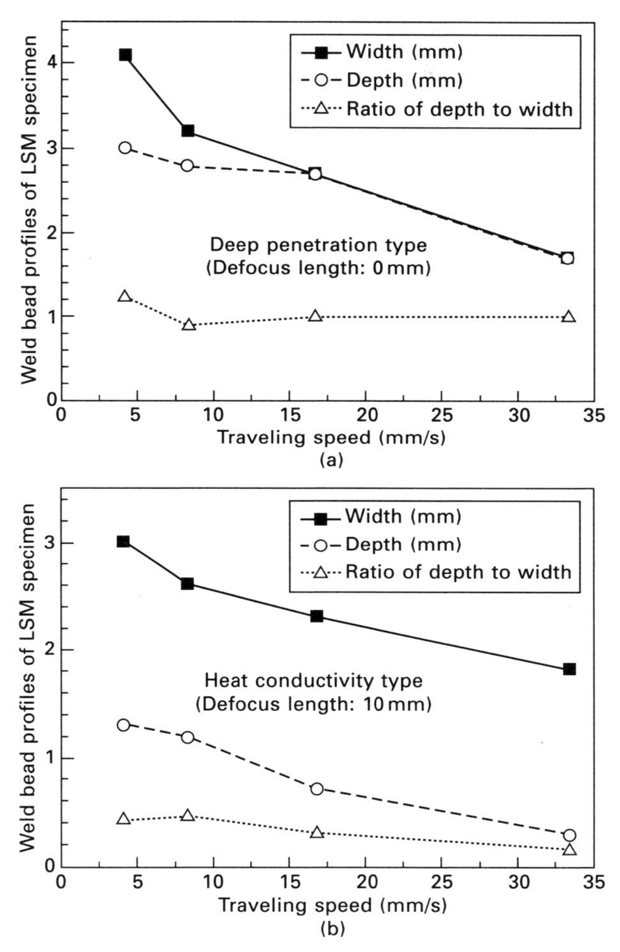

2.7 Relationship between weld bead shape and LSM parameters (1.92% Nb Inconel 182): (a) deep penetration type; (b) heat conductivity type.

penetration depth, a heat conduction weld bead, free of defects and with a shallower penetration depth is more practical. To obtain a repair depth that is as deep as possible, a heat conduction LSM treatment with slower traveling speed and a suitable lapping rate was chosen for the repair process in this study.

After the LSM treatment, the cross-section of the specimen was also observed to check the status of the SCC which had been sealed (see Fig. 2.10). Although the pre-existing SCC on the specimen surface was sealed by LSM treatment, some cracking was observed. The factors causing SCC are considered to be material, environment and stress, in the present study.

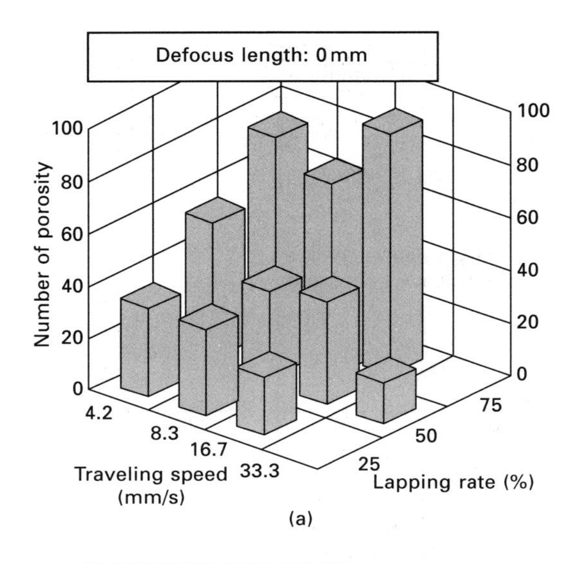

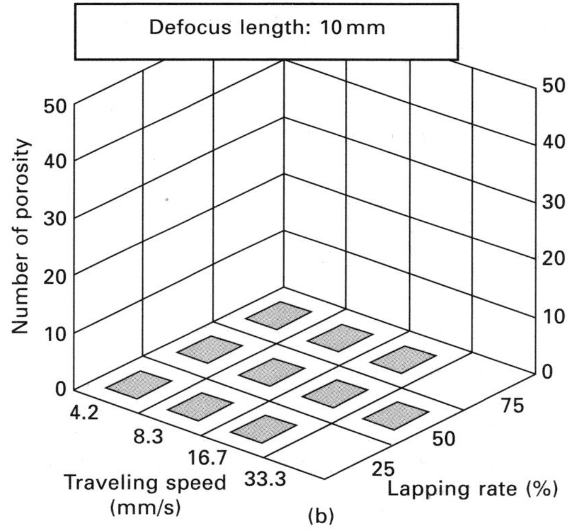

2.8 Relationship between LSM bead shape and porosity (1.92% Nb Inconel 182): (a) deep penetration type; (b) heat conductivity type.

Since the pre-existing SCC was sealed by the laser melted metal and not in contact with the caustic environment, the pre-existing SCC was sealed successfully by the LSM treatment.[14]

The other issue to be considered is the extension of cracking after the sealing process. As shown in Fig. 2.10, two types of sealed cracks were distinguished: Type A and Type B. For Type A, cracking showed no extension after the sealing process. Close examination of Type B specimens shows that the extended cracking immediately next to an original crack is probably

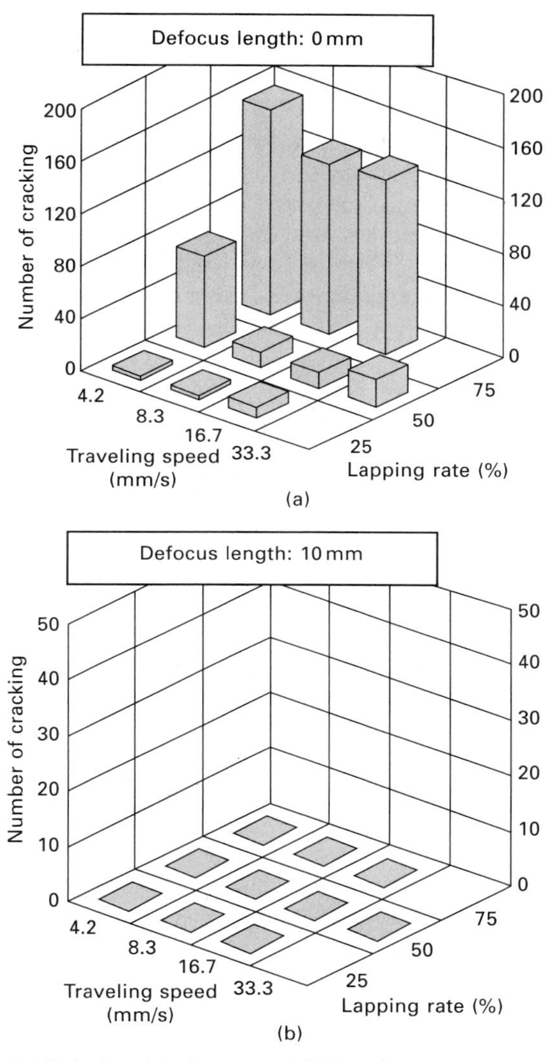

2.9 Relationship between LSM bead shape and cracking (1.92% Nb Inconel 182): (a) deep penetration type; (b) heat conductivity type.

induced by the escape of air existing in the cracking before LSM treatment rather than as a result of an extension of the original cracking. Since most specimens were Type A (82.7%, observed in six cross-sections), it can be considered that direct LSM sealing treatment is a useful and valuable method for SCC repair. And since the laser power used in the present study is only 1.5 kW, it can also be expected that any SCC could be totally melted and sealed if a higher laser power is used.

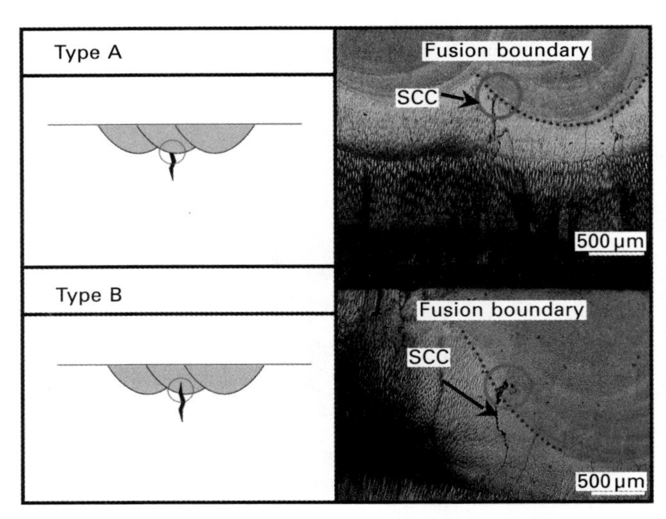

2.10 Cross-section of specimen after SCC sealing by LSM process (1.92% Nb Inconel 182).

2.4 Corrosion resistance of the laser surface melting (LSM) treatment zone

2.4.1 Effect of microstructure on IGC/IGSCC susceptibility during heat treatment

Corrosion test

The IGC/IGSCC susceptibility of 1.92% Nb Inconel 182 during repair procedure heat treatments was investigated. Figure 2.11 shows the appearance of a specimen after a Streicher test. The maximum IGC depth of the specimens measured after a Streicher test was used to evaluate IGC/IGSCC susceptibility.

According to the results of the Streicher tests, the maximum IGC depth decreases (see Fig. 2.12) in the following order: SMAW + SR, SMAW + SR + LTS, SMAW, SMAW + SR + LTS + LSM + LTS, SMAW + SR + LTS + LSM. After LSM treatment, the specimen shows excellent IGC/IGSCC resistance. Even after LTS treatment, the LSM treatment specimen still retains good corrosion resistance to IGC or IGSCC. IGC or IGSCC susceptibility is worse after SMAW + SR treatment, while the specimen undergoing SMAW + SR + LTS heat treatment is better than the specimen after SMAW + SR treatment.[15]

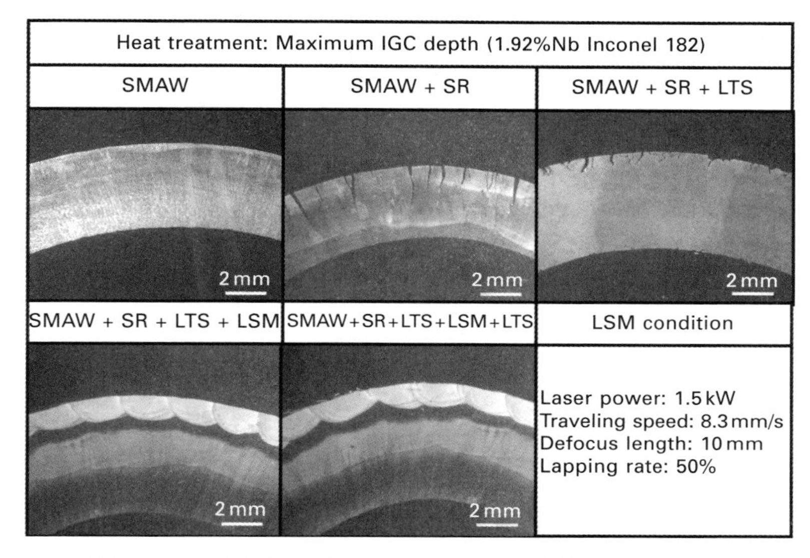

2.11 IGC of Inconel 182 for heat treatments and LSM process by Streicher test.

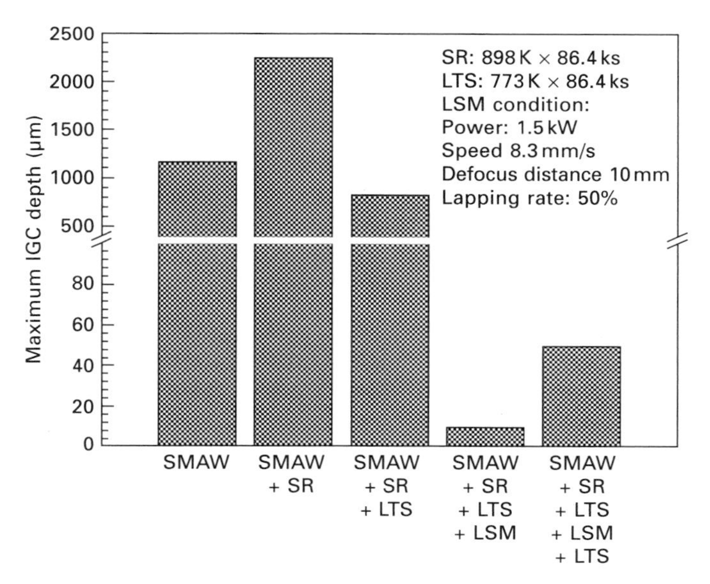

2.12 IGC susceptibility for heat treatments and LSM process (1.92% Nb Inconel 182).

Microstructures

There is almost no change in grain morphology before LSM treatment. As shown in Figs 2.13(a), (b) and (c), the SMAW, SMAW + SR and SMAW

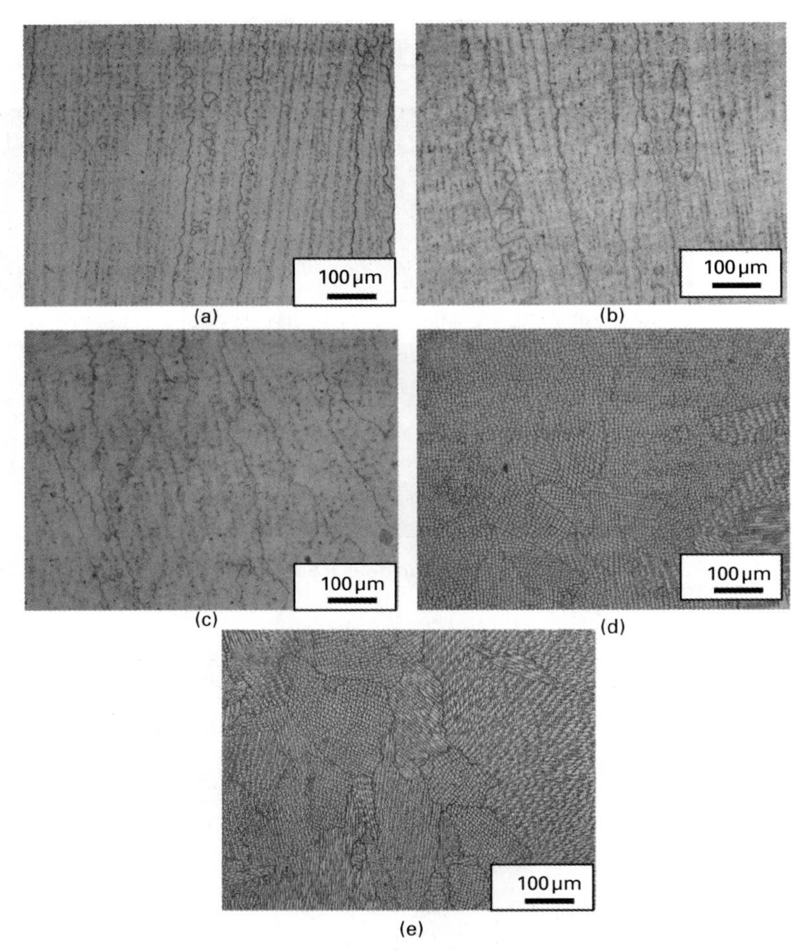

2.13 Optical micrographs showing various microstructures under heat treatments: (a) SMAW; (b) SMAW + SR; (c) SMAW + SR + LTS; (d) SMAW + SR + LTS + LSM; (e) SMAW + SR + LTS + LSM + LTS.

+ SR + LTS specimens show similar crystallization morphology; coarse columnar crystals formed from the SMAW process grow in the epitaxial growth direction. A more refined grain with fine cellular morphology is obtained after laser beam irradiation due to the high heating and cooling rate of the LSM treatment (Fig. 2.13(d)). And the morphology of the microstructure following LTS also seems to be similar to that seen after LSM (Fig. 2.13(e)). A more refined grain with a fine sub-grain has two beneficial effects: the suppression of micro segregation of the chemical composition, and a reduction of precipitates because of an increase in the grain boundary area. Both of these improve the corrosion resistance. However, the cell boundary of the LSM treatment sample was rich in Cr, which can be attributed to the micro segregation or coring that occurred even during the rapid solidification procedure.[16–18]

Intergranular film like micro constituent

Figure 2.14 shows a feature of the grain boundary under various treatments such as SMAW, SMAW + SR, SMAW + SR + LTS, SMAW + SR + LTS + LSM and SMAW + SR + LTS + LSM+LTS.[15] As shown in Fig. 2.14 (a), for the SMAW specimen, only discontinuous micro constituents were found

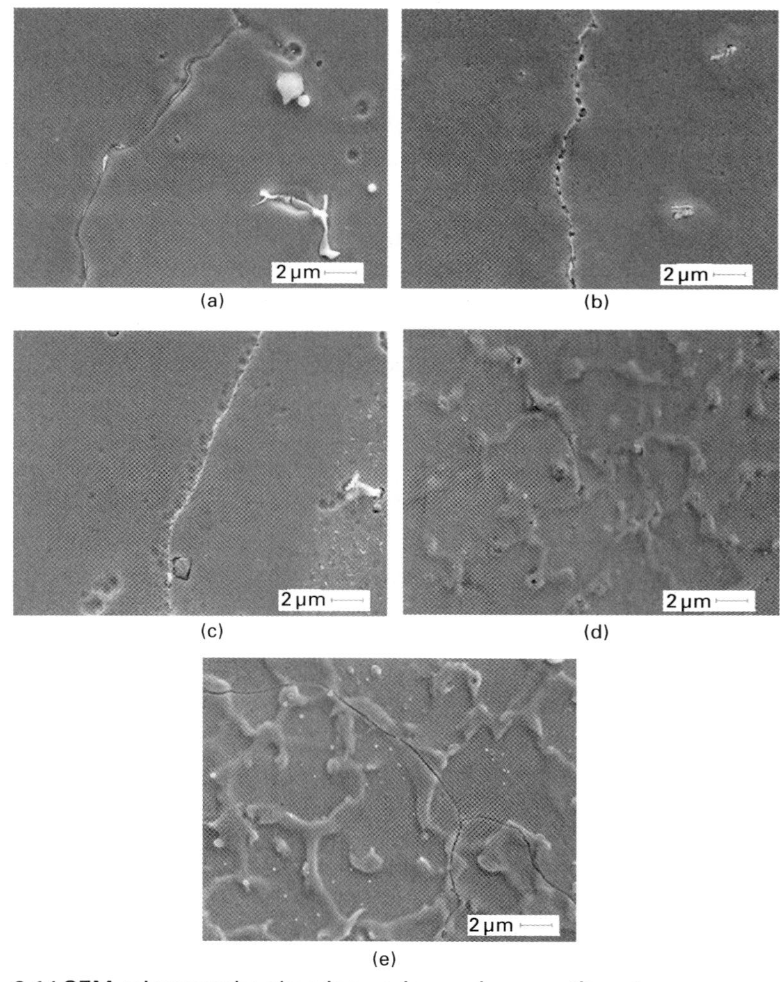

2.14 SEM micrographs showing various microconstituent distributions under heat treatments: (a) SMAW; (b) SMAW + SR; (c) SMAW + SR + LTS; (d) SMAW + SR + LTS + LSM; (e) SMAW + SR + LTS + LSM + LTS.

along the grain boundary. The specimen after SR heat treatment shows a semi continuous morphology along the grain boundary, as shown in Fig. 2.14(b), while Fig. 2.14(c) indicates that the micro constituent distribution on the grain boundary after LTS heat treatment is nearly continuous. Figure 2.14(d) shows the microstructure after LSM treatment, with no micro constituent found at the grain boundary. This result indicates that the pre-existing continuous micro constituent has been completely dissolved into the matrix after the sensitized treatment due to the high energy density of the laser beam and that it did not have enough time to re-precipitate in the LSM treatment region because of the extremely rapid solidification velocity of the LSM treatment. As shown in Fig. 2.14(e), almost no obvious changes happened in the LSM specimens following LTS, with only a few micro constituents found at the grain boundary.

Morphology and composition of interdendritic and intergranular micro constituents

As well as the film-like intergranular micro constituents, two other kinds of larger particles are also found, namely, white particles with complex shape and blocky particles. Figure 2.15 shows the distribution of interdendritic and intergranular micro constituents before LSM treatment.[15] The white particles are present mainly in the interdendritic spaces. Most of the blocky particles are distributed in the intragranular area. As shown in Fig. 2.15, a typical

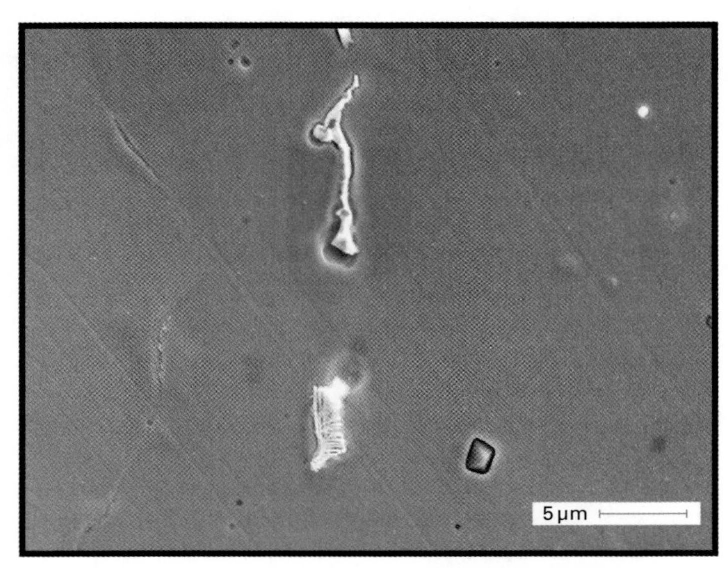

2.15 Primary and eutectic carbides formed during SMAW process (1.92% Nb Inconel 182).

eutectic morphology is formed during the final stage of solidification. Eutectic micro constituents are formed between dendrite arms and follow the shape of the dendritic arms, branching out frequently. These branches grow in the form of rods together with the γ phase into the remaining liquid.

As shown in Figs 2.14(d) and (e), no micro constituent was found on the grain boundary after LSM treatment. After LTS treatment, the pre-existing micro constituents in the interdendritic and intergranular area in the sensitized alloy had been completely melted and/or dissolved during cooling due to the high cooling rate during the LSM treatment. There were no obvious changes in the LSM treatment zone of the specimen following LTS and few micro constituents were found on the grain boundary. The refined grain microstructure formed during LSM treatment was preserved and remained almost unchanged during the following LTS treatment.

Identification of micro-constituents

As mentioned above, white particles with a complex shape were mainly present in the interdendritic spaces; their distribution in the interdendritic area indicates that these particles were formed after the crystallization of the γ matrix. According to a thermo dynamic calculation by thermo-calc, they were probably primary and eutectic MC carbides formed during SMAW, and remained almost unchanged during the following heat treatments.

Since there was no difference between the cooling rates at the core and in the interdendritic spaces, the difference in particle sizes could only be attributed to a difference in composition arising from microsegregation. The crystallization of γ dendrites quickly accumulated the micro constituent-forming elements in the residual liquids contained in the narrow interdendritic passages, and the segregation could not be alleviated efficiently by diffusion. In addition, the higher solidification rate also supplied a larger driving force for the carbide growth by a deeper supercooling. Therefore, once the carbide nucleated, it grew quickly along the residual liquid passages.

The result of energy dispersive X-ray (EDX) spectrum analysis of the white particles proves that these particles are rich in Nb compared with the matrix, as shown in Fig. 2.16. In the case of the chemical composition of Inconel 182, the content of Nb is 1.98%. Because Nb is an element which is easily rejected into interdendritic regions during solidification, it can be concluded that the white particles are probably primary and eutectic niobium carbides formed at the later stages of solidification. However, as Fig. 2.16 shows, the EDX spectrum also indicated that the chemical composition of the blocky particles is rich in Cr.

Because the heating and cooling speeds are extremely fast during laser beam irradiation, precipitation behavior of micro constituents deviates significantly from the equilibrium state. As shown in Fig. 2.14(d) and (e),

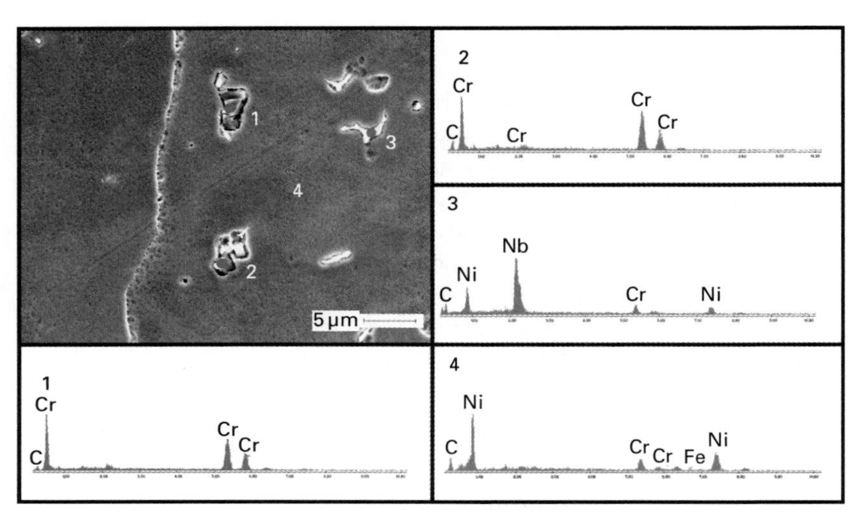

2.16 EDS spectrum of microconstituent after SMAW + SR treatment (1.92% Nb Inconel 182).

the sub-grains in the LSM treatment zone are predominantly of a refined cellular morphology.

According to thermo dynamic calculations, there are three possible carbides for Inconel 182: MC, M_7C_3 and $M_{23}C_6$. To identify the phase and crystal structure of the particles precipitated before LSM treatment, transmission electron microscopy (TEM) was used by means of an extraction replica method performed on the specimen before LSM treatment. One large precipitate is rich in Nb and identified as NbC (a = 0.441 nm) with a face-centered cubic (fcc) structure as shown in Fig. 2.17; the small granular precipitate is rich in Cr and identified as $M_{23}C_6$, as shown in Fig. 2.18.

According to the morphology of the micro constituents in Figs 2.14 and 2.15, the interdendritic particles with a complex shape should be primary or eutectic niobium carbide formed during the later stage of solidification. The small quantity of Cr-rich $M_{23}C_6$ should be the film-like intergranular precipitates. Therefore, for Inconel 182, the precipitates at the grain boundary should be Cr-rich $M_{23}C_6$ after various heat treatments.

Relationship between IGC/IGSCC susceptibility and precipitate coverage on the grain boundary

Because the precipitates at the grain boundary have a significant influence on the IGC/IGSCC susceptibility of the alloys, the precipitate coverage (density and distribution) after different heat treatments have been analyzed. Precipitate coverage at the grain boundary was evaluated from 10 scanning electron microscopy (SEM) micrographs selected randomly for observation

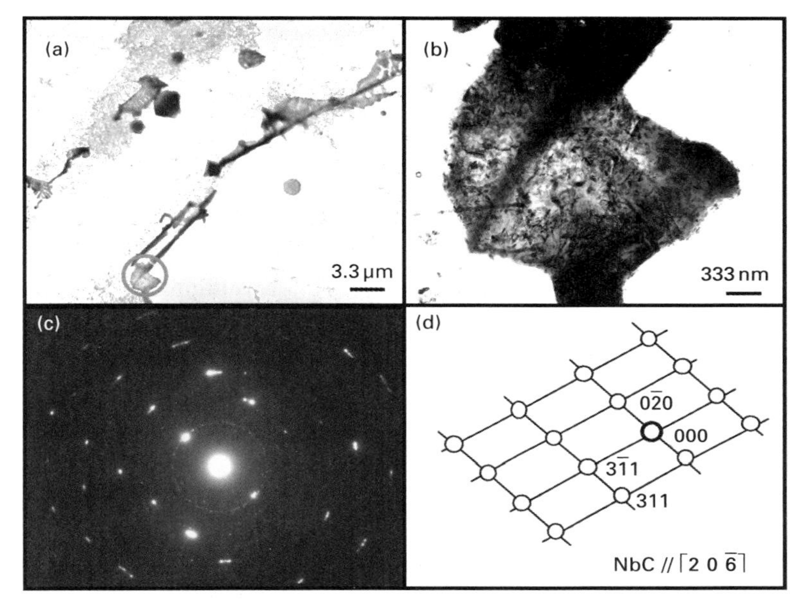

2.17 TEM micrographs of NbC from specimen of SMAW + SR + LTS (1.92% Nb Inconel 182): (a) bright field TEM image; (b) EDS spectrum; (c) selected area diffraction pattern (SADP); (d) key diagram.

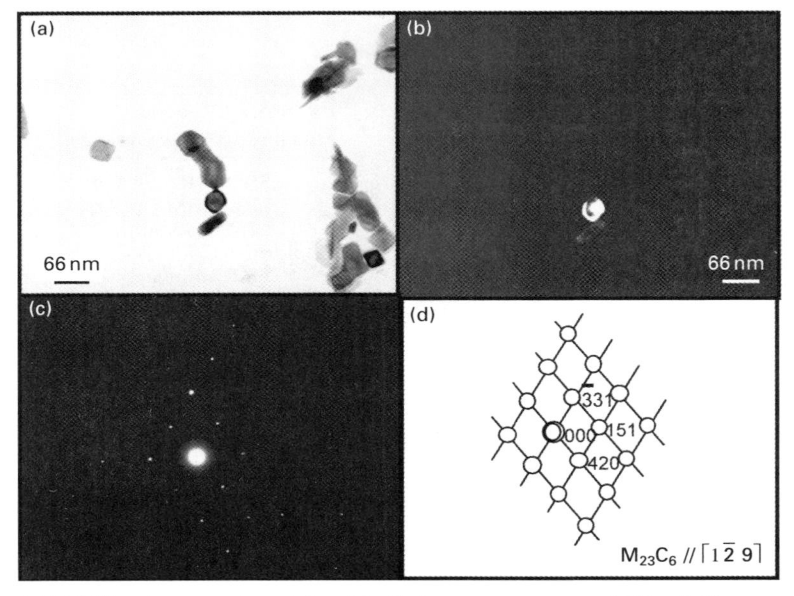

2.18 TEM micrographs of tiny $M_{23}C_6$ from specimen of SR + LTS (1.92% Nb Inconel 182): (a) bright field TEM image; (b) dark field TEM image; (c) SADP; (d) key diagram.

(taken at a magnification of × 5000). Linear coverage of carbides over these boundaries was measured.

A comparison of average precipitate coverage on the grain boundary with maximum IGC depth for samples of SMAW, SMAW + SR, SMAW + SR + LTS, SMAW + SR + LTS + LSM, SMAW + SR + LTS + LSM + LTS is shown in Fig. 2.19. It can be seen that a semi continuous precipitate distribution is most harmful for the ICC/IGSCC resistance of materials. This result means that SR heat treatment also degraded the SCC resistance at the same time as reducing the residual stress after welding. Since residual stress become very low after SR treatment, it can be concluded that microstructure plays a more important role than the residual stress in this situation.

2.4.2 Effect of niobium and carbon contents on IGC/IGSCC susceptibility

Streicher test

Yamauchi[13] has long argued that the Nb/C ratio is the significant parameter for IGC/IGSCC susceptibility of a material. Therefore, in the present study, the Nb/C ratio is also used as the evaluation parameter to study the influence of Nb and C content on IGC/IGSCC susceptibility of materials before and after LSM treatment. 1.92% Nb Inconel 182, 1.10% Nb Inconel 182 and Inconel 600 were used as the experimental materials. The maximum IGC

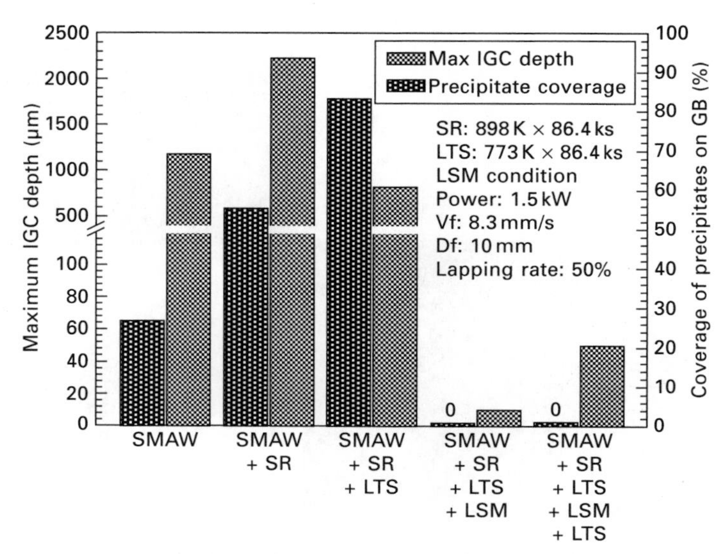

2.19 Coverage of precipitate on grain boundary (GB) after different heat treatments and LSM process.

depth of specimen measured after a Streicher test was used to evaluate the IGC/IGSCC susceptibility of materials.

Figure 2.20 shows the cross-sections of specimens under different heat treatment conditions after Streicher tests for all three materials. In all three cases, corrosion cracking is found to be of the intergranular type. For all three materials, the samples after LTS treatment show high IGC/IGSCC susceptibility. Both Inconel 182s (with different Nb content) have a similar IGC/IGSCC susceptibility, both of them being better than that of Inconel 600.

The results of the Streicher tests are shown in Fig. 2.21. The relationship between the IGC/IGSCC susceptibility and Nb/C ratio of materials after LSM treatment is also studied. For all of the materials, IGC/IGSCC resistance is improved after LSM treatment. According to the results from the Streicher tests, the material with a lower Nb/C ratio after LSM shows the worst IGC/IGSCC resistance. The specimens after LSM treatment following LTS treatment show the same tendency. The Inconel 600 after LSM treatment following LTS treatment, and with the lowest Nb/C ratio, shows the worst IGC/IGSCC resistance. The results imply that the Nb/C ratio should have a very significant effect on the IGC/IGSCC susceptibility of a material. The Inconel 182 with Nb/C ratio of 18.3 shows excellent IGC/IGSCC resistance after LSM treatment. Since the Nb/C ratio necessary to obtain high IGC/IGSCC resistance commonly needs to be higher than about 92,[13] it can be

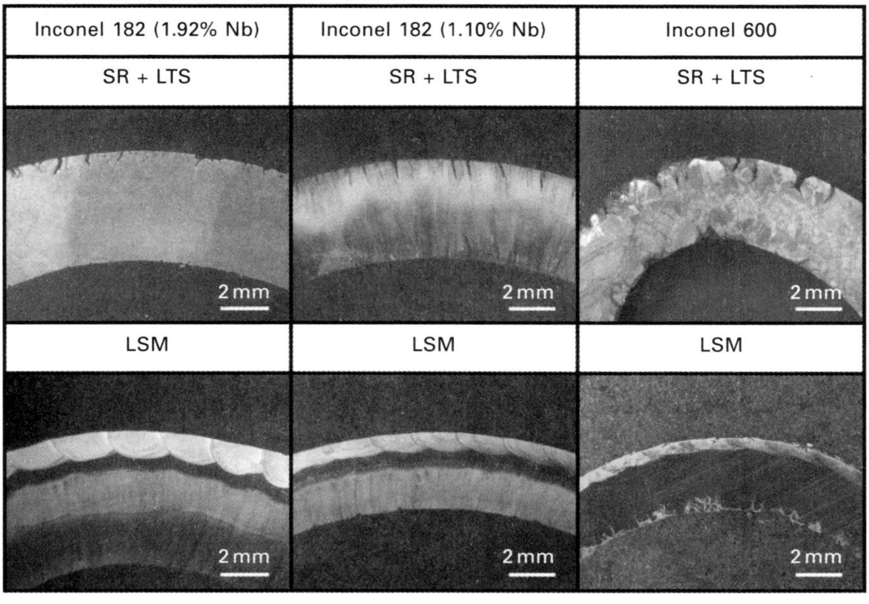

Inconel 182 (1.92% Nb)	Inconel 182 (1.10% Nb)	Inconel 600
SR + LTS	SR + LTS	SR + LTS
2 mm	2 mm	2 mm
LSM	LSM	LSM
2 mm	2 mm	2 mm

2.20 IGC of 1.92%Nb Inconel 182, 1.10%Nb Inconel 182 and Inconel 600 for heat treatments and LSM process by Streicher test.

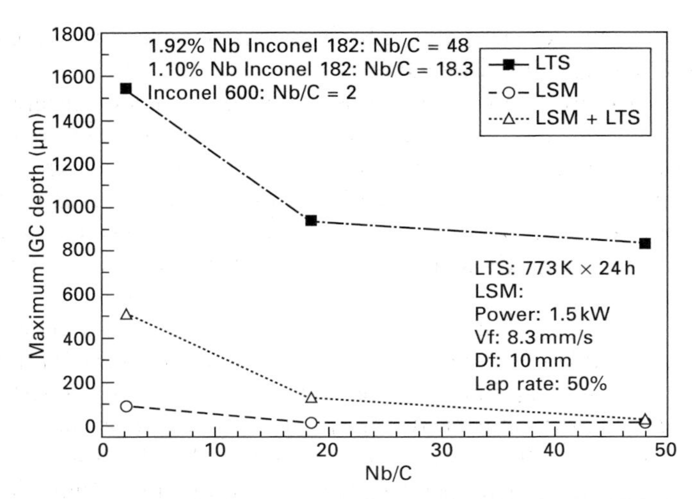

2.21 Relationship between Nb/C ratio and Maximum IGC depth with different heat treatment procedures.

concluded that LSM treatment can greatly improve IGC/IGSCC resistance even with a lower Nb/C ratio.

OBB test

The SCC susceptibility of three materials after LSM treatment following LTS treatment were also evaluated by an OBB test. For Inconel 600 with almost no Nb addition, SCC was found to occur earlier, as shown in Table 2.3. By contrast, the two kinds of Inconel 182 containing 1.92% Nb and 1.10% Nb show good SCC resistance, with no SCC being found until 3.6 Ms.

Intergranular carbide distribution before and after LSM treatment

The composition of grain boundaries of an alloy has a remarkable effect on IGC/IGSCC susceptibility, especially the precipitation of Cr-rich carbide at the grain boundary, as shown in Fig. 2.19. To test this, the intergranular precipitation of three materials after LTS treatment was examined carefully (Fig. 2.22). For all of the three materials, the precipitate distribution at the grain boundary after LTS treatment is nearly continuous. According to TEM results, for Inconel 182, precipitation at the grain boundary was verified as Cr-rich $M_{23}C_6$. According to the results from the literature,[6, 11] the precipitates of Inconel 600 at the grain boundary are probably Cr-rich $M_{23}C_6$ or M_7C_3.

As shown in Fig. 2.23(c), for the Inconel 600 LSM treatment specimen, no precipitate can be found on both the cellular dendrite and grain boundaries. But it should be noted that some tiny white particles are found on the grain

Table 2.3 SCC susceptibilities of Inconel 600 and Inconel 182s containing 1.92% Nb and 1.10% Nb with LSM treatment evaluated by OBB test

Materials	Heat treatments	Loading strain (%)	Time duration before SCC occurrence (h) 200 400 600 800 1000
Inconel 600	SR + LTS + LSM + LTS	1.0	IGSCC < 30 h
1.92% Nb Inconel 182	SR + LTS + LSM + LTS	1.0	No SCC > 1000 h
1.10% Nb Inconel 182	SR + LTS + LSM + LTS	1.0	No SCC > 1000 h

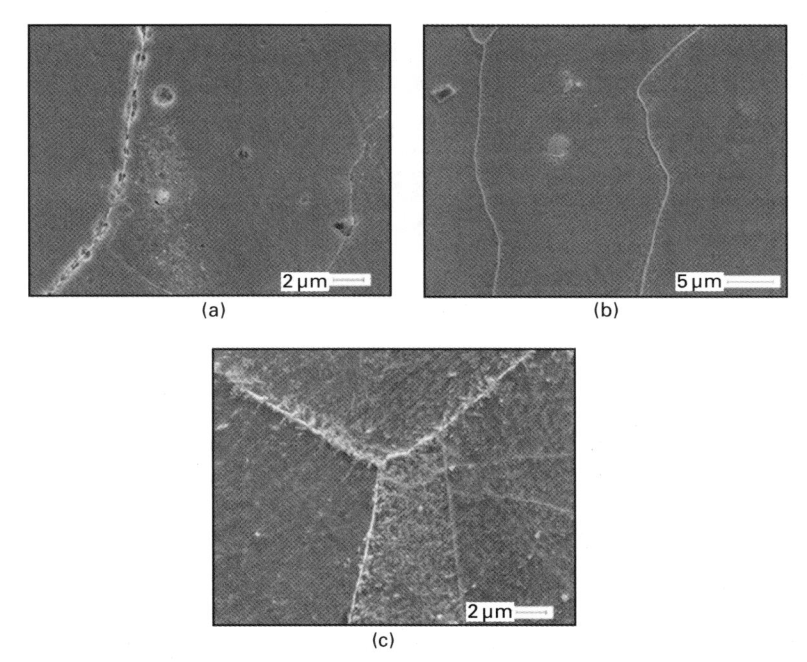

2.22 Intergranular carbide distribution along the grain boundary after LTS treatment: (a) 1.92% Nb Inconel 182; (b) 1.10% Nb Inconel 182; (c) Inconel 600.

boundary in the LSM treatment zone. Their size is smaller than 100 nm and they are difficult to identify by EDX using an SEM. Much research has been undertaken referring to the microstructure of the treatment zone of Inconel 600.[3-6, 11] However, in this case, no intergranular carbide was found at the grain boundary after LSM treatment and this is probably related to the higher heat input used in this present study. The tiny intergranular precipitation can be considered to be the reason why the Inconel 600 sample after LSM treatment showed worse IGC/IGSCC susceptibility.

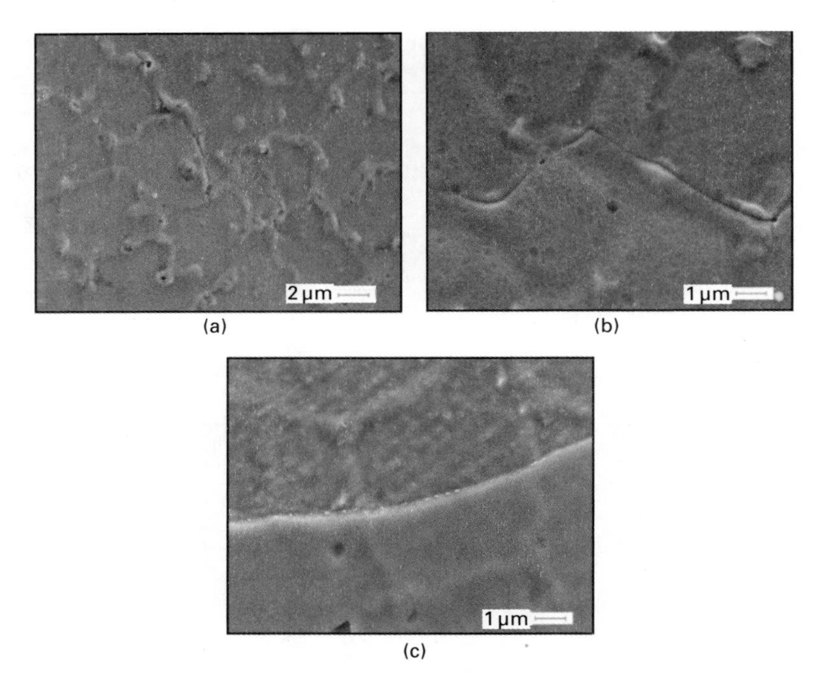

2.23 Microstructure comparison of LMZ after LSM processing: (a) 1.92% Nb Inconel 182; (b) 1.10% Nb Inconel 182; (c) Inconel 600.

However, for the two kinds of Inconel 182 with different Nb addition, no intergranular precipitate was found, as shown in Figs 2.23(a) and (b). The pre-existing continuous intergranular precipitates were melted by laser beam irradiation and did not have enough time to re-precipitate again during the extremely fast solidification procedure.

Figure 2.24 shows the microstructure of the LSM treatment zone following LTS treatment. Almost no variation could be found for the two kinds of Inconel 182 with different Nb content. No precipitate was found at the grain boundary. On the other hand, some intergranular precipitates were found in the LSM treatment zone for Inconel 600 following LTS treatment. These results indicate that the tiny precipitate found in the LSM treatment zone of Inconel 600 showed growth to some extent.

Relationship between IGC/IGSCC susceptibility and precipitation coverage at grain boundary

Because carbide precipitation at the grain boundary has a significant influence on the IGC/IGSCC resistance of alloys, the precipitate coverage on the grain boundary of three materials with different Nb/C ratio was studied, as shown in Fig. 2.25. For all three materials, almost continuous precipitate coverage

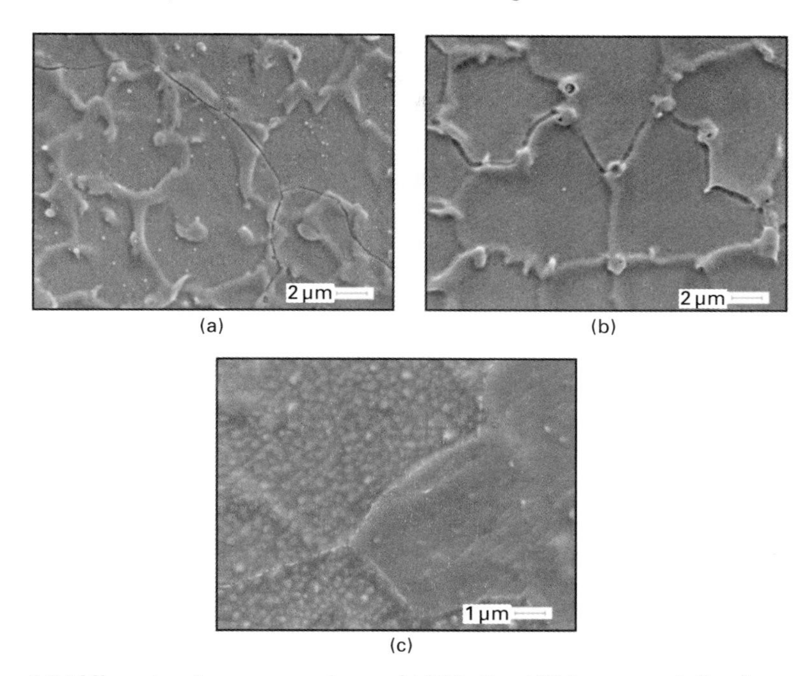

2.24 Microstructure comparison of LMZ after LSM process following LTS treatment: (a) 1.92% Nb Inconel 182; (b) 1.10% Nb Inconel 182; (c) Inconel 600.

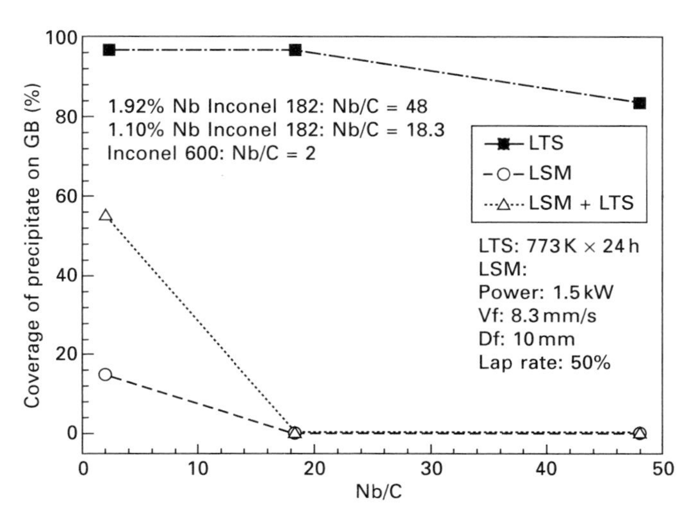

2.25 Comparison of coverage of precipitate on grain boundary with different Nb/C ratios.

is formed on the grain boundary after LTS treatment. Both kinds of Inconel 182 with different Nb content are free of intergranular precipitates after SMAW + SR + LSM and SMAW + SR + LSM + LTS. For Inconel 600, a higher intergranular precipitation coverage is found after SR + LSM + LTS following LSM treatment. Comparing the Streicher test results shown in Fig. 2.21 and the OBB test results shown in Table 2.3, it can be found that the higher Nb/C ratio induced lower precipitation coverage on the grain boundary resulting in higher IGC/IGSCC resistance.

2.4.3 Modeling of Cr depletion profiles near grain boundaries during heat treatment and LSM treatment

Modeling

In the case of the intergranular precipitation of Inconel 182 weld metal observed in the previous sections, the intergranular carbides commonly show semi continuous or continuous morphology. In our model, the carbide type was assumed to be Cr-rich $M_{23}C_6$, as shown in Fig. 2.18. Generally, Cr-rich $M_{23}C_6$ at the grain boundary results in Cr depletion near the grain boundary; therefore the Cr-rich carbide was assumed to precipitate uniformly at the grain boundaries and the diffusion problem was treated over a wide temperature range as a one-dimensional model and the Cr depletion phenomena was estimated by calculation.

The precipitation calculation was very complex and time consuming, and was therefore restricted to an alloy with the following four major components: Fe, C, Cr and Ni, which are the most important elements for $M_{23}C_6$ precipitation. Additionally, only two phases (matrix and carbide) were considered. The calculated area was near the grain boundary because the width of carbide is supposed to be very thin at the grain boundary. The precipitation process can commonly be divided into three stages: nucleation, growth and coarsening. The coarsening process requires a long time close to equilibrium conditions, and so is not considered in this study. Similarly, since the nucleation process is quite complex and difficult to be simulated, no attempt has been made to account for the carbide nucleation or incubation time; instead, the carbide is assumed to nucleate instantaneously. In the present model, only the growth process is considered. Rather than considering the growth of carbide from size zero, a finite size (3 nm) was assumed as the initial carbide size for calculation. The matrix size was assumed to be 3 and 0.5 μm for the specimens, before and after the LSM treatment, respectively. The simulation was carried out using the DICTRA package, linked with Thermo-Calc, and it was solved by applying a numerical procedure for a system of coupled parabolic partial differential equations.[19, 20] The numerical procedure is based on the Galerkin method for space discretization and on the Gaussian elimination technique with incomplete factorization to solve the algebraic equations. Time integration can be chosen to be implicit, explicit or trapezoidal.

The following simplifying assumptions were used in the model. The simulation was based on local equilibrium at the moving phase interface, which meant the compositions of both the matrix and carbide adjacent to the phase interface comply with the thermodynamic equilibrium. Carbon was considered as the interstitial diffusing species and the other elements were taken into account as substitutional species. Carbide growth was controlled by the diffusion of the substitutional components. The rate of transformation was controlled only by the component transport near the interface, thus we treated the diffusion problem for the heat treatments and the LSM process as a one-dimensional case.

In a multicomponent system, the diffusion of species was controlled by the diffusion equation shown below:

$$\frac{\partial c_k}{\partial t} = \frac{\partial}{\partial x}(-J_k) \qquad [2.1]$$

where c_k and J_k are the concentration and the diffusive flux of component k, respectively. The flux of species J_k corresponded to the Fick–Onsager law (as shown in Equation 2.2), which meant that the flux of species was the function of the diffusion coefficient and concentration gradient:

$$J_k = -\sum_{j=1}^{n-1} D_{kj}^{n} \nabla_{cj} \qquad [2.2]$$

where n is the species number and species n is dependent; D_{kj}^{n} is the diffusion coefficient matrix, and ∇_{cj} is the concentration gradient for species j. The diffusion coefficient matrix D_{kj}^{n}, which was used for the simulations, was obtained by both kinetic and thermodynamic databases. The thermodynamic matrix was obtained from the Thermo-Calc with Nickel database, which was based on the minimization of the total Gibbs free energies of the individual phases in the system and on the calculation of phase diagrams approach (CALPHAD). The diffusion data were taken from the DITRA databank of MOB2. After setting up the initial conditions for simulation (carbide and matrix compositions, temperature), the diffusion coefficient matrix varied during the simulation, which was determined by DICTRA according to the corresponding variation of compositions of matrix and temperature during the calculation. The extent of diffusion of various species in the carbides were still unknown and thus the carbides were therefore treated as non-diffusion phases.

Calculation of Cr concentration distribution near the intergranular carbide

Since the results of experimental data from the literature and those calculated using the current model showed good agreement, the current model was applied to our practical conditions. Using the thermal cycle shown in Fig. 2.2, Cr depletion profiles during practical heat treatments and LSM treatment were calculated.

For the diffusion calculation of the LSM treatment, a thermal cycle calculated using a finite element method (FEM) analyzing model[21] was used. In a similar way to SMAW, only a single pass LSM treatment was taken into account, without considering the reheat effect by sequential LSM treatment. As is the case with SMAW, LSM treatment induces melting and solidification procedures which are very complex and difficult to calculate by numerical simulation. It was also assumed that the precipitation of Cr-rich carbide usually occurs from the equilibrium precipitation temperature of $M_{23}C_6$ (calculated by Thermo-Calc) during the cooling procedure for LSM treatment, and no consideration was made for the microsegregation which occurred during the treatment. Since almost no NbC was found in the LSM treatment zone, no consideration of NbC was made for the LSM treatment. The lowest precipitation temperature selected for the LSM treatment was 773 K, which was the same as the LTS temperature used. The calculation results are shown in Fig. 2.26. For specimens after LSM treatment, the Cr depletion zone is very narrow, whereas it becomes very large after SR treatment.

Evaluation of Cr depletion profiles

A depletion parameter was adopted to quantitatively characterize the amount of material surrounding a grain boundary depleted of Cr. Since the minimum Cr concentration near the grain boundary is very important for the IGC/IGSCC susceptibility of a material and the Cr depletion zone must possess a minimum width for IGC/IGSCC occurrence, the depletion area[22] under critical Cr concentration (with consideration to both minimum Cr content and Cr depletion width) was used to evaluate the Cr depletion of Inconel 182 under different heat treatments and under LSM treatment (see Fig. 2.27). It has been reported that Inconel 182 is susceptible to IGSCC/IGC when the grain boundary Cr level is below approximately 12 mass%.[23] Thus, in the present study, the critical Cr concentration of the depletion area for Inconel 182 was assumed to be 12 mass%. The depletion area for the different heat treatments and LSM treatment was calculated from the simulation results above, and the evaluation results are summarized in Fig. 2.28. According to the calculation results, the Cr depletion area is largest after SR treatment,

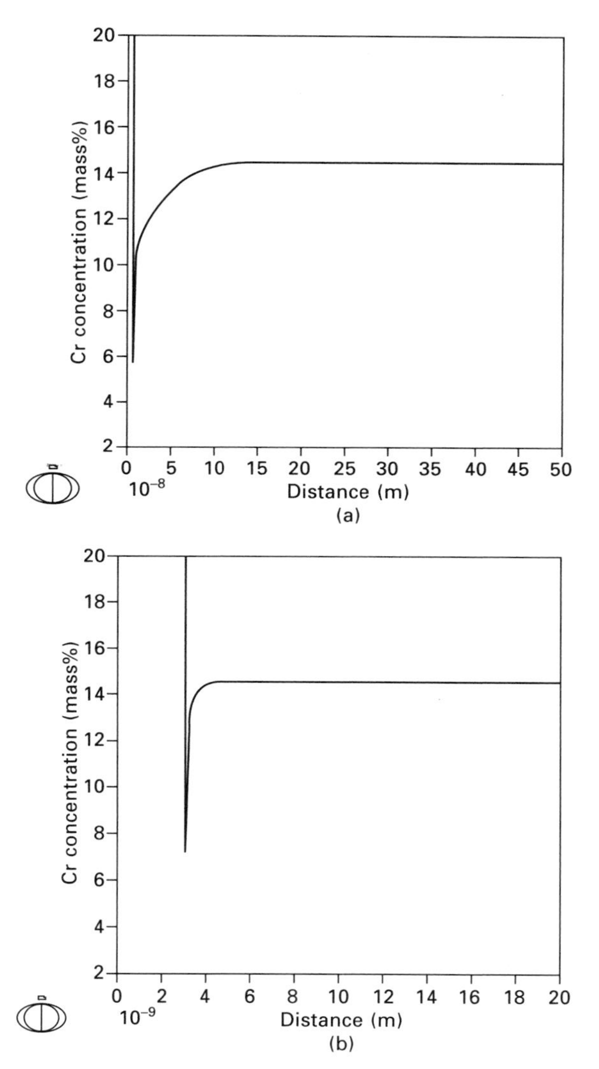

2.26 Cr depletion profiles of Inconel 182 for heat treatment and LSM process: (a) SMAW + SR; (b) LSM.

while it becomes smaller during subsequent LTS treatment. By contrast, the Cr depletion areas for SMAW, LSM and LSM + LTS treatments are quite small.

Discussion

Taking the effects of NbC into account, the Cr depletion profiles of Inconel 182 were calculated and its IGC/IGSCC resistance under the different heat

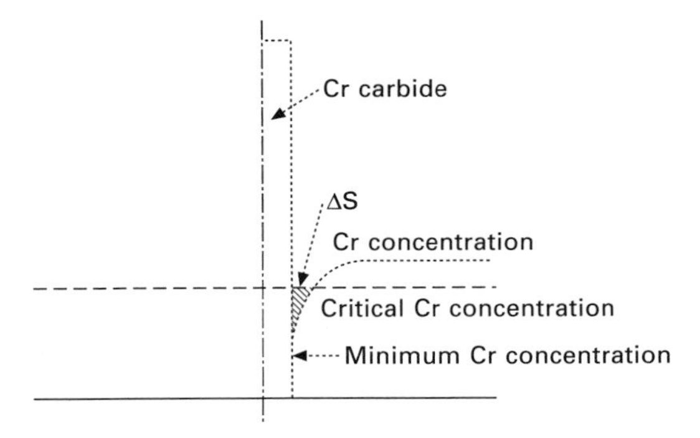

2.27 Evaluation Cr depletion zone around the Cr carbide.

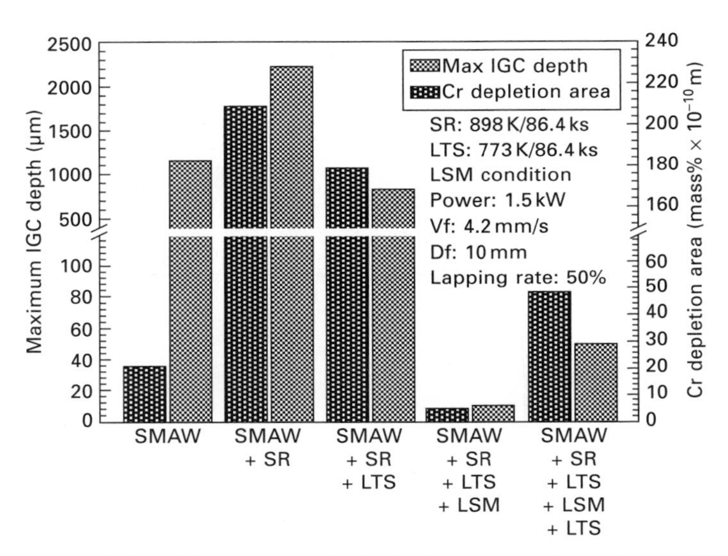

2.28 Relationship between Cr depletion area and maximum IGC depth.

treatments and LSM treatment could be predicted very well. Since the complex influences of microstructural and metallurgical factors (carbide nucleation, dislocations, etc.) were not included in the current model, the composition of the matrix and the heat treatment or processing parameters (temperature and time duration) were the most significant factors which determined the final Cr depletion profiles.

The SMAW and LSM treatments show a small Cr depletion area due to the rapid cooling rate and short time duration in the sensitization temperature range. In particular, LSM treatment shows the smallest Cr depletion area due to its extremely fast cooling rate during laser processing, although the C concentration in the matrix for the SMAW is lower. The results for SR treatment provides the largest Cr depletion area, most likely due to the longest duration at the sensitization temperature of the material. The Cr depletion area after LTS treatment is smaller than that after SR treatment, which implies that there is a recovery of Cr concentration in the Cr depletion zone near the Cr-rich carbide precipitates; Cr depletion is reduced during the LTS treatment following the SR treatment.

These results agree with the IGC/IGSCC susceptibilities evaluated by the Streicher test as shown in Fig. 2.28. Since the Streicher test is sensitive to the extent of the Cr depletion region, we can conclude that the depletion areas calculated and the IGC/IGSCC susceptibilities show good agreement. Heat treatments inducing larger Cr depletion areas always give poor IGC/IGSCC susceptibility. The analysis and calculations above also demonstrate that, with the present model, it is possible to predict Cr depletion as well as the corresponding IGC/IGSCC susceptibility of Inconel 182 for different heat treatment conditions.

2.4.4 SCC susceptibility evaluation and SCC sealing by LSM treatment

The LSM procedure was carried out using the optimal parameters achieved above. The SCC susceptibility of specimens before and after the LSM process was evaluated by an OBB test. As shown in Table 2.4, for the specimen after LTS heat treatment, SCC was found to occur earlier. The specimen after LSM treatment, however, showed good SCC resistance, and no SCC was found until 3.6 Ms.

The specimen with SCC was prepared, and the LSM sealing treatment was carried out on the specimen surface; a penetration test was then carried out to check for the occurrence of SCC. Figure 2.29 shows the results of the penetration test before and after LSM treatment. Pre-existing SCC before

Table 2.4 Comparison of SCC susceptibilities of Inconel 182s containing 1.92% Nb with and without LSM and repaired sample by LSM treatment evaluated by OBB test

Materials	Heat treatments	Loading strain (%)	Time duration before SCC occurrence (ks)				
			720	1440	2160	2880	3600
1.92% Nb Inconel 182	SR + LTS	1.0	IGSCC < 172.8				
1.92% Nb Inconel 182	SR + LTS + LSM	1.0			No SCC > 3600		
1.92% Nb Inconel 182	SCC sealing	1.0			No SCC > 3600		

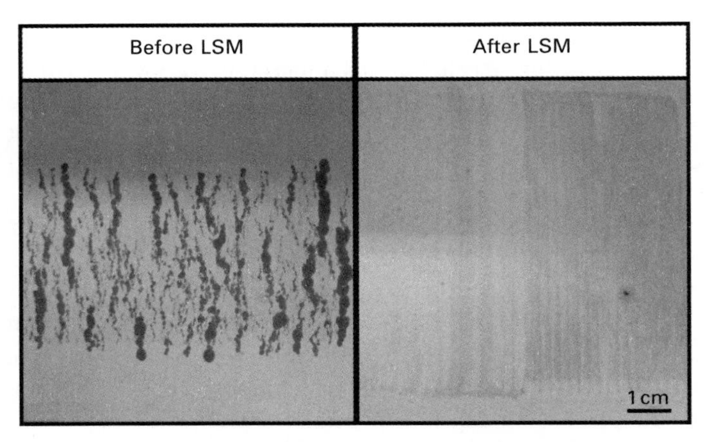

2.29 Results of penetration test before and after LSM sealing process in parallel direction of SCC occurrence (1.92% Nb Inconel 182).

LSM treatment was melted and sealed by the laser irradiation. No SCC were found in the LSM treatment zone.

An OBB test was also conducted to check the effect of SCC sealing by the LSM procedure. As shown in Table 2.4, no SCC was found on the specimen surface after an LSM sealing treatment up to 3.6 Ms. The sample after SCC sealing by LSM treatment still showed good SCC resistance, which also proves the possibility of SCC sealing by LSM treatment.

2.5 Effect of residual stress on stress corrosion cracking (SCC) susceptibility

2.5.1 FEM modeling

To consider the effect of residual stress, FEM analysis was used to simulate the residual stress distribution of samples after LSM treatment. A sequential thermal-stress couple calculation was carried out using an ANSYS FEM code. A 2-dimensional cross-section model of generalized plane strain elements was established to compute the local residual stress distribution after LSM treatment.[21]

Since no data concerning the material properties of Inconel 182 could be obtained, the specimen material was assumed to be Inconel 600 (since the material properties of Inconel 182 are quite similar to Inconel 600). Both thermal analysis and mechanical analysis were conducted using temperature-dependent material properties. Under thermal analysis, both convection and radiation were considered as an initial boundary condition, whilst the Von-Mises yield criterion and associated flow rules and linear kinematic hardening rule were assumed in the structural analysis. Both the thermal and stress analyses were verified by experimental measurements. Figure 2.30 presents a schematic diagram of the laser surface melting experimental system and the direction of the stress distribution.

Figure 2.31 shows a comparison between the analytically and experimentally obtained temperature distributions of a single pass LSM treatment. The black lines are fusion boundary lines measured by experiment. The fusion boundary curve calculated demonstrates good correlation with the welding pool shape measured, which indicates that the temperature distribution of the LSM treatment can be simulated very well using the current model.

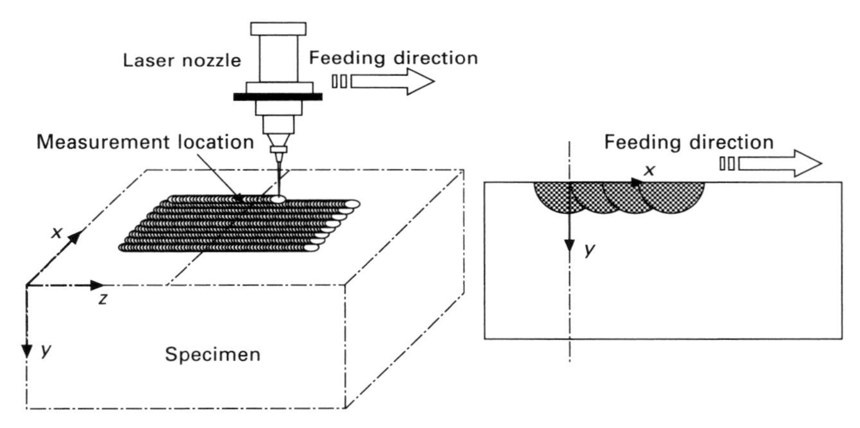

2.30 Schematic of weld specimen and LSM process, location of residual stress measurement.

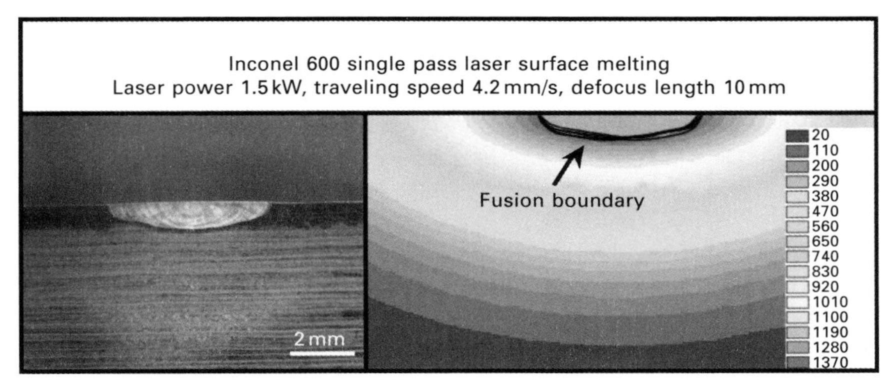

2.31 Molten pool comparison between experimental measurement and FEM calculation.

2.5.2 Factors affecting residual stress distribution

Figure 2.32 shows a comparison between calculated and measured residual stress distribution on the specimen surface of after multi-pass LSM treatment (11 passes). According to the simulation results, maximum transverse residual stress is about 460 MPa, and maximum longitudinal residual stress is about 510 MPa. Neither of these exist in the weld bead itself but can be found in the base metal near the weld bead. Whilst the value calculated is bigger

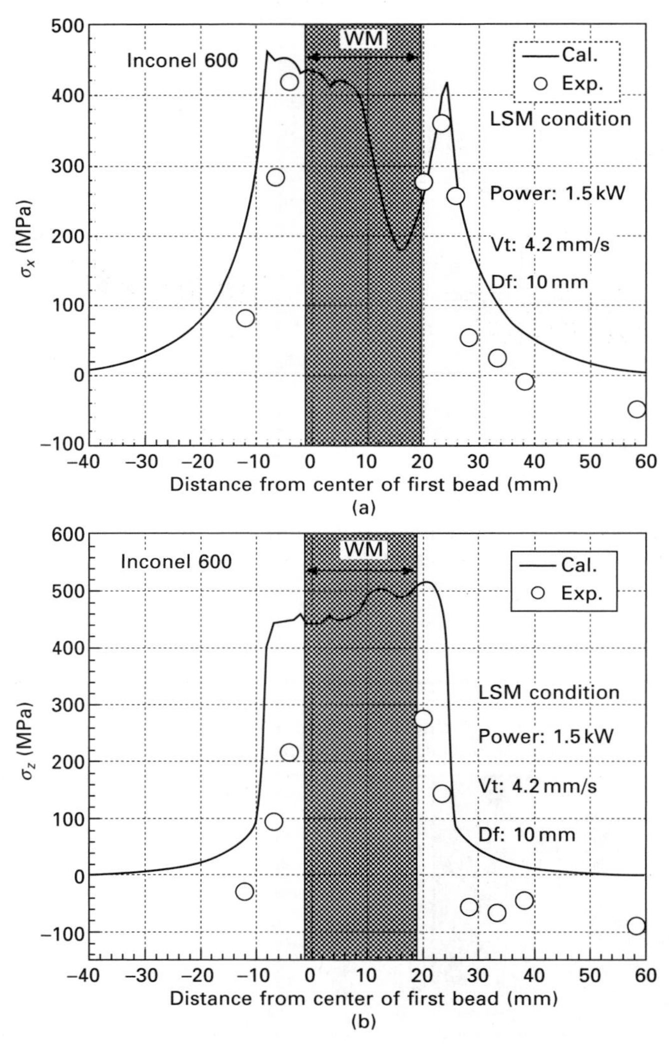

2.32 Residual stress distribution comparison between experimental data and FEM simulation.

than that found from experimental measurement (because of the assumption of plane strain used which results in a strict constraint in the direction of welding in the 2-dimensional model), the tendency of it generally agrees with the measured data; the stress calculated shows good coherence with the experimental data. This good agreement suggests that the current thermal-stress coupled model could be used as a tool for a parametric analysis and study of LSM treatment.

The calculation results in Fig. 2.32 prove that the maximum transverse residual stress σ_x is about 460 MPa, and the maximum longitudinal residual stress σ_z is about 510 MPa on the specimen surface for Inconel 182, and neither of them exists in the weld bead but rather in the base metal near the weld bead. Kim and Moon[7] suggest that the higher stress resulted in deeper SCC penetration and that the tensile residual stress would seriously weaken the ability of material corrosion resistance; thus, the high residual stress existing in the material in the heat-affected zone (HAZ) or base metal near the weld bead would make it quite susceptible to SCC.

Figure 2.33 shows SCC occurrence in 1.10% Nb Inconel 182 after immersing it in a caustic solution for 86.4 ks with no external loads applied. No SCC occurrence is found in the LSM treatment zone; all the SCCs are located in the HAZ or base metal near the LSM treatment zone. Additionally, almost all the SCCs extend in the direction parallel to welding, as a consequence of the tensile residual stress acting in the vertical direction. This implies that SCC occurrence should have some significant relationship with transverse residual stress and with strain distribution. Therefore, we must consider the method by which the tensile stress in the vertical direction in the HAZ or the base metal is reduced.

2.6 Conclusions

The purpose of this study was to develop a practical technology for repairing the occurrence of SCC after LSM treatment of Inconel 182 weld metal, which

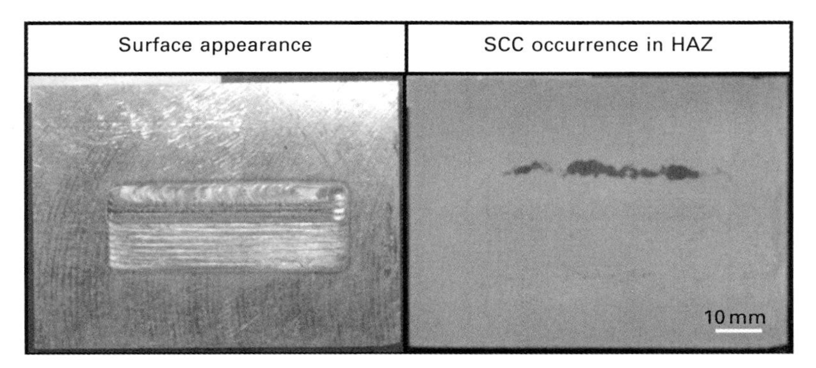

2.33 SCC occurrence in HAZ of LSM specimen.

has been used for a number of decades as a weld material in parts of nuclear power plants. In order to achieve optimized LSM treatment conditions for the repairing procedure, the effects of LSM treatment conditions on weld bead shape, penetration depth and welding defects were examined. Based on experimental results, optimum LSM treatment conditions were obtained and used for the sealing procedure of SCC occurrence.

The corrosion resistances of the LSM treatment zone were evaluated. As a consequence, research was undertaken to analyze the factors affecting the IGC/IGSCC susceptibility from both the microstructural and metallurgical viewpoints. The effects of different heat treatments during the repairing procedure on microstructure evolutions, and the relationship between microstructure and IGC/IGSCC susceptibility were discussed. Heat treatment conditions have significant influence on IGC/IGSCC susceptibility of materials. For Inconel 182, the SR treatment induced the worst IGC/IGSCC susceptibility. The specimen after SMAW + SR + LTS treatment was a bit better than that after SR treatment alone. The specimens undergoing LTS treatment following LSM treatment showed excellent IGC/IGSCC resistance.

A study was then carried out to analyze the effect of chemical composition on microstructure and corresponding IGC/IGSCC resistance for materials with different Nb/C ratios. Two Inconel 182s with higher Nb/C ratio showed better IGC/IGSCC resistance than that of Inconel 600. Since it is well known that the precipitation kinetics of chromium carbide plays a very important role in determining the IGC/IGSCC susceptibility of materials, a model to simulate the Cr concentration profile of Inconel 182 for the heat treatments and LSM treatment were presented, and the Cr concentration profiles adjacent to the grain boundary based on conventional diffusion equations were calculated by numerical methods. Finally, the IGC/IGSCC susceptibility of Inconel 182 during the heat treatments and LSM treatment was evaluated and compared by depletion parameter. As for the calculation, the sample after SMAW + SR treatment exhibited the maximum Cr depletion area, whilst that following LTS treatment with recovery of IGC/IGSCC resistance showed a slightly smaller Cr depletion area; the sample after LTS treatment following LSM treatment showed a smaller Cr depletion area, whilst the sample after LSM treatment showed the smallest Cr depletion area.

However, the presence of tensile residual stresses in the weld area has been identified as another significant factor leading to the occurrence of SCC, and SCC mainly occurs in the region where residual stresses due to manufacturing are highest. A 2-dimensional FEM model was created to predict the temperature and residual stress distribution of the LSM treatment of Inconel 600 and Inconel 182 on the basis of the model's good agreement with experimental measurements. According to the calculation of residual stress, the high residual stress existing in HAZ or the base metal near the LSM treatment zone make it quite susceptible to SCC. From the corrosion test, it was apparent that there was no SCC occurrence found in the LSM treatment zone; all the SCCs were located in the HAZ or base metal near the LSM treatment zone.

Given good IGC or IGSCC resistance in the LSM treatment zone, future research could concentrate on other methods which are expected to improve SCC behavior, especially for material in the HAZ or base metal near the weld bead, using a combination of high amplitude and deep compressive residual stress fields.

2.7 References

1. R.M. Horn, G.M. Gordon, F.P. Ford and R.L. Cowan: Experience and assessment of stress corrosion cracking in L-grade stainless steel BWR internals, *Nuclear Engineering and Design* **174** (1997) 313–325.
2. P.H. Hoang, A.G. Gangadharan and S.C. Ramalingam: Primary water stress corrosion cracking inspection ranking scheme for alloy 600 components, *Nuclear Engineering and Design* **181** (1998) 209–219.
3. Y.S. Lim, H.P. Kim, J.H. Han, J.S. Kim and H.S. Kwon: Influence of laser surface melting on the susceptibility to intergranular corrosion of sensitized Alloy 600, *Corrosion Science* **43** (2001) 1321–1335.
4. J.K. Shin, J.H. Suh, J.S. Kim and S.J. Kang: Effect of laser surface modification on the corrosion resistance of Alloy 600, *Surface and Coating Technology* **107** (1998) 94–100.
5. J.H. Suh, J.K. Shin, S.J. Kang, Y.S. Lim, I.H. Kuk and J.S. Kim: Investigation of IGSCC behavior of sensitized and laser-surface-melted Alloy 600, *Materials Science and Engineering* **A254** (1998) 67–75.
6. Y.S. Lim, J.H. Suh, I.H. Kuk and J.S. Kim: Microscopic investigation of sensitized Ni-base Alloy 600 after laser surface melting, *Metallurgical and Materials Transactions A* **28A** (1997) 1223–1231.
7. J.D. Kim and J.H. Moon: C-ring stress corrosion test for Inconel 600 and Inconel 690 sleeve joint welded by Nd:YAG laser, *Corrosion Science* **46** (2004) 807–818.
8. W. Kono, S. Kimura, S. Kawano and R. Sumiya: Laser de-sensitization heat treatment for inside surface of SUS304 stainless steel pipe welds, 7th International Conference on Nuclear Engineering, Tokyo, Japan, 1999.
9. K.P. Cooper, P. Slebodnick and E.D. Thomas: Seawater corrosion behavior of laser surface modified Inconel 625 alloy, *Materials Science and Engineering* **A206** (1996) 138–149.
10. J.D. Kim, C.J. Kim and C.M. Chung: Repair welding of etched tubular components of nuclear power plant by Nd:YAG laser, *Journal of Materials Processing Technology* 114 (2001) 51–56.
11. Y.S. Lim, J.S. Kim, and H.S. Kwon: Effects of sensitization treatment on the evolution of Cr carbides in rapidly solidified Ni-base Alloy 600 by a CO_2 laser beam, *Materials Science and Engineering* **A279** (2000) 192–200.
12. G. Posch and J. Tosch: Corrosion behavior of high alloyed stainless steel weld metals, joints and overlays in standardized corrosion tests with some examples, IIW Doc. II-C-283-04 (2004).
13. K. Yamauchi: Improvement of stress corrosion cracking resistance of Ni–Cr–Fe

alloy 600 and its weld metal for nuclear power plants, PhD thesis, Osaka University, Osaka, Japan, 1988 (in Japanese).

14. G. Bao, K. Shinozaki, S. Iguro, M. Inkyo, M. Yamamoto, Y. Mahara and H. Watanabe: Stress corrosion cracking sealing in overlaying of Inconel 182 by laser surface melting, *Journal of materials Processing Technology* **173** (2006) 330–336.
15. G. Bao, K. Shinozaki, S. Iguro, M. Inkyo, Y. Mahara and H. Watanabe: Influence of heat treatments and chemical composition on SCC susceptibility during repairing procedure of overlaying of Inconel 182, *Science and Technology of Welding and Joining* **10**(6) (2005) 706–716.
16. Y.S. Lim, H.P. Kim, J.H. Han, J.S. Kim and H.S. Kwon: Influence of laser surface melting on the susceptibility to intergranular corrosion of sensitized Alloy 600, *Corrosion Science* **43** (2001) 1321–1335.
17. J.H. Suh, J.K. Shin, S.J. Kang, Y.S. Lim, I.H. Kuk and J.S. Kim: Investigation of IGSCC behavior of sensitized and laser-surface-melted Alloy 600, *Materials Science and Engineering* **A254** (1998) 67–75.
18. T.Y. Kuo and H.T. Lee: Effects of filler metal composition on joining properties of alloy 690 weldments, *Materials Science and Engineering* **A338** (2002) 202–212.
19. A. Borgenstam, A. Engstorm, L. Hoglund and J. Agren: DICTR, a tool for simulation of diffusional transformations in alloys, *Journal of Phase Equilibria* **21** (2000) 269–280.
20. G. Bao, M. Yamamoto and K. Shinozaki: Precipitation and Cr depletion profiles of Inconel 182 during heat treatments and laser surface melting, *Journal of Materials Processing Technology*, **209** (2009), 416–425.
21. G. Bao, S. Iguro, M. Inkyo, K. Shinozaki, M. Yamamoto, Y. Mahara and H. Watanabe: Repair of stress corrosion cracking in overlaying of Inconel 182 by laser surface melting, *Welding in the World* **49** (7/8) (2005) 37–44.
22. K. Nishimoto, H. Mori and Y. Nakao: Consideration on accelerated heat-treating condition for low temperature sensitization in weld metals of austenitic stainless steels, *Quarterly Journal of the Japanese Welding Society*, 14 (4) (1996) 703–708.
23. C.L. Briant and E.L. Hall: The microstructural causes of intergranular corrosion of Alloy-82 and Alloy-182, *Corrosion* **43** (9) (1987) 539–548.

3 Laser surface melting (LSM) of stainless steels for mitigating intergranular corrosion (IGC)

W. K. CHAN, C. T. KWOK and K.H. LO,
University of Macau, China

Abstract: This chapter reviews laser surface melting (LSM) of aged stainless steels for mitigating intergranular corrosion (IGC) by various researchers. In addition, the effect of LSM on IGC behavior of aged austenitic stainless steels (UNS S30400, S34700 and FeCrMn) and aged duplex stainless steels (UNS S31803 and S32760) is investigated.

Key words: intergranular corrosion (IGC), laser surface melting (LSM), Ni-free manganese stainless steel, Nb-stabilized stainless steel, super duplex stainless steel (SDSS).

3.1 Introduction

Austenitic stainless steels (ASSs) and duplex stainless steels (DSSs) are FeCrNi alloys which have been extensively applied in various industrial sectors including marine, chemical and petrochemical industries, paper production, desalination and nuclear plants, civil engineering and architecture because of their ease of weldability and formability, high corrosion resistance, high mechanical strength and ductility. However, sensitization renders these stainless steels susceptible to intergranular corrosion (IGC) attack, which leads to premature failure of the engineering components. The sensitization temperature (in the range of 450 to 850 °C) can be reached during fabrication processes of hot-worked parts, welding, isothermal treatment of components for stress relief, and also slow cooling from higher temperatures (for example, solution annealing or shut down of a plant operated at high temperatures) [1].

For single-phase ASSs, IGC is caused by sensitization due to the formation of Cr-rich carbides along the grain boundaries and the concurrent depletion of Cr in the immediate vicinity. If the local Cr content drops below 12 wt%, the Cr-depleted zones become prone to localized corrosion in chloride and caustic environments [2]. Stabilized ASSs have been widely used in applications where IGC has been a concern. Stabilization is achieved by Ti in UNS S32100, while Nb is used in UNS S34700. Ti and Nb are both known to be strong carbide formers and can prevent the formation of Cr carbides at and near grain boundaries with Cr depletion. They also play an important role in precipitation reactions, contributing to creep strength at elevated temperature. However, the high cost of ASSs reduces their applications and stems from their relatively high Ni content. Ferritic stainless steels (FSSs) do not contain Ni and are thus cheaper than ASSs; however, they suffer from poor weldability.

The combination of austenite (γ) and ferrite (δ) results in duplex stainless steels (DSSs) which combine the corrosion resistance and weldability of ASSs with the lower cost of FSSs. DSSs consist of a dual-phase microstructure with a weight fraction of about 50% δ and 50% γ. Owing to the presence of δ, the mechanical strength of DSSs is about twice that of ASSs. Moreover, DSSs possess excellent corrosion resistance to chloride-containing environments by virtue of their high Cr and Mo contents. Super duplex stainless steels (SDSSs), a sub-group of DSSs, have been developed recently with pitting resistance equivalent number (PREN) greater than 40 and excellent mechanical properties. Nevertheless, DSSs and SDSSs are susceptible to sensitization at 600 to 950 °C arising from the precipitation of the Cr-rich and Mo-rich intermetallic sigma phase (σ). The presence of a σ phase in DSSs and SDSSs results in the deterioration of corrosion and mechanical properties. Deng and his co workers reported that a drastic reduction in critical pitting temperature and impact energy for S31803 aged at 850 °C for 10 minutes is mainly attributed to σ phase precipitation, while the hardness is almost unchanged [3]. However, DSSs and SDSSs do contain some Ni and are thus still quite expensive. Replacing Ni with Mn would reduce the cost of stainless steels. In fact, the Ni-free FeCrMn stainless steels are inferior to the 300 series with respect to general corrosion resistance especially under reducing conditions [4]. Cr and Ni have beneficial effects on the corrosion resistance of stainless steels, while Mn is generally seen to have a detrimental effect on their general corrosion resistance [5]. Reports on IGC of the FeCrMn stainless steels are rare in the literature.

3.2 Merits of laser surface melting (LSM)

Conventionally, sensitization can be avoided by high-temperature solutionizing and rapid cooling through the sensitization range. However, such a treatment is not always possible and is impracticable for bulk components used in the nuclear and chemical industries. Also, large thermal stresses may be introduced due to such rapid quenching. In this case, an *in situ* method is necessary to selectively remove the sensitized microstructure at critical locations. Laser surface melting (LSM) is a feasible method for eliminating sensitization and sealing cracks for technical applications [6, 7]. Compared with other surfacing methods, LSM has derived its attractiveness and advantages in engineering applications:

- It is a simple, economical and efficient rapid solidification process for the surface of materials leading to extended solid solution of the alloy system, formation of metastable phases, homogenization and refinement of microstructure, and dissolution/redistribution of precipitates or inclusions while the bulk properties can be preserved.
- The hardness, corrosion, wear and fatigue resistance can be enhanced by refinement and homogenization of microstructure.
- After LSM, the heat-affected zone (HAZ) formed is small; it leaves the bulk properties unchanged and minimizes the effect of distortion.
- No additional elements need to be on the surface and the final surface obtained has a good chemical cleanliness.
- The laser beam does not physically touch the workpiece, there is no force exerted on the part, and also the magnetic field, radiation field and air will not affect the beam.
- It is relatively easy to control the processing parameters (such as power, speed and beam size) by automation.
- After laser surfacing, the post-machining process can be minimized.

3.3 Laser surface modifcation of stainless steels for mitigating intergranular corrosion (IGC)

To solve the problem caused by sensitization, LSM of stainless steels for mitigating IGC has been investigated by several researchers in the three decades as shown in Table 3.1. The earliest work on LSM of sensitized 304 ASS was carried out by Anthony and Cline [8]. Strauss tests proved that the sensitized 304 after LSM was completely resistant to IGC. However, it was observed that intergranular attack spread rapidly through a gap which was accidentally made between the overlapping laser passes on the surface of the sensitized 304. In addition, mechanical testing at strains less than 15% showed LSM to indefinitely extend specimen life in a stress corrosion environment. However, at strains greater than 15%, the laser-melted layer was breached by cracks. By transmission electron microscopy (TEM) analysis, it was confirmed that the laser-melted zone contained acicular bands of martensite which will only elongate 15% before fracture. Maximum and minimum critical laser-scanning velocities for normalization and avoiding resensitization of the surface layer were determined. Later, assorted work on the IGC of laser-modified stainless steels has been undertaken, classified under the following themes.

3.3.1 Effect of LSM, alloying and cladding

LSM of as-received 18/13/Nb ASS was shown to improve the resistance to the propagation of IGC (end grain attack) [9]. LSM was also used to

Table 3.1 A summary of research on laser surface modification of stainless steels for combating IGC

Year	Sensitization of stainless steel	Laser parameters*	IGC test conditions	Findings	Ref.
1978	304 (650 °C for 24 h)	CW CO_2 laser P = 70–200 W v = 5 & 10 mm/s d = 0.25 mm	Strauss test (ASTM A262 practice E) – boiling in H_2SO_4 (10%) and $CuSO_4$ (10%) for 72 h	LSM of sensitized 304 reduced the susceptibility to IGC	[8]
1993	As-received 18/13/Nb steel, 304L (675 °C for 120 h)	CW CO_2 laser P = 2 kW	Huey test (ASTM A262 Practice C) – boiling in HNO_3 for 240 h	LSM effectively prevented propagation of end grain attack in 18/13/Nb in HNO_3 and also eliminated sensitization of 304L, alloying of the 304L surface by injecting extra Cr or cladding with 310 steel had little extra benefit compared with LSM	[9]
1995	TIG welded 308 with various carbon content (400–900 °C)	CW CO_2 laser with Ar shielding P = 2.5 kW v = 1.7–166.7 mm/s	DL-EPR test in 0.5 M H_2SO_4 + 0.01 M KSCN	Low temperature sensitization behavior in the weld 308 and IGC resistance of the sensitized weldment of 308 was improved by LSM	[10]
1998	321 (650 °C)	CW CO_2 laser with Ar, He shielding P = 1–3 kW v = 5–100 mm/s	DL-EPR test in 0.5 M H_2SO_4 + 0.01 M KSCN	LSM can effectively eliminate the carbides formed during sensitizing treatment and homogenize the sensitized microstructures, leading to a remarkable improvement in IGC resistance of 321	[11]

Table 3.1 Continued

Year	Sensitization of stainless steel	Laser parameters*	IGC test conditions	Findings	Ref.
2001	316 (cold-worked to 5 to 25%, aged at 625 °C for 10–50 h)	CW CO_2 laser P = 150–260 W, 5 kW t = 5–30 s (150–260 W) v = 20 mm/s (5 kW)	ASTM A262 Practice A, ASTM G108 – Single loop EPR in 0.5 M H_2SO_4 + 0.01 M NH_4SCN	Under identical LSM conditions, the extent of desensitization decreases with an increase in the degree of cold work	[13]
2008	304 (aged at 650 °C for 0.5–15 h)	CW & pulsed CO_2 laser P = 2 kW (CW) P = 0.5 kW (base) + 3.2 kW (peak) v=5–10 mm/s	DL-EPR test in 0.5M H_2SO_4 + 0.01 M KSCN	LSM of 304 resulted in a large improvement in its resistance against sensitization and IGC	[14]
2008	304 (aged at 650 °C for 20 h)	CW Nd:YAG laser with Ar shielding P = 721–1415 W v = 5–50 mm/s d = 3 mm	DL-EPR test in 0.5 M H_2SO_4 + 0.01 M KSCN	Effects of LSM on IGC resistance of 304 was reassessed from viewpoint of GBE. LSM could make the sensitized microstructures locally desensitize, and could improve the IGC resistance	[15]
2008	316LN (SMAW welded)	CW Nd:YAG laser + pulsed or CW CO_2 laser with Ar shielding P = 150 kW (average) v = 0.3 33 mm/s d = 1.4 3.5 mm (post-annealing at 1100 °C)	ASTM A262 Practice E – Boiling in 10% $CuSO_4$ + 16% H_2SO_4 for 24 h	LSM parameters have been found to have a profound effect on the IGC resistance of 316LN and the resultant microstructure after subsequent solution annealing	[16]
2009	304 (GTAW welded)	Pulsed CO_2 laser P = 3.1 kW (peak) v = 5 mm/s d = 4 mm	DL-EPR test in 0.5 M H_2SO_4 + 0.01 M KSCN	LSM of HAZ of 304 weldment exhibited a significantly lower DOS and susceptibility to IGC than that of unmelted HAZ	[6]

(Contd)

Table 3.1 Continued

Year	Sensitization of stainless steel	Laser parameters*	IGC test conditions	Findings	Ref.
2011	304, 316L, 321, 347 (aged at 600 °C for 40 h), 2205, 7MoPlus (aged at 800 °C for 40 h)	CW Nd:YAG laser with Ar shielding P = 1 kW v = 25 mm/s d = 4 mm	DL-EPR test in 0.5 M H_2SO_4 + 0.01 M KSCN for ASSs and 2 M H_2SO_4 + 0.01 M KSCN + 0.5 M NaCl for DSSs	After LSM, the IGC of various ASSs and DSSs are improved as reflected by the reduction in DOS	[17]

**P* = power; *v* = scanning speed; *d* = beam size; *t* = interaction time

eliminate sensitized 304L either preventing the initiation of IGC or halting corrosion which has already started. Moreover, laser surface alloying of the 304L surface by injecting extra Cr or laser cladding with grade 310 steel was found to provide little extra benefit, as compared with LSM. It was probably due to the limited coverage of the laser-treated samples since the clad and alloyed surfaces showed very little evidence of corrosion. A high carbon content in the 310 clad and increased ferrite content in the Cr-alloyed layer also limited any improvement.

3.3.2 LSM of weldments

Several research groups have reported that IGC resistance of the weldments of stainless steel 308 [10], 316 LN [16] and 304 [6] can be enhanced by LSM. Nishimoto and co workers found that LSM improved the IGC resistance of sensitized 308 weldment, especially under conditions with fast cooling rates (if the laser scanning speed was higher than 83.3 mm/s then the size and amount of ferrite decreased), because of the dissolution of chromium carbides resulting in annihilation of the chromium depletion region adjacent to grain boundaries [10]. However, Parvathavathini's group reported that LSM of 316(N) weldments with a high repetition rate, pulse-modulated CO_2 laser successfully developed immunity against IGC after solution annealing involving cooling at the rate of 65 K/h [16]. The results have been explained with the help of electron backscattered diffraction measurement (EBSD) and numerical simulation studies. The direct implication of these results is that the 316(N) weldments can be cooled at a slower rate during subsequent solution annealing for stress relief without the risk of sensitization, thus minimizing distortion and reintroduction of thermal stresses. After that, Parvathavathini's group applied the similar LSM method on 304 stainless steel sheets by a CO_2

laser for effectively suppressing HAZ sensitization during subsequent gas tungsten arc welding (GTAW) [6]. The laser-melted HAZ of the weldment exhibited a significantly lower degree of sensitization (DOS) and susceptibility to IGC than those of untreated HAZ. This is attributed to the higher fraction of $\Sigma 1$ subgrain boundaries introduced by LSM and resolidification.

3.3.3 Effect of cooling rate (δ-ferrite content)

Pan *et al.* reported that the microstructure and the solidification mode in the laser-melted layer of 321 ASS were very sensitive to the cooling rate [11]. With the increase of cooling rate, the solidification mode changed from primary δ-ferrite to primary austenite. LSM can effectively eliminate the carbides formed during sensitizing treatment and homogenize the sensitized microstructures, leading to a remarkable improvement in the IGC and pitting corrosion resistance of 321. Moreover, an artificial neural network (ANN) method was used to model the non-linear relation between laser-processing parameters (such as laser power, scanning speed and shielding gas) and the corrosion resistance (DOS and pitting potential), and a genetic algorithm was further introduced to optimize the LSM process for different property demand. The verifying experiment indicated that experimental results agreed well with the optimized ones. ANN combined with a genetic algorithm offers a new effective means for optimization of laser-processing technology.

3.3.4 Effect of cold work

Mudali *et al.* firstly attempted LSM for eliminating the sensitized microstructure and the IGC resistance of the cold-worked 316 [12]. The results of ASTM A262 practice A showed that a cellular-dendritic structure was present in the melt zone while ASTM A262 practice E showed no intergranular crack was detected in the laser-melted specimens. After that, Parvathavarthini's group reported that under identical LSM conditions, the extent of desensitization decreased with an increase in the degree of cold work (5 to 25%) of 316, and hence higher power levels and an extended interaction time were adopted to homogenize the sensitized microstructure with prior cold work [13].

3.3.5 Effect of processing parameters

Kaul and co workers demonstrated that LSM of 304 with a CO_2 laser used in continuous wave (CW) and pulsed modes has resulted in a large improvement to its resistance against sensitization and IGC [14]. Double-loop electrochemical potentiokinetic reactivation (DL-EPR) tests demonstrated that the DOS of laser-melted 304 remained low even in the post-heating condition in contrast to the large increase in DOS of heat-treated base metal at 650 °C. The specimens treated with the pulsed laser exhibited lower DOS than those treated with CW laser. However, the change in the direction of laser scanning and degree of overlapping between two successive melt tracks did not bring any systematic change in DOS. It was found that under the best conditions, the DOS of the laser-melted surface remained unchanged even exposed to severe re-sensitization heat treatment. DOS of the re-sensitized specimens were comparable and even better than that of the untreated specimen. Development of a $\gamma + \delta$ duplex microstructure in the laser-melted surface with higher fraction of low-angle grain boundaries are believed to be the reasons for enhancement in IGC resistance. The approach suggested by Kaul *et al.* can find application as a prewelding laser surface treatment to suppress HAZ sensitization in unstabilized grade ASS weldments.

3.3.6 Effect of resensitization and grain boundary engineering (GBE) study

The effects of LSM on the IGC resistance of 304 were reassessed from the viewpoint of grain boundary engineering (GBE) by Yang *et al.* [15]. LSM could make the sensitized microstructures locally desensitize, and could improve the IGC resistance. The improved IGC resistance of the laser-melted 304 could be attributed in part to Cr redistribution at the boundaries of the cells and grains and in part to the existence of a large amount of low energy $\Sigma(1 \leq \Sigma \leq 29)$ boundaries and the formation of $\langle 0\ 0\ 1 \rangle (1\ 0\ 0)$ texture. However, the laser-melted specimens were more susceptible to IGC in the sensitization temperature region, and the corrosion rate of the resensitized specimen was even higher than that of the base materials under the same sensitization condition. It is inconsistent with the findings of Kaul and his co workers [14]. By a subsequent annealing treatment, the grain boundary character distribution was remarkably changed and the IGC resistance of the processed specimens was then improved.

3.3.7 Effect of LSM on various stainless steels with different compositions

Kwok and co workers have studied the effect of LSM on various stainless steels (ASSs and DSSs) with different chemical compositions [17]. The final microstructure of the laser-melted stainless steels depends on their compositions. It was found that the aged ASSs 304, 316L, 321 and 347 were essentially austenitic with some δ-ferrite but the Cr carbides were completely removed after LSM. For the aged DSSs steels 2205 and 7MoPlus after LSM, δ became the major phase and the δ/γ phase balance was redisturbed but the σ and γ_2 phases were eliminated. The IGC of 304, 316L and 321 was found to be considerably improved as reflected by the reduction in DOS due to a more homogeneous microstructure and the redissolution of chromium carbides. For 2205 and 7MoPlus, the IGC was also significantly enhanced as indicated by the decrease in DOS owing to a more homogeneous microstructure and to the redissolution of σ and γ_2 phases.

Although several studies have been reported on LSM of some ASSs and DSSs for improving IGC resistance, systematic analysis and comparison of more stainless steels with different compositions have not yet been done. In this chapter, besides the common ASS (S30400) and DSS (S31803) selected for investigation, more types of stainless steels such as Ni-free FeCrMn, Nb-stabilized S34700 and SDSS S32760 after aging followed by LSM will be discussed. In addition, LSM is likely to result in the formation of γ–δ in ASSs and redistribute the γ–δ phase balance in DSSs; their IGC behavior will also be affected.

3.4 Experimental details

3.4.1 Materials and isothermal aging

Three ASSs (UNS S30400, UNS S34700, FeCrMn), one DSS (UNS S31803) and one SDSS (UNS S32760) in the form of plates with the dimensions of $20 \times 20 \times 5\,mm^3$ were selected for the present study. The nominal compositions of the stainless steels are shown in Table 2.2. Prior to aging treatment, the stainless steels were subjected to solution annealing at 1100 °C in a furnace for an hour followed by water quenching. The volume fractions of δ-ferrite in the annealed S31803 and S32960 were found to be about 40% by image analyzing. Then isothermal aging treatments were conducted at 600 °C for ASSs [18] and at 800 °C for the DSS and SDSS [19] for 40 h, which are known to be the most suitable heat treatment conditions for the two different grades of stainless steels most susceptible to sensitization. For ASS 347, longer aging time (720 h) was needed to achieve sensitization. Practically, the short exposure time at high sensitization temperatures simulates the HAZ of weldments, while the long aging time simulates the long-term thermal components such as a turbine operated at high temperature. For the aged ASSs, carbide phases were detected by X-ray diffractometry as reported in a previous study [17].

3.4.2 LSM

LSM of the aged stainless steels were carried out by a CW Nd:YAG (yttrium aluminum granet) laser with a power of 1 kW, a beam diameter of 4 mm in and a scanning speed of 25 mm/s. Argon flowing at 20 l/min was used as the shielding gas. The surface was achieved by overlapping the melt tracks, with degree of overlap of 50 ± 5%. The sensitized specimens attained by

Table 3.2 Compositions of various stainless steels in wt%

Stainless steel	UNS Designation	Fe	Cr	Ni	Mn	Mo	Nb	Si	C	N	S	P	Cu	W
Austenitic	S30400	Bal.	18.4	8.7	1.6	–	–	0.5	0.08	–	0.03	0.045	–	–
	S34700	Bal.	17	10	1.5	–	0.8	0.5	0.03	–	0.03	0.045	–	–
	Fe-Cr-Mn	Bal.	15.6	–	10.4	–	–	0.47	0.08	–	–	–	1.08	–
Duplex	S31803	Bal.	22.3	5.6	1.5	2.9	–	0.5	0.03	0.18	0.02	0.030	–	–
Super duplex	S32760	Bal.	25.6	7.2	0.6	4			0.03	0.2			0.7	0.8

aging were designated as XXXXXX-S and the aged specimens followed by LSM were designated as LM-XXXXXX-S (XXXXXX represents the stainless steel type).

3.4.3 DL-EPR test

To investigate the IGC behavior of the aged specimens before and after LSM, the specimens were embedded in and the edges were sealed with cold curing epoxy resin, exposing a surface area of 1 cm^2 for avoiding crevice corrosion. Prior to the corrosion test, the specimens were mechanically ground with progressive grits of SiC papers up to 800 grit and then rinsed with ethanol and distilled water. The DOS of the specimens was determined by the DL-EPR method in test solution naturally aerated at 25 ± 1 °C by means of a potentiostat (VoltaLab 10) according to ISO Standard 12732-2006 [20]. The standard test solution (0.5 M H_2SO_4 and 0.01 M KSCN) was used for ASSs. However, a more aggressive solution (2 M H_2SO_4, 0.01 M KSCN and 0.5 M NaCl) was used for the more corrosion-resistant DSS and SDSS [21]. All potentials were measured with respect to a saturated calomel electrode (SCE, 0.244 V versus SHE at 25 °C) as the reference electrode. Two parallel graphite rods served as the counter electrode for current measurement. During the DL-EPR test as schematically shown in Fig. 3.1, the specimen was first polarized anodically from the open-circuit potential through the active region to the passive region at a scan rate of 6 V/h to form a passive layer on the surface. Then the reactivation scan in the reverse direction was carried out from a potential of +300 mV (SCE) at the same scan rate, leading to the breakdown of the passive film on the Cr-depleted regions. So the anodic loop and the reactivation loop were generated. The DOS was evaluated by measuring the

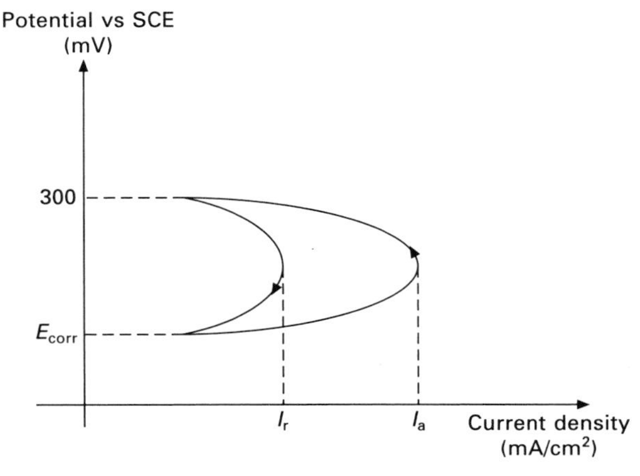

3.1 DL-EPR test criteria for stainless steels.

ratio $(I_r/I_a) \times 100\%$, where I_r and I_a are the peak reactivation current density and the peak activation current density, respectively. The larger the DOS value, the higher the susceptibility of the stainless steel to IGC.

3.4.4 Microstructural analysis

The microstructural characteristics, corrosion morphologies and chemical compositions of the aged stainless steels before and after LSM were studied by optical microscopy (OM) and scanning electron microscopy (SEM) equipped with energy-dispersive X-ray spectroscopy (EDX). The microstructure of the aged DSS and SDSS was revealed by electrolytic etching at 3 V in 10 M NaOH solution for 10 s. Microstructural observation was conducted by backscattered electron (BSE) imaging, on the basis of the atomic number contrast effect: the δ appears slightly darker than γ, while the σ phase is the brightest. The volume fraction of the σ phase was determined using an image analysis software (ImageJ 1.43) with the BSE micrographs. Acidified ferric chloride was used to etch the aged stainless steels after LSM for microstructural observation. Analysis of volume fraction of δ was also performed by image analyzing.

3.4.5 Hardness measurement

The Vickers hardness of the stainless steels in annealed, aged and laser melted conditions was evaluated by means of a microhardness tester, applying a load of 9.8 N for 10 s. At least five indentations were taken on each specimen, and the average hardness values were reported with errors of ±4% approximately.

3.5 Metallographic and microstructural analysis

The transverse cross-sections of the aged stainless steels after LSM at a low magnification and the microstructure of their melt pools at a high magnification are shown in Figs 3.2 to 3.6. The compositions of various phases in the aged stainless steels before and after LSM are also shown in Figs 3.5 and 3.6. LSM caused melting and rapid solidification of the surface layers but no cracks and pores exist. Melt depths of 0.2 to 0.3 mm were obtained for various laser-melted specimens.

For all aged ASSs after LSM, the large sensitized grains were refined and the carbide phases were completely dissolved and did not reprecipitate. More homogeneous compositions were obtained after LSM. The different compositions rate may lead to the change in solidification mode and hence different microstructure. Compared with the substrate, the δ-ferrite content in aged ASSs after LSM increased significantly. When the melt pools of the

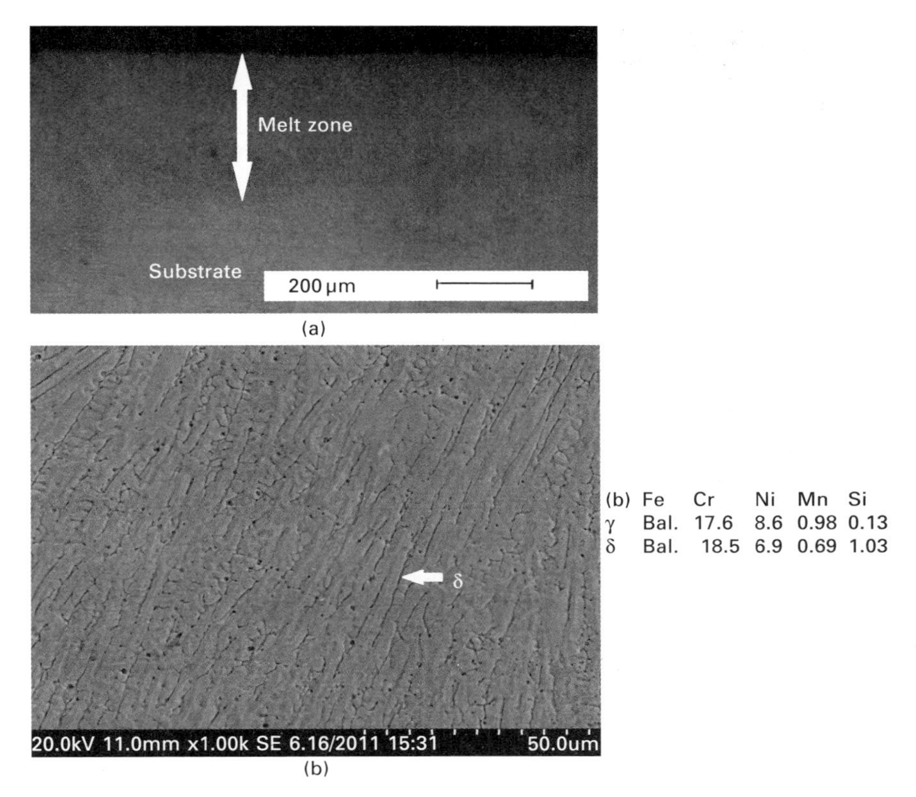

(b)	Fe	Cr	Ni	Mn	Si
γ	Bal.	17.6	8.6	0.98	0.13
δ	Bal.	18.5	6.9	0.69	1.03

3.2 SEM micrographs of LM-30400-S: (a) cross-sectional view (OM) and (b) microstructure of melt pool (SE image with EDX results in wt%).

laser-melted specimens solidify, the possible phase transformation sequence upon cooling is represented as:

$$\text{Liquid} \rightarrow \text{Liquid} + \delta \rightarrow \delta \rightarrow \gamma\ (\text{major}) + \delta\ (\text{minor})$$

Due to primary γ with secondary δ solidification (AF) mode, residual interdendritic vermicular δ is observed in LM-S30400-S and LM-S34700-S and makes the γ grain network discontinuous while only γ is observed in LM-FeCrMn-S.

For the aged DSS and SDSS, eutectoid decomposition of δ causes transformation of the sigma phase (σ) and the secondary austenite (γ_2) through the eutectoid reaction:

$$\delta \rightarrow \sigma + \gamma_2$$

After LSM, the melt layer is primarily composed of bulky δ with massive γ as the minor phase formed by primary δ with secondary γ solidification (FA)

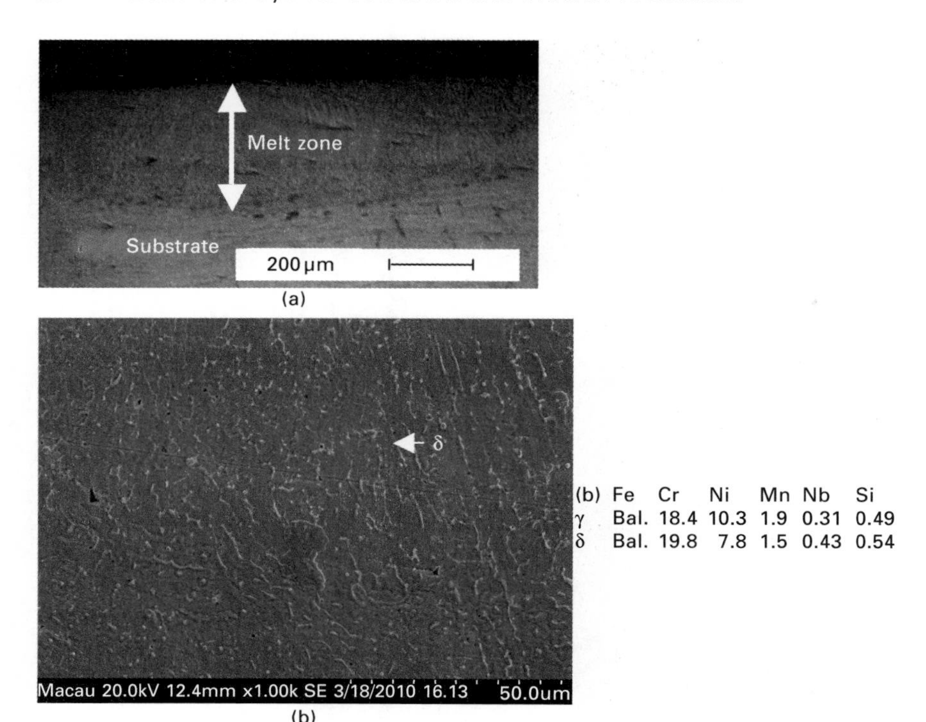

3.3 SEM micrographs of LM-34700-S: (a) cross-sectional view (OM) and (b) microstructure of melt pool (SE image with EDX results in wt%).

mode. Widmanstatten structure of semi-continuous dendritic γ (needle-shape) is present in the columnar grain boundaries of δ. When the melt pools of the laser-melted specimens solidify, the possible phase transformation sequence upon cooling may be represented as:

$$\text{Liquid} \rightarrow \text{Liquid} + \delta \rightarrow \delta \rightarrow \delta\ (\text{major}) + \gamma\ (\text{minor})$$

The degree of completion of the transformation and hence the final microstructure depend on compositions of the stainless steels and laser processing parameters [22].

From the BSE images of S31803-S and S32760-S (Figs 3.5(c) and 3.6(c)), the σ phase has the brightest contrast reflecting the presence of elements of high atomic mass compared with that in the δ and γ phases. The σ phase depleted the neighboring δ matrix of Cr and Mo and enriched it with Ni, causing the formation of γ_2. For S31803-S, σ and γ_2 phases were found to occupy the whole of original δ region at prolonged aging time while partial δ transformation to σ and γ_2 was observed in S32760-S. For the DSS S31803, saturation of σ phase can be achieved at 800 °C for longer than 10 h [23].

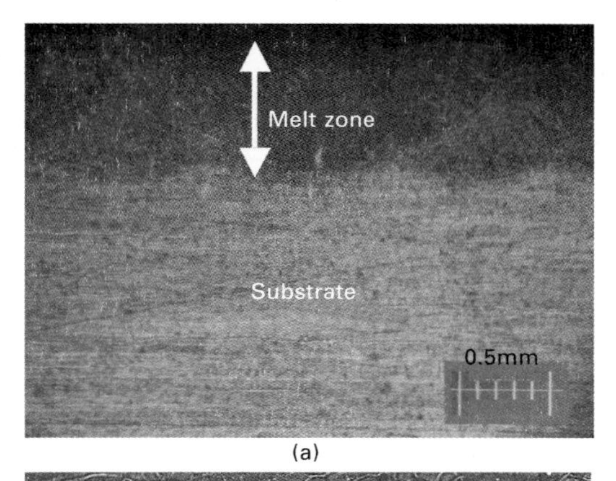

(a)

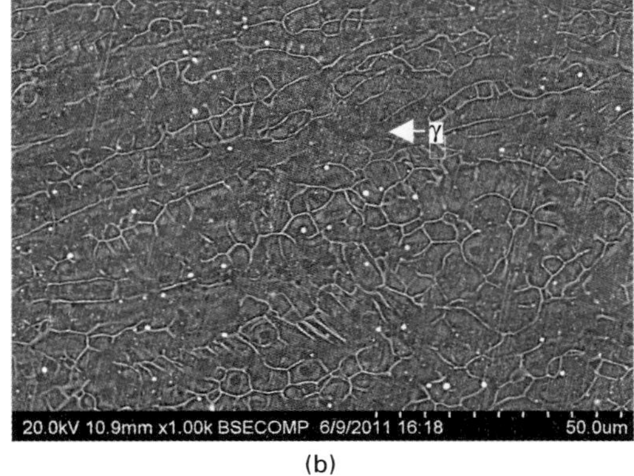

(b)

(b)	Fe	Cr	Mn	Cu	Si
γ	Bal.	15.0	10.0	1.47	0.56
IB	Bal.	15.1	10.2	1.47	0.58

3.4 SEM micrographs of LM-FeCrMn-S: (a) cross-sectional view (OM) and (b) microstructure of melt pool (SE image with EDX results in wt%, IB = interdendritic boundary).

It has been reported that the volume fraction of σ of aged SDSS SAF 2507 increases with increasing aging time and temperature, reaching about 36% for specimens aged for 168 h at 900 °C [24]. In the present study, the volume fraction of σ of aged S32760-S is 21% and is slightly lower than that of S31803-S (30%).

Besides the σ phase, there may be another intermetallic phase (χ) precipitated in the DSS and SDSS during aging. However, the χ content was too low in comparison with the σ phase and it had far less effect on the DSS and SDSS than the σ phase [25]. Carbides are also not detected because of the low carbon content of the DSS and SDSS. On the other hand, aging of SDSS at 800 °C can cause the precipitation of chromium nitrides inside γ, as observed also by other authors [24, 26]. However, the role of chromium nitrides is less harmful because of the low amount of these precipitates [24].

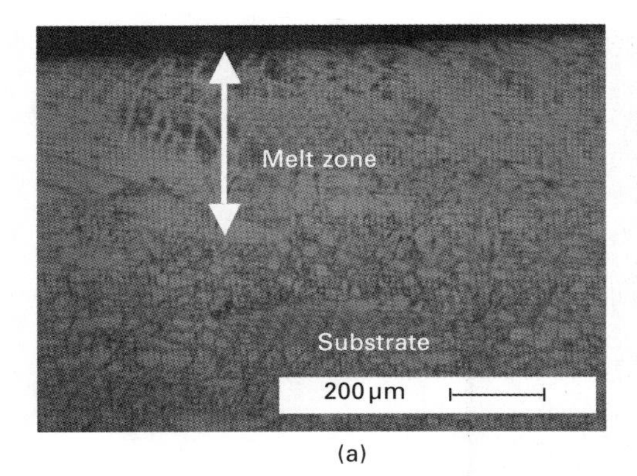

(a)

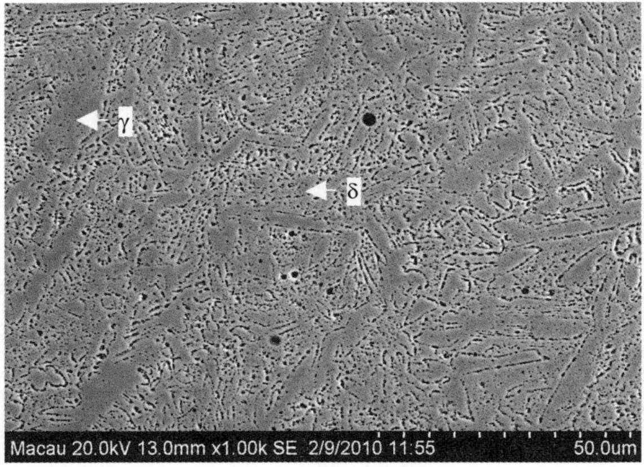

(b)	Fe	Cr	Mo	Ni	Mn	Si
γ	Bal.	21.7	3.8	5.7	1.3	0.31
δ	Bal.	22.2	3.6	6.0	1.1	0.33

(b)

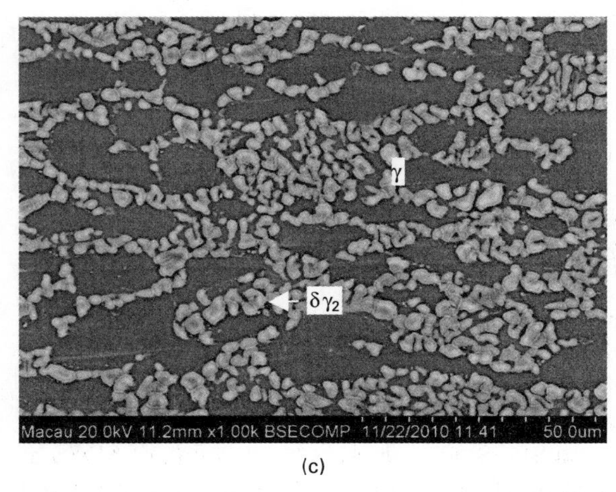

(c)	Fe	Cr	Mo	Ni	Mn	Si
γ	Bal.	20.9	1.4	6.7	1.7	0.34
γ_2	Bal.	18.9	1.3	6.2	1.6	0.55
σ	Bal.	27.1	6.4	2.8	1.5	0.74

(c)

3.5 SEM micrographs of LM-S31803: (a) cross-sectional view (OM), (b) microstructure of melt pool (SE image with EDX results in wt%), and (c) microstructure of sensitized zone (BSE image with EDX results in wt%).

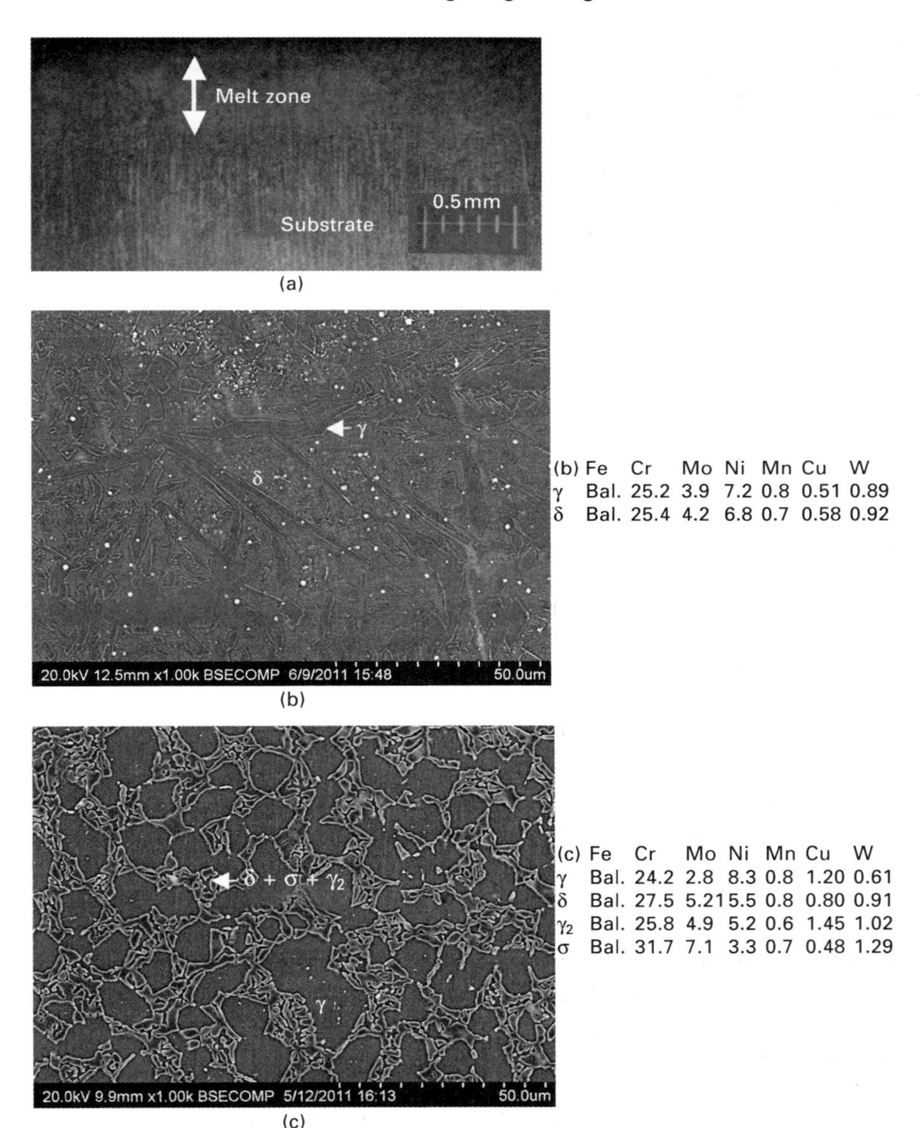

3.6 SEM micrographs of LM-S32760-S: (a) cross-sectional view (OM), (b) microstructure of melt pool (BSE image with EDX results in wt%), and (c) microstructure of aged substrate (BSE image with EDX results in wt%).

The ratios of the Cr equivalent (Cr_{eq}) and the Ni equivalent (Ni_{eq}) of various stainless steels are shown in Table 3.3. As the Cr_{eq}/Ni_{eq} ratio increases, the δ-forming tendency of the stainless steels increases. The highest δ content is observed in LM-S31803-S, which has the highest ratio ($C_{\delta} = 86\%$), followed

Table 3.3 Cr_{eq}/Ni_{eq} ratio, phase present and volume fraction of ferrite (C_δ) in aged stainless steels before and after LSM

Stainless steels	$Cr^{*}_{eq}/Ni^{\#}_{eq}$ ratio	Phase present in aged specimens before LSM	Phase present in aged specimens after LSM	C_δ in aged specimens after LSM (%)
S30400	1.60	γ, $M_{23}C_6$, M_7C_3	γ, δ	11.6
S34700	1.46	γ, NbC	γ, δ	10.0
FeCrMn	1.68	γ, $M_{23}C_6$, M_7C_3	γ	0.0
S31803	2.17	γ, σ, γ_2, $M_{23}C_6$	δ, γ	86
S32760	2.06	γ, δ, σ, γ_2, $M_{23}C_6$	δ, γ	72

*Cr_{eq} = [Cr] + [Mo] + 1.5[Si] + 0.5[Nb].
#Ni_{eq} = [Ni] + 30[C] + 30[N] + 0.5[Mn].

by LM-S32760-S ($C_\delta = 72\%$). In addition, the solid-state transformation of δ to γ is considered to be diffusional. Thus, the high solidification rate typical in LSM would also suppress the δ-to-γ transformation, leading to high δ content in LM-S31803-S and LM-S32760-S.

The average hardness of aged stainless steels before and after LSM is shown in Fig. 3.7. The hardness of the annealed specimens is also shown for comparison. The hardness values within the melt zones of the aged ASSs after LSM were nearly constant and higher than those of the annealed ones. LM-S30400-S shows the most significant increase in hardness (increased by 23%) as compared with that of the annealed specimens. The hardness of FeCrMn in annealed and laser-melted conditions is high due to the high Mn content for solid-solution hardening of γ. The increase in hardness of the aged ASSs after LSM could be attributed to the refinement of grains, presence of δ, solid-solution hardening of γ [27].

It was reported that σ precipitation and δ content in DSSs can increase the hardness but drastically decrease the toughness properties [28–30]. However, various investigators reported that a small amount of the σ phase does not increase the hardness of DSSs [31, 32]. The hardness values of the annealed S31803 and S32760 are 250 and 290 HV respectively. From the results obtained, the hardness of the aged S31803 and S32760 before LSM (355 and 368 HV respectively) is higher than that of the annealed specimens and aged specimens after LSM (322 and 282 HV respectively) due to the presence of a large amount of σ in the aged specimens. The change in hardness of the aged DSSs after LSM could be attributed to the remarkable increase in δ content and elimination of the σ phase.

3.6 Intergranular corrosion (IGC) behavior

3.6.1 Aged ASSs before LSM

For the aged ASSs, due to the effect of segregation at the vicinity of Cr carbides, corrosion attack in the forward scan occurs at the locations anodic

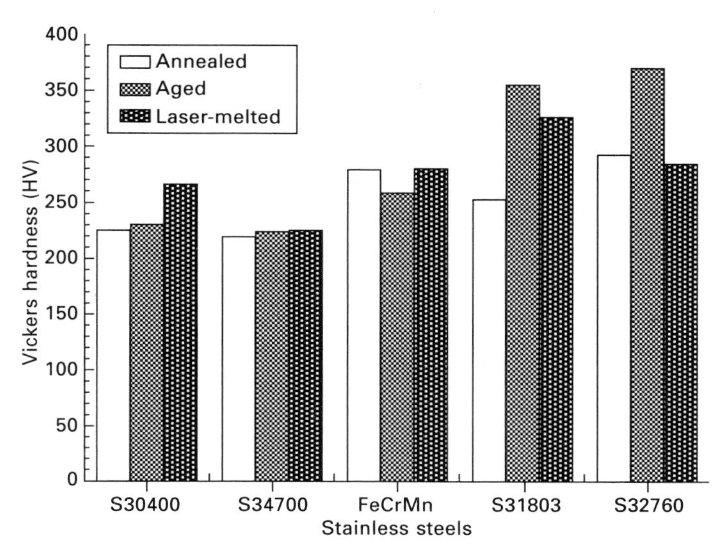

3.7 Hardness of various stainless steels in annealed, aged and laser-melted conditions.

to the γ, where the Cr-depleted regions are adjacent to the grain boundaries. From a previous study by the present authors, aged S34700 was found to be unsensitized for 40 h and the DOS value is only 0.18% [17] since Nb was added to enhance the formation of stabilized NbC instead of Cr carbides. A longer aging period (720 h) was attempted to sensitize S34700 in the present study. The DL-EPR curves of S30400-S and FeCrMn-S (aged for 40 h) and S34700-S (aged for 720 h) are shown in Fig. 3.8(a). For S30400-S, the relatively high I_a (58.9 mA/cm^2) is attributed to the active dissolution of the alloying elements. Before reaching the preset reverse potential of +300 mV, it falls to a relatively low passive current density (0.06 mA/cm^2). Also, a wide passive range between –55 and +300 mV was recorded due to the formation of a thin chromium oxide film on the specimen surface. This film acts as a barrier to protect the surface from the environment and hence reduces the dissolution rate. Moreover, IGC attack occurs at the reverse scan due to the incomplete passivation during the forward scan. It is mainly due to the Cr depletion resulting from the precipitation of Cr carbides.

From Table 3.4, the DOS value of S30400-S (36.6%) is quite large compared with aged S34700-S (9.5%) and aged FeCrMn-S (11.7%), indicating that severe IGC attack occurs in S30400-S after DL-EPR test. From the corrosion morphologies of the aged specimens after DL-EPR test, S30400-S also shows the largest intergranular susceptibility among the aged ASSs. Continuous severe ditches along the grain boundaries are clearly observed in S30400-S (Fig. 3.9(a)(i)). Narrow discontinuous corrosion ditches along the grain boundaries

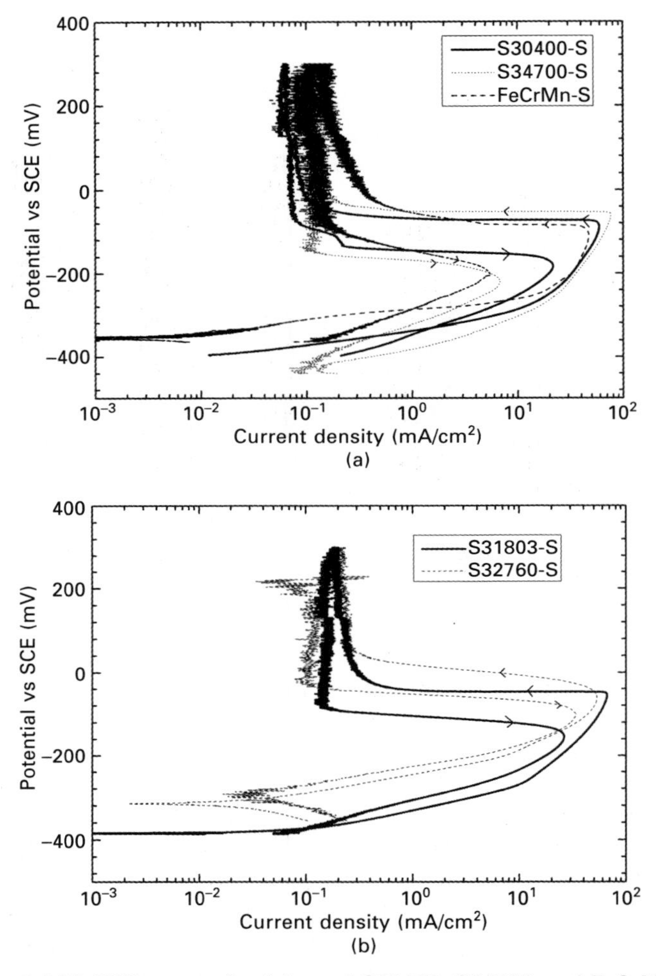

3.8 DL-EPR curves for (a) aged S30400, S34700 and FeCrMn before LSM in 0.5 M H_2SO_4 + 0.01 M KSCN solution at 25 °C and (b) aged S31803 and S32760 before LSM in 2 M H_2SO_4 + 0.01 M KSCN + 0.5 M NaCl solution at 25 °C.

and some corrosion cavities inside the grains are observed in FeCrMn-S (Fig. 3.9(c)(i)). It can be noticed that no intergranular attack but individual corrosion cavities appeared near the boundaries of S34700-S (Fig. 3.9(b)(i)) and the phase with high Nb content (probably is NbC) can be detected as shown in Fig. 3.10. Recently, Kim and his co workers proposed that IGC occurring in the stabilized FSS and ASS is induced by Cr-depletion due to segregation of unreacted Cr atoms around carbides of stabilizer elements (Ti or Nb) along the grain boundary, but not due to formation of Cr-carbide as

Table 3.4 *DOS* of various stainless steels after DL-EPR test

Specimens	I_r (mA/cm^2)	I_a (mA/cm^2)	DOS = 100 × I_r/I_a (%)	DOS_S/DOS_{LM}*
S30400-S	21.60	58.9	36.6	30.5
LM-S30400-S	0.86	67.9	1.2	
S34700-S	7.41	77.9	9.5	23.8
LM-S34700-S	0.28	70.6	0.4	
FeCrMn-S	5.53	47.2	11.7	2.5
LM-FeCrMn-S	2.29	48.9	4.7	
S31803-S	26.90	66.2	40.6	23.9
LM-S31803-S	0.71	41.5	1.7	
S32760-S	33.80	53.3	63.3	45.2
LM-S32760-S	0.16	11.3	1.4	

*DOS_S = DOS for sensitized specimen; DOS_{LM} = DOS for laser-melted specimen.

evidenced by TEM investigation [33]. IGC developed much more slowly in ASS than in FSS because of its slow kinetics of diffusion and precipitation. The stabilized NbC precipitated at the grain boundaries instead of Cr carbide resulted in the highest IGC resistance of S34700 among the ASSs. In short, the ranking of IGC resistance of the aged ASSs is:

S34700-S > FeCrMn-S > S30400-S

It has been reported that the Ni-free stainless steels with high Mn content of more than 10 wt% are less resistant to crevice and pitting corrosion, suggesting that high Mn content adversely affects the corrosion properties of stainless steels [34]. From a study of Wu *et al.* [35], the DL-EPR tests proved that the Ni-free and Mn alloyed high N stainless steels (HNSSs) were rather susceptible to the sensitization treatment at 650 °C for 2 h compared to 316L because Mn possesses a strong chemical activity. The relatively weak resistance of the HNSSs to uniform and IGC is due to the high Mn content leading to anodic dissolution. However, the DOS value for FeCrMn-S has been found to be lower than that of S30400-S, probably because of the lower Cr content in the former resulting in fewer Cr-depleted regions as the initiation sites for IGC attack.

3.6.2 Aged DSS and SDSS before LSM

A more aggressive solution (2 M H_2SO_4, 0.01 MSCN and 0.5 M NaCl) was used in the DL-EPR study for S31803-S and S32760-S. The result shows that higher I_r is obtained for both specimens (Fig. 3.8(b)). The DOS of S31803-S and S32760-S are 40.6 % and 63.3% respectively. The ranking of IGC resistance is:

S31803-S > S32760-S

The susceptibility to IGC of the aged DSS and SDSS depends on the degree

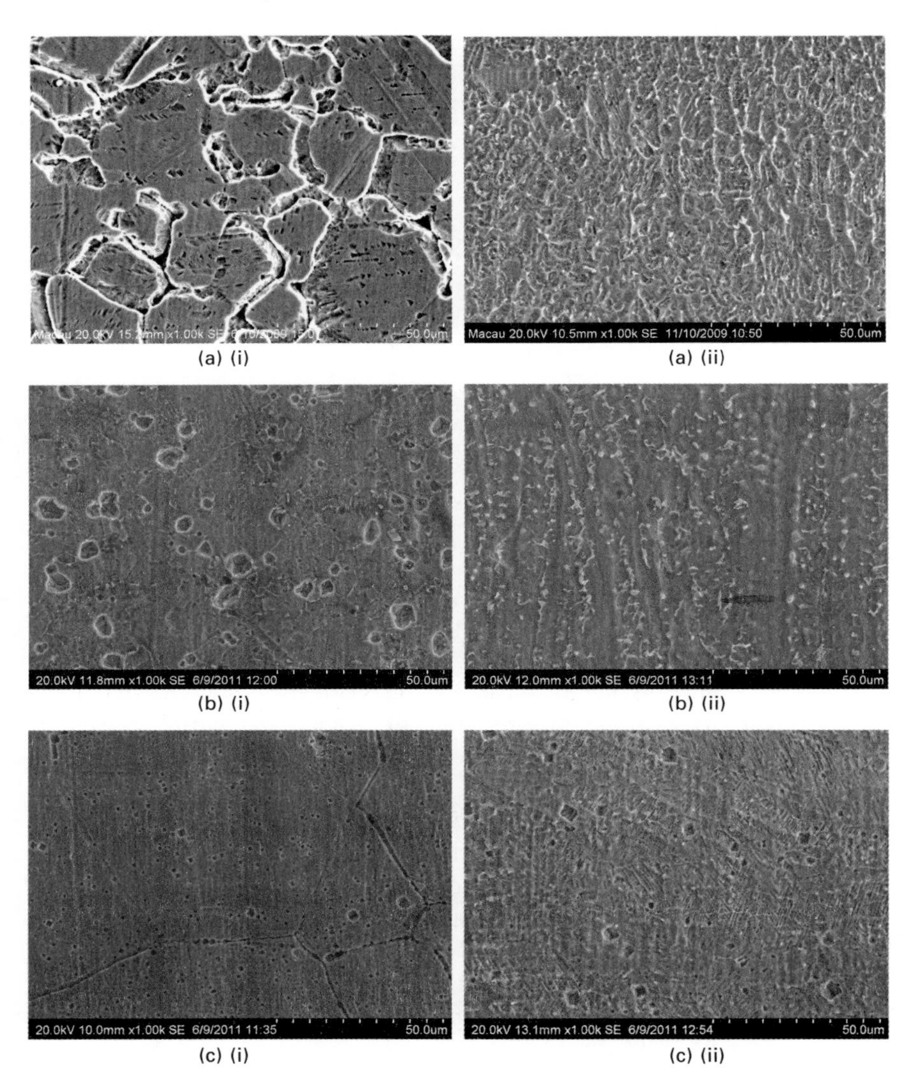

3.9 Corrosion morphologies of aged ASSs after DL-EPR test (a) S30400, (b) S34700, and (c) FeCrMn (i) before LSM and (ii) after LSM.

of depletion of the compositions, in particular the Cr and Mo contents. The numbers of phases (σ, δ, γ and γ_2) in S32760-S are higher than in S31803-S (σ, γ and γ_2). The degree of depletion of Cr and Mo for S31803-S and S32760-S are represented by the DOS. The DOS for S32760-S was lower than that for S31803-S when they were equivalently aged, which demonstrated that the degree of depletion of Cr and Mo in S32760-S is higher due to the faster precipitation rate of σ phase. SEM examination on S31803-S and S32760-S after DL-EPR tests (Fig. 3.13(a)(ii) and (b)(i)) shows that IGC attack takes

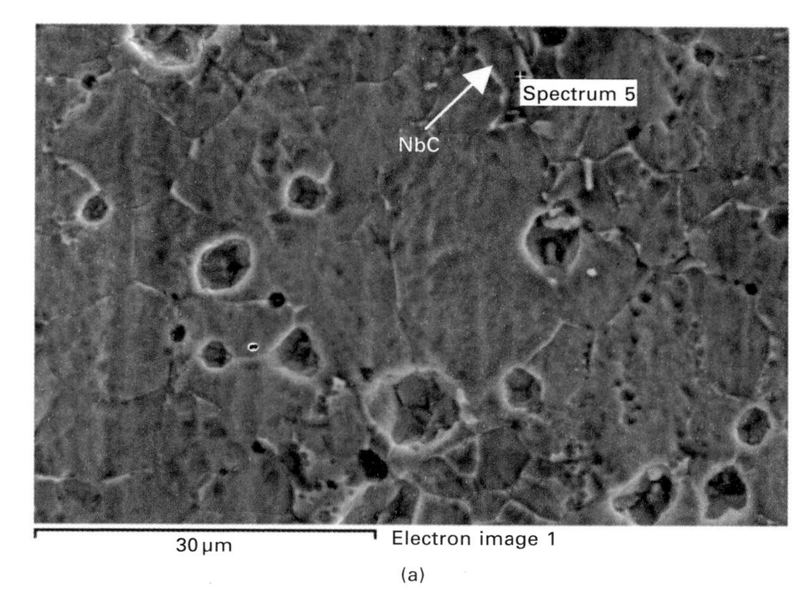

(a)

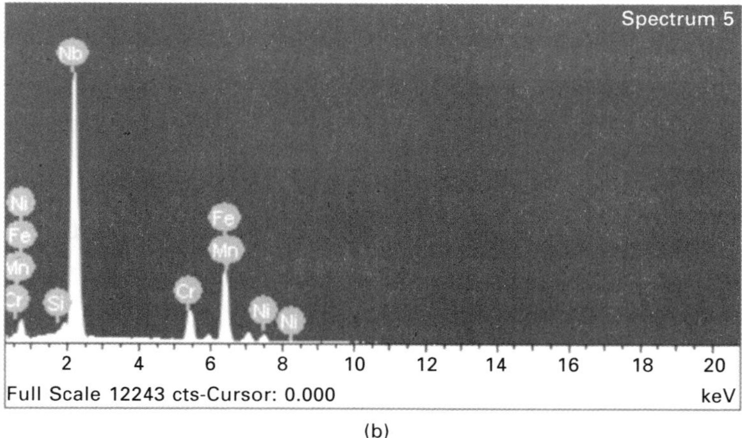

(b)

3.10 (a) Corrosion morphology of aged S34700 after DL-EPR test and (b) EDX spectrum of Nb-rich phase.

place at the phase boundaries of σ phase and the metallic phases (γ and δ) and propagation of corrosion occurs more easily along their boundaries. The result indicates that Cr and Mo depletion at the phase boundaries is crucial since they are the main contents in σ phase. From the EDX results as depicted in Figs 3.5(c) and 3.6(c), the σ phase is partitioned with the highest contents of Cr, Mo and Si, and then followed by δ, while primary γ and γ_2 are partitioned with higher Ni and Mn contents. The σ phase originates from the

δ phase because Cr diffusion is much faster in δ than in γ. Cr concentrates in the neighboring σ phase [36] and causes the σ phase to grow fast within the δ region. As a result, the γ_2 phase with lower Cr content degrades the stability of the passive film [37].

3.6.3 Aged ASSs after LSM

All three types of ASSs possess reduced DOS values after LSM. LM-S34700-S possesses much smaller DOS value (0.4%) than those of LM-S30400-S (1.2%) and LM-FeCrMn-S (4.7%). The latter two laser-melted ASSs are slightly sensitized. For LM-S30400-S, the DOS value is 30.5 times that of the S30400-S (Table 3.4). Corrosion attack in the forward scan takes place at locations anodic to the γ or δ, that is, at the γ-δ interdentritic regions due to small segregation effects [14] for LM-S30400-S with duplex-phase structure (Fig. 3.2b). Attack takes place at locations anodic to γ, that is, at interdentritic boundaries due to even smaller segregation effects for LM-FeCrMn-S with single-phase structure (Fig. 3.4b). Corrosion attack for LM-S30400-S in the reverse scan arises from incomplete passivation during a forward scan and is probably attributed to a high δ content (13.9%) resulting in a galvanic effect between γ and δ. A similar finding was reported by Kina and co workers [27]. A small I_r is observed in as-received S30400 containing Cr-rich δ islands but no such peak is observed after annealing.

Among the aged ASSs after LSM, the DOS value of LM-FeCrMn-S is the largest. The DOS value of LM-FeCrMn-S is only 2.5 times that of FeCrMn-S (Table 3.3). The possible reason for the high DOS is the absence of Ni in FeCrMn stainless steel. From the Poubaix diagram of Mn, the stable range of pH value and potential for Mn^{2+} is wider than that of Fe^{2+} and Ni^{2+}, which indicates that Mn is more active than Fe and Ni, and tends to become Mn^{2+} in a corrosive environment. Moreover, Mn-rich oxides are less protective [35]. Furthermore, high Mn alloying was detrimental to pitting corrosion of stainless steels [34, 38].

The DOS value of S34700-S is 23.8 times that of the LM-S34700-S (Table 3.4). From Table 3.2, LM-S34700-S wita lower δ content than S30400-S and redissolution of NbC is can be considered as desensitized with DOS values smaller than 1 and has high IGC resistance. It shows noticeable anodic loops but no obvious reactivation loop (Fig. 3.11(b)). According to the DOS, the ranking of IGC resistance of the aged ASSs after LSM becomes:

LM-S34700-S > LM-S30400-S > LM-FeCrMn-S

Compared with the ranking before LSM, it can be noticed that the ranking of the IGC resistance of aged S30400 and aged FeCrMn after LSM are swapped. From the corrosion morphologies of all laser-melted ASSs, no continuous ditch is observed at the grain boundaries as depicted in Fig. 3.11. In fact,

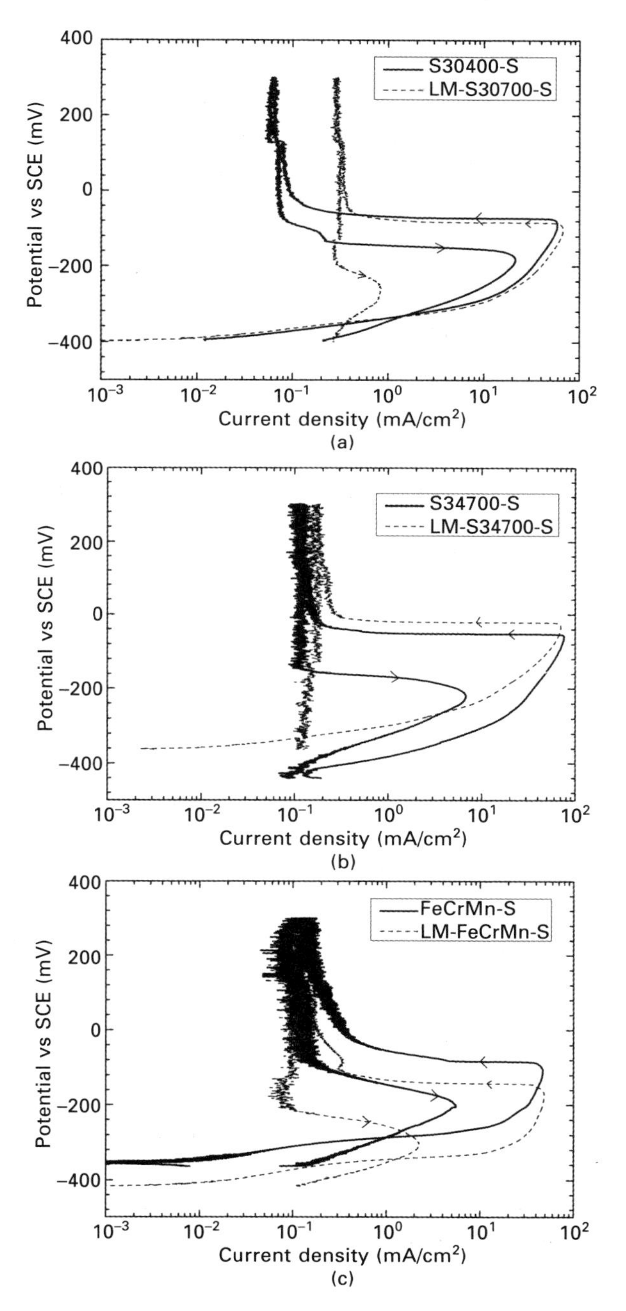

3.11 DL-EPR curves for aged (a) S30400, (b) S34700 and (c) FeCrMn before and after LSM in 0.5 M H_2SO_4 + 0.01 M KSCN solution at 25 °C.

these specimens exhibit only a small corrosion attack at the interdendritic phase boundaries.

There are two factors responsible for the improvement of corrosion resistance of ASSs after LSM. The first one is the redistribution and/or complete removal of Cr carbides, since they deplete the Cr content nearby and act as the corrosion initiation, hence degrading the IGC resistance. The second factor is due to the formation of δ after LSM. It makes the γ grain network discontinuous and hence forms a γ–δ duplex structure [14]. Enhanced IGC resistance of the duplex microstructure is attributed to increased grain boundary area and rapid healing of Cr depletion due to enhanced diffusion kinetics in δ.

3.6.4 Aged DSS and SDSS after LSM

LM-S31800-S and LM-S32760-S show low I_r (0.71 and 0.16 mA/cm^2) respectively (Fig. 3.12 and Table 3.4). The value of I_a of LM-S32760-S is dramatically decreased to 11.3 mA/cm^2. After LSM, the DOS values are significantly reduced mainly due to the redissolution of σ and γ_2 phases. The DOS value of LM-S32760-S (1.4%) is slightly lower than that of LM-S31803-S (1.7%). According to the DOS, the ranking of IGC resistance of the aged DSS and SDSS after LSM is:

LM-S32760-S > LM-S31803-S

Compared with the ranking before LSM, it can be also noticed that the rankings of the IGC resistance of aged S31803-S and aged S32760-S after LSM are swapped.

The formation of σ in the aged DSS and SDSS leads to Cr and Mo depletion in the adjacent zone and γ_2, and degrades the IGC resistance. The main factor responsible for improving the IGC resistance of aged DSS and SDSS after LSM is the redistribution of γ and δ phases, and/or complete elimination of σ and γ_2 phases. The second factor is the presence of high δ content in DSS and SDSS after LSM. Although δ is more prone to corrosion than γ, the galvanic effect is insignificant because the δ content (anodic) is much higher than the γ content.

The main reason for the dramatic decrease in DOS of LM-S32760-S (1.4%) after LSM is due to its high content of alloying elements. Although a higher alloying content in S32760-S may lead to easier formation of the σ phase at high temperature, leading to a high DOS value (63.3%), it will enhance IGC after LSM due to the redistribution of the alloying elements in the solid solution.

For the aged DSS and SDSS after LSM, corrosion attack initiates at the interdendritic boundaries with slight segregation of alloying elements. There is a large increase in the δ content in the specimens but the chemical

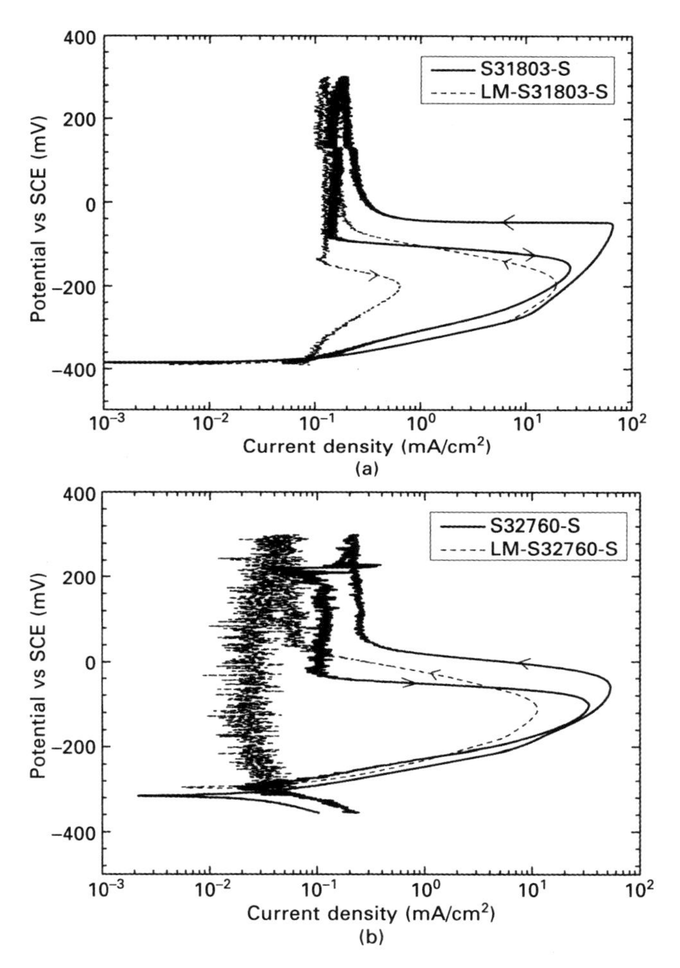

3.12 DL-EPR curves for aged (a) S31803 and (b) S32760 before and after LSM in 2 M H_2SO_4 + 0.01 M KSCN + 0.5 M NaCl solution at 25 °C.

compositions (Cr, Mo and Ni) of δ and γ are more homogeneous as depicted in Figs 3.5(b) and 3.6(b). From the corrosion morphology of the specimens, no continuous grain boundaries attack is observed. On the contrary, they exhibit mild corrosion attack at the interdendritic boundaries (Fig. 3.13). It might be concluded that a more homogeneous microstructure and redissolution of σ and γ_2 phases tend to improve the IGC resistance.

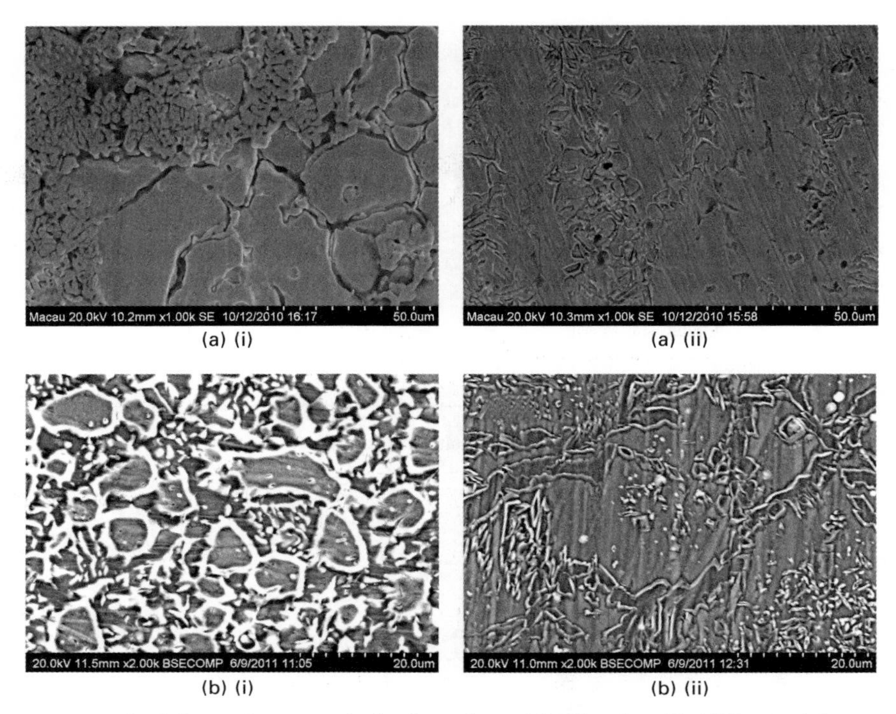

3.13 Corrosion morphologies of aged DSSs after DL-EPR test (a) S31803 and (b) S32760 (i) before LSM and (ii) after LSM.

3.7 Conclusions

LSM was attempted for desensitization of different austenitic and duplex grade stainless steels and the IGC behaviour was investigated. The following conclusions can be drawn:

- After LSM, the aged austenitic stainless steels S30400, S34700 and FeCrMn were essentially austenitic with some δ (except FeCrMn) but the chromium carbides were completely removed.
- For the aged duplex grade stainless steels S31803 and S32760 after LSM, δ became the major phase, the δ/γ phase balance was disturbed but the σ and γ_2 phases were eliminated.
- After LSM, the intergranular corrosion resistance of the aged austenitic stainless steels in 0.5 M H_2SO_4 and 0.01 M KSCN solution was found to be considerably improved as reflected by the reduction in DOS. This could be attributed to a more homogeneous microstructure and the redissolution of chromium carbides.
- The intergranular corrosion resistance of aged S31803 and S32760 after LSM in 2 M H_2SO_4, 0.01 M KSCN and 0.5 M NaCl solution was

also significantly enhanced as indicated by the decrease in DOS. This could be attributed to a more homogeneous microstructure and to the redissolution of σ and γ_2 phases.

- For the aged specimens before LSM, SEM examination revealed that intergranular attack occurs at the Cr-depleted grain boundaries for the austenitic stainless steels while attack takes place at the phase boundaries of the intermetallic σ phase and metallic phase (γ, δ, γ_2) for the duplex grade stainless steels.
- For the aged specimens after LSM, SEM examination revealed that intergranular attack occurs at the interdentritic boundaries of γ–δ or γ–γ for both austenitic and duplex grade stainless steels.

3.8 Acknowledgments

The work described in this paper was fully supported by a research grant from the Science and Technology Development Fund (FDCT) of Macau SAR (Grant no. 019/2009/A1).

3.9 References

[1] R.K. Dayal, J.B Gnanamoorthy and G. Srinivasan: in *ASM handbook of cases histories in failure analysis*, Vol. 2, Ed. K.A. Esaklul, ASM International, 1993, 506–508.

[2] H.S. Khatak and B. Raj, *Corrosion of austenitic stainless steel: mechanism, mitigation and monitoring*, Woodhead Publishing Limited, Cambridge, UK, 2002, p.117.

[3] B. Deng, Z. Wang, Y. Jiang, T. Sun, J. Xu and J. Li, *Corrosion Science*, 2009, **51**, 2969–2975.

[4] R.A. Lula and W.G, Renshaw, *Metal Progress*, 1965, 69.

[5] J.S. Stanko and D. Wellbeloved, Manganese in corrosion resistant steels, SAMANCOR, 1991.

[6] R. Kaul, N. Parvathavarthin. P. Ganesh, S.V. Mulki and I. Samajdar, *Welding Journal*, 2009, **88**, 233s–242s.

[7] G. Bao, K. Shinozaki, S. Iguro, M. Inkyo, M. Yamamoto, Y. Mahara and H. Watanabe, *Journal of Materials Processing Technology*, 2006, **173**, 330–336.

[8] T.R. Anthony and H.E. Cline, *Journal of Applied Physics*, 1978, **49**, 3, 1248–1255.

[9] J.Y. Jeng, B.E. Quayle, P.J. Modern, W.M. Steen and B.D. Bastow, *Corrosion Science*, 1993, **35**, 5-8, 1289–1296.

[10] K. Nishimoto, H. Mori and Y. Nakao, *ISIJ International*, 1995, **35**, 10, 1265–1271.

[11] Q.Y. Pan, W.D. Huang, R.G. Song, Y.H. Zhou and G.L. Zhang, *Surface and Coatings Technol.*, 1998, **102**, 245–255.

[12] U.K. Mudali, R.K. Dayal and G.L. Goswami, *Anti-corrosion Methods and Materials*, 1998, **45**, 3, 181–188.

[13] N. Parvathavarthini, R.V. Subbarao, Sanjay Kumar, R.K. Dayal and H.S. Khatak, *JMEPEG*, 2001, **10**, 5–13.

[14] R. Kaul, S. Mahajan, V. Kain, P. Ganesh, K. Chandra, A.K. Nath and R.C. Prasad, *Corrosion*, 2008, **64**, 10, 755–763.
[15] S. Yang, Z. Wang, H. Kokawa and Y.S. Sato, *Materials Science and Engineering A*, 2008, **474**, 112–119.
[16] N. Parvathavarthini, R.K. Dayal, R. Kaul, P. Ganesh, J. Khare, A.K. Nath, S.K. Mishra and I. Samajdar, *Science and Technology of Welding and Joining*, 2008, **13**, 4, 335–343.
[17] C.T. Kwok, K.H. Lo, W.K. Chan, F.T. Cheng and H.C. Man, *Corrosion Science* 2011, **53**, 1581–1591.
[18] S. Yang, Z.J. Wang, H. Kokawa and Y.S. Sato, *Journal of Materials Science*, 2007, **42**, 847–853.
[19] C. Huang and C. Shih, *Materials Science and Engineering A*, 2005, **402**, 66–75.
[20] International Standard, ISO 12732 (2006), Corrosion of metals and alloys – electrochemical potentiokinetic reactivation measurement using the double loop method (based on Cihal's method).
[20] T. Amadou, C. Braham and H. Sidhom, *Metallurgical and Materials Transactions A*, 2004, **30A**, 3499–3513.
[21] L. Karlsson, *Welding in the World*, 1990, **43.5**, 20–40.
[22] S.A. David, J.M. Vitek and T.L. Hebble, *Welding Journal*, 1987, **10**, 289s–300s.
[23] C. Huang and C. Shih, *Materials Science and Engineering* A, 2005, **402**, 66–75.
[24] E. Angelini, B. de Benedetti and F. Rosalbino, *Corrosion Science* 2004, **46**, 1351–1367.
[25] B. Deng, Z. Wang, Y. Jiang, H. Wang, J. Gao and J. Li, *Electrochimica Acta*, 2009, **54** 2790–2794.
[26] J.O. Nilsson and A. Wilson, *Materials Science and Technology*, 1993, **9**, 545.
[27] A. Yae Kina, V.M. Souza, S.S.M. Tavares, J.M. Pardal and J.A. Souza, *Materials Characterization* 2008, **59**, 651–655.
[28] Y.S. Ahn and J.P. Kang, *Materials Science and Technology* 2000, **16**, 382–388.
[29] T.H. Chen, K.L. Weng and J.R. Yang, *Materials Science and Engineering* A 2002, **338**, 259–270.
[30] E. Angelini, B. de Benedetti and F. Rosalbino, *Corrosion Science* 2004, **46**, 1351–1367.
[31] K.M. Lee, H.S. Cho and D.C. Choi, *Journal of Alloys and Compounds* 1999, **285**, 156–161.
[32] J.O. Nilsson, P. Kangas, T. Karlsson and A. Wilson, *Metallurgy and Materials Transaction*, 2000, **31A** 35–45.
[33] J.K. Kim, Y.H. Kim, B.H. Lee and K.Y. Kim, *Electrochimica Acta*, 2011, **56**, 1701–1710.
[34] P.J. Uggowitzer, R. Magdowski and M.O. Speodel, *ISIJ International*, 1996, **36**, 901.
[35] X. Q. Wu, S. Xu, J. B. Huang, E. H. Han, W. Ke, K. Yang and Z. H. Jiang, *Materials and Corrosion*, 2008, **59**, 675–684.
[36] C.-J. Park and H.S. Kwon, *Corrosion Science*, 2002, **44**, 2817–2830.
[37] M. Pohl, O. Storz and T. Glogowski, *Materials Charactization*, 2007, **58**, 65–71.
[38] Y.S. Lim, J.S. Kim, S.J. Ahn, H.S. Kwon and Y. Katada, *Corrosion Science*, 2001, **43**, 53.

4

Pulsed laser surface treatment of multilayer gold–nickel–copper (Au/Ni/Cu) coatings to improve the corrosion resistance of components in electronics

N. SEMMAR and C. BOULMER-LEBORGNE,
University of Orléans, France

Abstract: This chapter deals with laser treatment of thin film materials for reducing the impact of corrosion phenomena in the electrical connector industry. Materials used in electrical contact applications usually consist of multilayered coatings (e.g. copper alloy electroplated with a nickel layer and finally by a gold layer). After the electrodeposition, microchannels and pores within the gold layer allow undesirable corrosion of the underlying layer. In order to modify the gold coating microstructure, a selective pulsed laser melting (SLM) is applied. The SLM treatment suppresses porosity and smooths the surface leading to low roughness and to good electrical contact. Corrosion tests in addition to SEM observations are correlated to SLM conditions to optimize the present process.

Key words: thin film laser treatment, laser selective melting (LSM), electric connectors, corrosion resistance by pulsed lasers, excimer lasers.

4.1 Introduction

The passive components industry plays a main part in the electronic procedures for the telecommunication and computing industries. In connectors, the electrical contacts mainly constitute a copper alloy (brass or bronze) covered by a nickel coating (diffusion barrier) and finally by a gold coating to increase simultaneously the corrosion resistance and the surface brightness at a reasonable cost. The study of the atmospheric corrosion mechanisms of nickel protected by gold has shown the existence of pores in the gold layer, inducing corrosion phenomena of the underlying layers [1, 2]. This porosity is probably linked to the columnar crystalline structure of the thin gold film. Nevertheless, the protection of nickel film requires a tight gold coating. To modify the structure of the gold coating, a selective laser melting (SLM) process has been undertaken. The laser treatment has to induce melting of the gold layer surface to suppress the pores. It must also smooth the surface as any roughness prevents the correct electrical contact. To determine the best laser processing conditions (pulse shape and pulse duration) the melting kinetics should be understood by way of thermal simulation. Thus, it is possible to calculate the melting depth as a function of laser pulse shape and pulse duration. Additionally the effect of thermal properties on the melting depth is discussed.

As indicated in Fig. 4.1, materials used in electrical contact applications are usually copper alloy (brass or bronze), for its excellent electrical conductivity and inexpensive cost. However, a major difficulty in the use of copper contacts is its high reactivity to the atmospheric environment. The classic solution is to use a protective coating of noble metals such as gold or palladium. In our study, the copper surface is firstly nickel-plated and then protected by a thin gold coating 0.4 to 1 μm thick. The nickel layer is used as a diffusion barrier between the copper and the gold used as a corrosion barrier. Electrodeposition of these metals usually leads to columnar structures that are favorable to the formation of micro channels. However, the thickness of the gold alloy coatings must be limited to less then 1 μm to keep the cost low. In order to enhance the protective role of the gold coatings, an additional surface treatment must be applied. Two positive effects are desired: reduction of porosity inside the thin gold layer, and smoothing of the surface to achieve a highly reflective surface, an efficient electrical contact, and hence a good product for industry. In this chapter a laser heat treatment is tested for this surface modification by selective melting of the surface.

The surface treatment for electrical contacts depends on various restraints closely linked to the manufacturing process. It must be fast and usable in air atmosphere. In considering the multilayer coatings at the micron level, one should treat only the gold layer; the nickel sublayer must not be damaged. Moreover, the surface of the material must be heated to the gold melting temperature (without reaching the boiling temperature) and the temperature should be lower than the nickel melting temperature. Thus, the process requires a shallow and a controlled penetration depth of the heat wave. During this last decade, SLM processes have been widely employed in several applications mainly for the improvement of resistance to corrosion and hardness in the continuous or pulsed regime [3-11]. To achieve the SLM on submicron multilayer coatings, pulsed laser re-melting technology, using nanosecond lasers, appears to be a soft and fast alternative method. Furthermore, the desired laser treatment could be achieved in a few pulses, thus being interesting and competitive for industrial applications. To check the

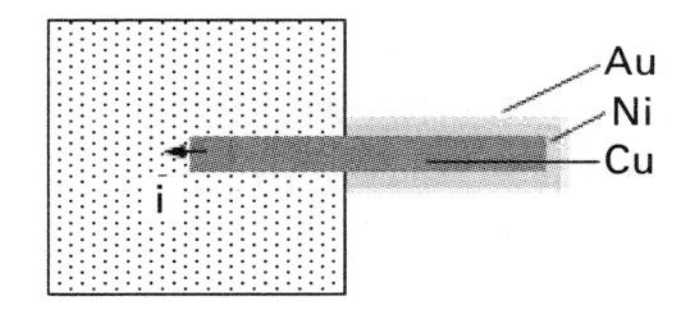

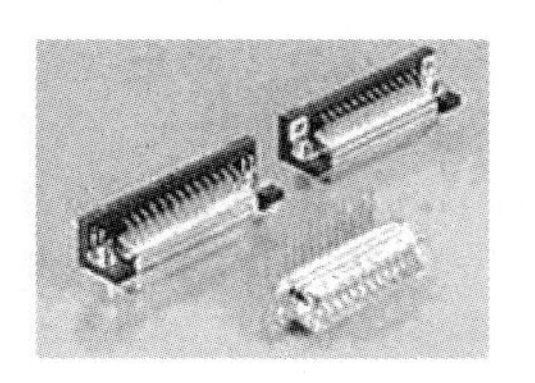

4.1 View of the multilayer coatings for electrical connectors.

efficiency of this laser treatment, samples were submitted to corrosion tests, microscopy characterization (scanning electron microscopy (SEM), optical microscopy) and chemical analyses (energy dispersive X-ray, Rutherford backscattering spectroscopy) for both treated and untreated samples. In turn, laser parameters were adjusted incrementally until the best improvement in corrosion resistance was achieved.

To understand the melting kinetics, the dynamic of laser–matter interaction is simulated via heat conduction equation numerical resolution considering phase changes and interface heat resistances. This simulation allows computation of the thermal fields (temperature and heat flux) versus thermal properties (thermal conductivity and thermal diffusivity). Because the thermophysics situation differs between bulk and thin film materials, the simulated thermal properties were altered with respect to ones from the literature [12–14]. The melting depth simulated in the case of gold coating is finally compared to the one observed by SEM analysis.

4.2 Experimental arrangements

Basically, the surface treatment is expected to melt the whole gold thin film volume without damaging the Ni sub-layer (Fig. 4.2). To achieve these optimal conditions, the laser parameters should be compared as illustrated in the following sections. The plot of reflectivity of 1 μm gold coating on Ni/Cu substrate analysed by Fourier transform infra-red (FTIR) spectroscopy extending to the UV range (250 nm) is shown in Fig. 4.3. There is a significant drop in reflectivity from 80% in the IR and visible ranges to 15% in the UV range. Thus, excimer lasers seem to be more efficient for photon/lattice coupling.

4.2.1 Selection of operating lasers

The laser treatments were performed with two different excimer lasers: a Lambda Physics Compex 205 in KrF gas mixing configuration (λ (wavelength) = 248 nm, τ (pulse duration) = 25 ns) and a Questek XeCl (λ = 308 nm, τ = 28 ns). A pulsed laser was chosen as the tool for surface treatment of

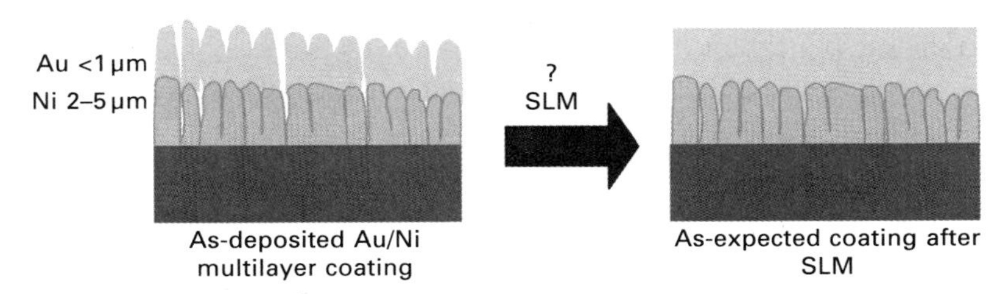

4.2 Schematic view of surface modification by SLM processing.

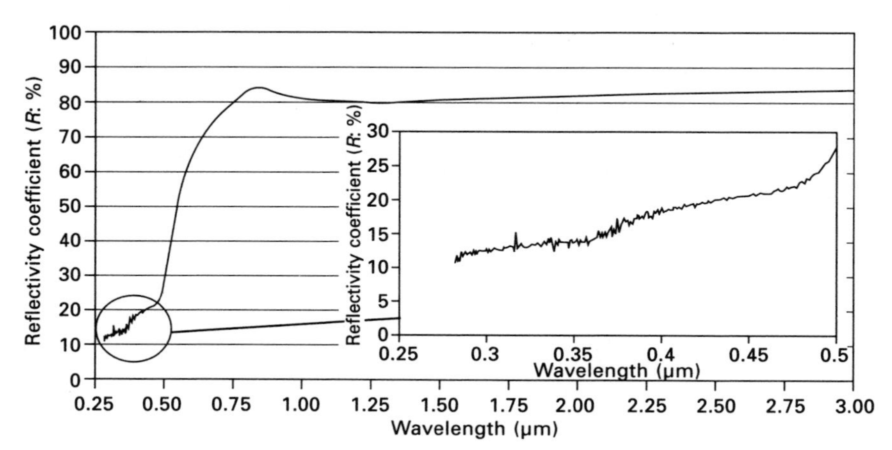

4.3 Spectral reflectivity of Au/Ni/Cu connector by FTIR spectroscopy extended to the UV range (250 nm).

gold layer for two reasons. The first is to introduce the SLM process in a production line of connectors. According to the line speed (6 m/min), this surface treatment must be fast. The second is concerned with the treatment process that has also a role to induce the gold layer melting without heating and damaging the nickel sub-layer. Thus, it needs a low penetration depth (δ_{th}) of heat which is proportional to the square root of the pulse time [$\delta_{th} = (d_{th} \times \tau_p)^{1/2}$]. Typically, in the case of nanosecond lasers the penetration depth is roughly 1 μm depending on the thermal diffusivity (d_{th}).

Among the available lasers, the choice of an excimer laser for surface treatment has been governed by two main parameters. First, the UV wavelength as the photon beam penetration depth is generally lower for UV range than for larger wavelengths and this corresponds to a better photon absorption coefficient. Second, the beam homogeneity of the excimer laser beam allows a homogeneous surface treatment. As illustrated in Fig. 4.4, the laser beam is focused onto a sample, the more homogeneous part of the beam being selected by a diaphragm. The sample is fixed on a holder supported by an automatic scanner allowing an *X*–*Y* movement of the samples in front of the laser beam. The laser fluence used for the surface treatment ranges from 300 to 1000 mJ/cm^2. In order to compare the influence of the laser pulse number at the same location, 1 to 50 laser pulses of the same impact were used. The approximate size of the samples is 1 cm × 1 cm and each laser beam size is roughly 3 × 5 mm^2.

4.2.2 Experimental set-up

As shown in Fig. 4.4, the SLM processing needs nanosecond lasers to perform the surface treatment of samples. A classic device is used here to

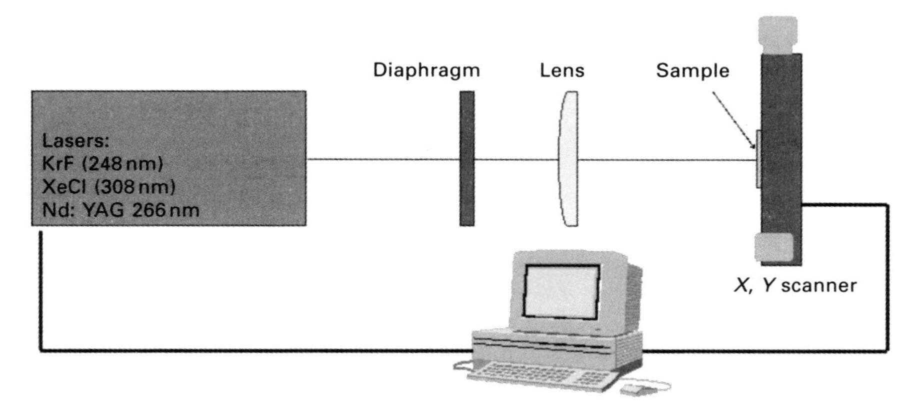

4.4 Experimental devices for Au/Ni/Cu connectors by SLM processing.

control both the laser fluence (by diaphragm and lenses) and position of the laser spot by the *X–Y* scanner.

4.2.3 Sample characteristics

The samples comprise a substrate (copper alloy: brass or bronze) of 0.2 mm thickness electroplated with two successive coatings, a nickel layer and a gold layer. The gold coating must be deposited on a nickel underlayer in order to limit the diffusion between copper and gold. The nickel and the gold deposits are performed by electrolytic process.

The samples have different roughnesses and thicknesses in order to study the influence of these parameters on the laser treatment efficiency in relation to the corrosion tests. The roughness of the substrate (mirror polished or laminate polished ($R_a < 0.1\,\mu m$) or laminate ($R_a \geq 0.3\,\mu m$)).

- The thickness of the nickel layer (2 μm or 5 μm)
- The thickness of the gold layer (0.75 μm or 1 μm)

After laser treatment, corrosion tests were carried out under observation by optical microscopy or SEM. Typical tests were systematically achieved on these substrates as reported in Fig. 4.5.

4.2.4 Corrosion test procedures

The corrosion tests were carried out at 25 °C, in wet synthetic air (relative humidity = 85%) containing three polluting gases: NO_2, SO_2 with concentrations of 0.2 vpm (volume per million, i.e. 0.2×10^{-6} liters per air liter) and Cl_2 with concentration of 0.01 vpm. The test duration varies from

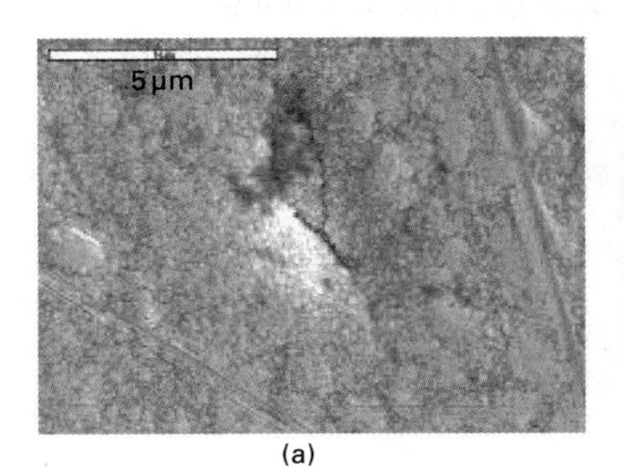

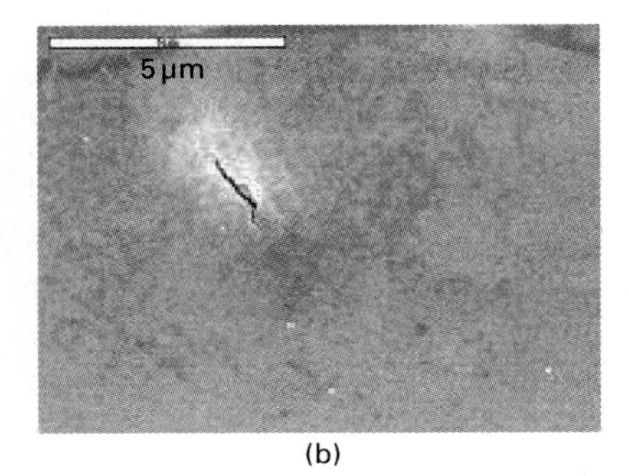

(a) (b)

4.5 SEM photos of the untreated (a) and treated (b) Au/Ni/Cu surface by KrF laser (248 nm, 27 ns).

a few days to a dozen days [15]. At regular intervals of time, the samples were observed to visualise the evolution of the corrosion.

4.3 Experimental results

For SLM processes, it is crucial to understand the role of laser parameters on the corrosion behaviour. The effect of the energy density by pulse (i.e. fluence) and number of pulses, and the space and time distributions of the laser beam are reported here and correlated at least in the analysis of the surface smoothing and melting.

4.3.1 Effect of laser fluence and number of shots

Different laser fluences were used to determine the energy range needed for gold melting in the case of rough substrates ($R_a > 0.15\,\mu m$ corresponding to gold thickness higher than $0.75\,\mu m$) commonly used in industry. For small fluences (lower than 400 mJ/cm^2), laser impacts were not observed. For fluences ranging between 400 and 700 mJ/cm^2 the laser impacts are gloss (as used in connector market), and over this fluence value, the laser impact induces a surface aspect modification (impact zones become white). In Table 4.1, the treatment results on laser impact appearance and corrosion behavior are classified. There was no significant improvement in corrosion resistance observed for samples treated with one laser shot at fluence lower than 700 mJ/cm^2. Corrosion aggregates were always very numerous in the low fluence treated zones (for more details, refer to Georges *et al.* [16]). Samples treated at fluences higher than 750 mJ/cm^2 exhibited more promising corrosion resistance. The tests have been achieved for few laser shots from one to five. Optical micrographs as shown in Fig. 4.6 (top) show the comparison between the untreated area (a) and treated areas respectively with one (b) and five laser shots (c) at 850 mJ/cm^2 after corrosion tests.

Table 4.1 Experimental laser conditions and corrosion behavior for samples treated between 400 and 950 mJ/cm^2

Excimer laser KrF fluence (mJ/cm^2)	Number of laser pulse/impact	Impact appearance	Corrosion behavior
400/450/500/550/600/650/700	1	Gloss	Poor
700/750/800/850/900/950/	1, 2, 5, 10	Opaque (fluences > 750 mJ/cm^2)	Good (number of pulses > 5)

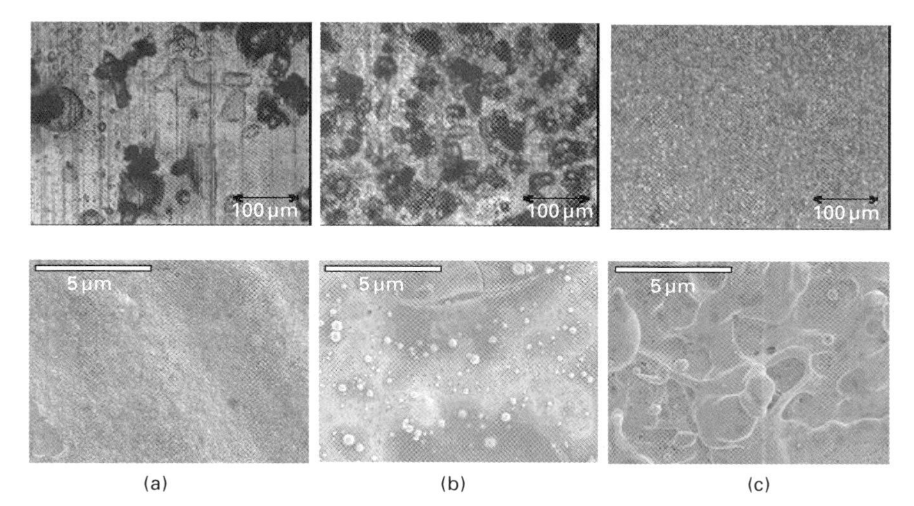

4.6 Optical microscopy (top) and SEM (bottom) images of samples after corrosion tests: (a) untreated zone, (b) treated with one laser pulse and (c) with five laser pulses at 850 mJ/cm^2.

The corrosion nuclei for a sample treated by one laser pulse (b) are smaller than the ones observed in the untreated zone (a) in Fig. 4.6. For samples treated by five laser pulses (Fig. 4.6(c)), the corrosion nuclei in the laser impact area are much less numerous. SEM observations (Fig. 4.6 bottom) at high magnification (× 10 000), respectively for the three zones (a, b, c), show a surface modification increasing the roughness versus the pulse number. This last observation has been verified using an optical profilometer. The roughness measurements corresponding to 900 mJ/cm^2 laser fluence are listed in Table 4.2.

With high fluences (>750 mJ/cm^2) and a number of laser pulses higher than five, an improvement in corrosion resistance was achieved. At the same time, roughness is increased, leading to poor electrical contacts undesirable for the industrial process. One should conclude that lower fluences (<700 mJ/

Table 4.2 Average (R_a) and maximum (R_{max}) measured roughness versus laser pulse number for 900 mJ/cm²

	Untreated zone	Laser impact with 1 laser pulse	Laser impact with 2 laser pulses	Laser impact with 5 laser pulses	Laser impact with 10 laser pulses
R_a (μm)	0.15	0.16	0.18	0.22	0.30
R_{max} (μm)	0.2	0.21	0.25	0.66	1.10

cm²) are necessary for treating these samples. However, in this case (high roughness and low fluence) no corrosion improvement was achieved as shown previously [6]. Hence polished samples (R_a < 0.1 μm) will be considered in the following.

4.3.2 Effect of the laser beam distribution

As the efficient fluence range necessary to treat the gold surface without damaging is very small (550 to 700 mJ/cm²) it is essential to have a spatially homogeneous laser beam to avoid any hot points. For the two excimer lasers used in this study, the results concerning the corrosion resistance are nearly similar but the use of the XeCl laser seems to give better results, perhaps because its time duration is slightly longer (28 ns compared with 25 ns for KrF excimer laser). To validate this, samples were irradiated by one pulse of a Cilas-XeCl (λ = 308 nm, τ = 50 ns) laser beam at 600 mJ/cm². Corrosion tests on these samples do not show any significant improvement of corrosion resistance, as this fluence was not high enough to melt the gold layer sufficiently. Indeed, the instantaneous power density is too weak (≈ 12 MW/cm² instead of ≈ 20 MW/cm² for Questek XeCl and Lambda KrF).

4.3.3 Effect of the surface layer thickness and interface roughness (1p)

Parameters of samples and particularly the roughness have also an important role on the efficiency of laser treatment. For a sample with a rough substrate (R_a ≥ 0.3 μm) the laser processing has no effect on the atmospheric corrosion prevention. Indeed, in Fig. 4.7(a), there are many nuclei inside the laser impact zone. On the photographs, each laser impact is represented in dotted frames. Conversely, for the sample presented in Fig. 4.7b, it can be observed that the laser treatment has a positive effect since the corrosion germs in the treated zone are much less numerous than in the untreated zone. Notice that the substrate roughness is the only different parameter between these two samples.

In fact, the roughness of the substrate is a limiting parameter for SLM processing because it could induce poor and inhomogeneous deposits of

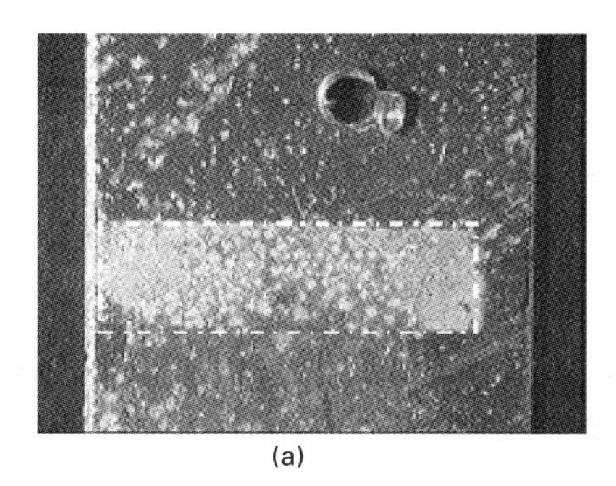

(a)

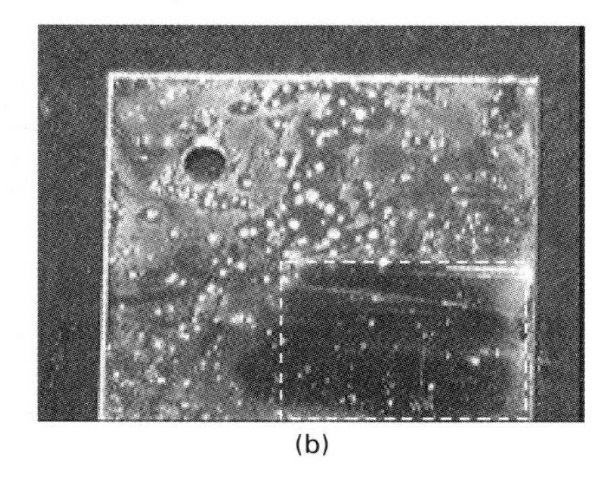

(b)

4.7 Optical microscopy observations after laser treatment and corrosion test: (a) Ni 2 μm, Au 0.75 μm, $R_a \geq 0.3$ μm; (b) Ni 2 μm, Au 0.75 μm, $R_a < 0.1$ μm.

nickel and gold. Figure 4.6 presents a SEM observation of the surface state of gold deposit before and after a laser treatment. It can be observed that the sample surface is covered by gold aggregates induced by the electrolytic process. Owing to these aggregates, the SLM process induces simultaneously melting and smoothing but also nanoporosities unfavorable to the corrosion limitation.

4.3.4 Surface smoothing and melting analysis (2p)

In Fig. 4.8, results of corrosion tests on a polished sample treated by 550 mJ/cm^2 are presented for one laser pulse. The laser impact is delimited by the white dots. The corrosion germs are much smaller than in the untreated area. In a previous study [17], it is shown that the laser treatment improved the corrosion resistance as reflected by the reduction in the size of the nuclei. As a general rule, the best results are obtained for a number of pulses ranging from one to five and laser treatment is more efficient on polished samples. These results are justified by the fact that the melting occurs on a small surface zone. When the roughness is high, some zones could not be fully recovered during the melting process.

To study the effect of laser treatment on the gold melting thickness, SEM characterizations on cross-sections of samples treated with the KrF laser at different fluences and with one laser pulse (Fig. 4.9) have been performed. For 430 mJ/cm^2 (Fig. 4.9a) there is no evidence of a melting zone in the gold thickness. For 750 mJ/cm^2 (Fig. 4.9b), the formation of cavities at 0.15 μm in depth can be observed, i.e. the laser melting treatment only acts on 0.15 μm. For 900 mJ/cm^2 the cavities are initiated at the gold/nickel interface with surface damage (Fig. 4.9c).

These last results on the melting depth, threshold of melting and surface damage are of great interest in the following section to establish a realistic

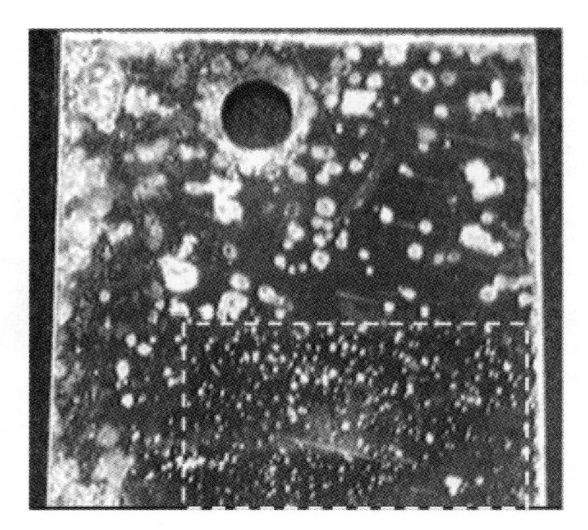

4.8 Optical observation of a polished sample [Au (0.75 μm)/Ni (2 μm)] after corrosion test (192 h), the treated area is surrounded by dotted lines (550 mJ/cm^2, 1 laser pulse).

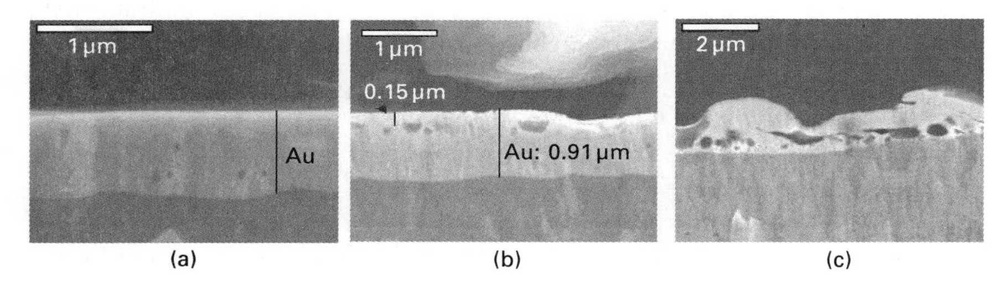

4.9 SEM observations of cross-sections of samples (substrate R_a > 0.15 μm) treated with different laser fluences, after one laser pulse: (a) 430 mJ/cm^2; (b) 750 mJ/cm^2; (c) 900 mJ/cm^2.

discussion on the melting kinetics in metallic thin films throughout the numerical simulation.

4.4 Numerical results

To understand how the melting occurs in depth in the gold layer, the heat-transfer problem with phase change is considered for both systems (i.e. thin film and substrate). The one-dimensional unsteady heat equation is resolved using the finite differences method in thin film and substrate domains. The related boundary conditions and the mathematical equations are given in the following:

$$\frac{1}{\alpha}\frac{\partial T}{\partial t} = \frac{\partial^2 T}{\partial x^2} \qquad [4.1]$$

$$\begin{cases} t = 0, & T_S(0, x) = T_0 \\ x = 0, & -k_S \dfrac{\partial T_S}{\partial x} = \varphi(t) \\ x = z_T & T_S(t, z_T) = T_0 \end{cases} \qquad [4.2]$$

and

$$\begin{cases} t = 0, & T_S(0, x) = T_0 \\ x = 0, & -k_L \dfrac{\partial T_L}{\partial x} = \varphi(t) \\ x = \xi(t), & T_S(t, \xi) = T_L(t, \xi) = T_M \\ x = z_T & T_S(t, z_T) = T_0 \end{cases} \qquad [4.3]$$

The boundary conditions for $t = 0$ and $x \to \infty$ (i.e. $x = z_T$) are expressed by room temperature (T_0). The condition at the immediate surface ($x = 0$) translates the absorption of heat flux density. At the melting position ($x = \xi(t)$), the continuity of the thermal field is considered, meaning both solid and liquid temperatures equal the melting temperature. At the same melting position, the heat flux balance expressed by relationship (4.4) must take the latent heat flux of melting into account. Expressly:

$$\varphi_L - \varphi_S \Big|_{x=\xi(t)} = \left(-k_L \frac{\partial T_L}{\partial x}\right) - \left(-k_S \frac{\partial T_S}{\partial x}\right)\Bigg|_{x=\xi(t)} \qquad [4.4]$$

In the case of a 'thermally thin' layer, the condition in (4.2) and (4.3) at $x = z_T$ should be replaced by $x = z$, and respectively $T_S(t, z) = T_0$ by

$$T_S(t, z) - T_0 = R_{TC} \times \varphi(t, z) = R_{TC} \times \left[-k_S \frac{\partial T_S(t, x)}{\partial x}\right]_{x=z}$$

Notice that the last expression that translates the presence of a thermal resistance at the interface position is equivalent to the *first Ohm's law* (i.e. $\Delta U = R \times i$).

4.4.1 Simulation of laser-induced melting

Considering the gold reflectivity, the optical depth is close to 10 nm, when the thermal penetration is about 1000 nm for 25 ns pulse duration. So, the absorbed laser beam is treated as a surface boundary condition regarding the optical depth, and not as a volume heat source [18, 19]. To ensure the convergence of the numerical calculation, the number of grids should be greater than 1000 in the two space directions (time and position).

The first simulation is employed to understand the effect of thermal properties, especially the thermal diffusivity on the melting depth time behavior (curves D). When the gold layer is electroplated on the substrate (see Fig. 4.2), the heat propagation is strongly modified by the presence of grains, unlike the bulk state. A few references [3, 20] have pointed out this crucial problem. In practice there is no data to correlate the physical properties at a macroscopic level (i.e. thermal or electrical conductivity, mechanical properties…) to the elaboration processing of thin layers. Thus to start this simulation of laser intensity shapes, KrF and XeCl are fitted by linear curves as shown on Fig. 4.10. The fitted KrF curve corresponding to 750 mJ/cm² laser fluence is considered in equation system (4.1) to (4.4). The same is done for the XeCl curve and then for the computation return melting curves D-KrF and D-XeCl. The melting kinetics are very different (at the same laser fluence 750 mJ/cm²) in terms of melting threshold time, melting duration and maxima. D-XeCl seems to be better for achieving soft (less intensity) and more progressive (higher melting duration ≈ 40 ns) melting of the gold thin film.

4.4.2 Effect of the multilayer thermal properties

The thermal diffusivity of gold at the bulk state is close to 1.0 cm²/s. As shown in Fig. 4.11, no melting occurs for this bulk value. Curves C1, C2 and C3 correspond to 0.4, 0.2 and 0.1 cm²/s. As shown previously, experiments with this laser fluence (750 mJ/cm²) are enough to initiate the surface melting. As expected, the strong change of the gold microstructure induces a change in the diffusivity value. As an example, to melt 0.25 μm of the layer (KrF

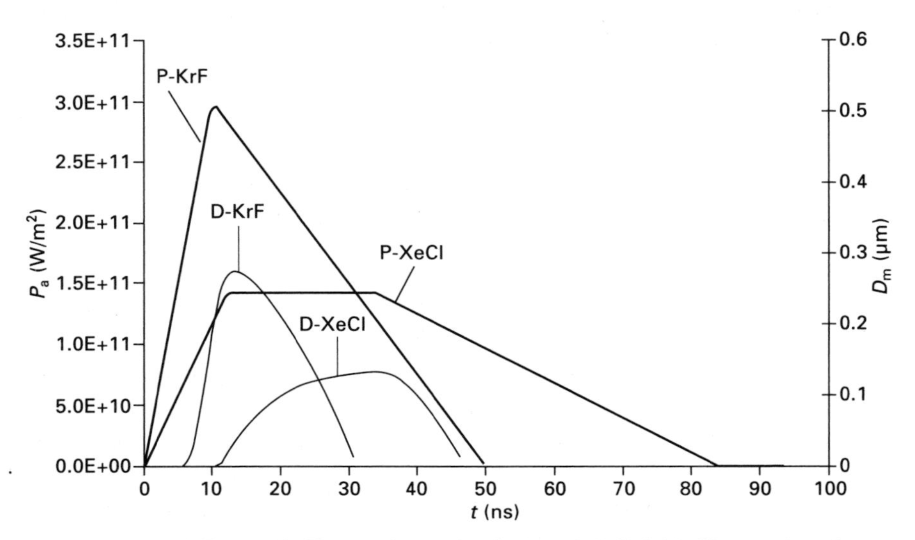

4.10 KrF and XeCl laser intensity fitting (P-KrF, P-XeCl) correlated to the induced melting depth (D-KrF, D-XeCl).

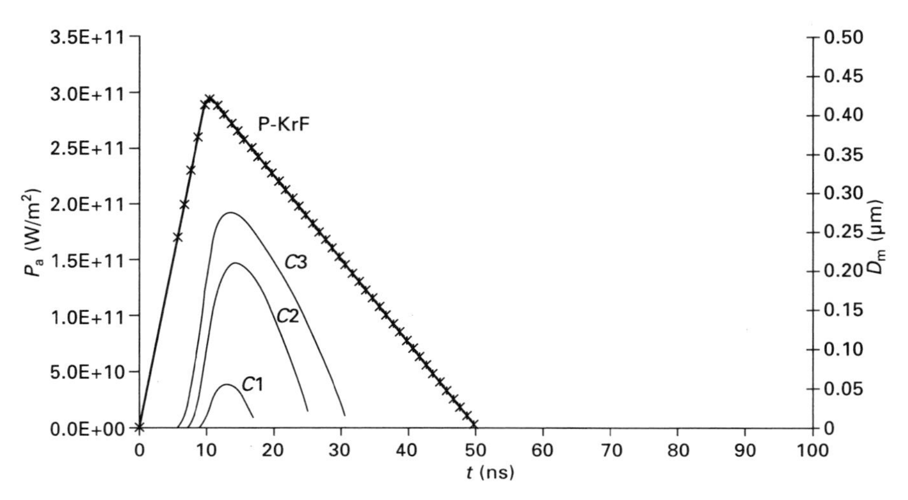

4.11 Melting depth versus time for different diffusivity values 4.E-01, 2.E-01, 1.E-01 cm²/s corresponding to melting curves C1 to C3 respectively. KrF laser fluence is 750 mJ/cm².

laser beam) the surface must absorb 750 mJ/cm² when thermal diffusivity is close to 0.1 cm²/s. The lifetime of this melting pool is then roughly 20 ns. The diffusivity value is lower; more important are the melting depth and the lifetime. In contrast, the curves C1, C2 and C3 show that the lower the thermal diffusivity, the shorter is the time to start the melting.

4.4.3 Effect of gold layer thickness and interface roughness

To understand the effect of film thickness on the melting kinetics, we have extended our modeling to study the 'thermally thin' metallic coatings [18]. The thermal contact resistance is introduced in the computation procedure when the heat wave propagation reaches the gold/nickel interface as reported in the numerical simulation section. It is then possible to calculate the critical temperatures and the melting threshold fluence for high and low contact resistance values (e.g. high and low substrate roughness). Under these conditions, the temperature profile and melting depth were plotted for different gold layer thickness (200, 500 and 800 nm). Considering a realistic thermal diffusivity a = 0.1 cm²/s and laser fluence 750 mJ/cm², the most important results are summarized in the following:

- The larger the layer thickness (200, 500 and 800 nm), the higher is the maximum melting zone (respectively 100, 220 and 340 nm), but the ratio of melting zone to layer thickness decreases steadily.

- The smaller the layer thickness, the higher is the effect of increasing thermal resistance at the interface on the melting zone. Typically, for 0.8 μm gold layer no change occurs on the melting zone (340 nm) whatever is the value of the thermal contact resistance in the range 1.7×10^{-11} (good thermal contact) to 1.7×10^{-7} K m²/W (bad contact) [20].

4.5 Conclusions

This study demonstrated that SLM processing could improve the corrosion resistance of gold-plated contacts for several laser conditions. Comparing the observed results with the modeling data, it appears that the most important laser parameter limiting the process efficiency is the space beam homogeneity. Concerning the material parameters, the roughness substrate is the most limiting. As a general rule, the best results for SLM treatment for corrosion limitation can be obtained using a pulsed laser emitting in the visible or UV wavelength range. It has been observed that the laser treatment is much more efficient on polished samples ($R_a < 0.1$ μm), nevertheless the treatment remains available for a substrate roughness reaching a value as high as 0.15 μm. The number of laser pulses has to be ranged from 1 to 5 at the same location on the sample and the laser fluence must be over 550 mJ/cm^2 to be efficient in corrosion tests but has to be limited to 750 mJ/cm^2 to prevent the sample from a roughness increase of its surface.

More appropriate modeling to simulate the thermal behavior of surfaces treated by pulsed lasers has been developed, especially for thin films by excimer lasers. This modeling is easily extended to the study of two situations: 'thermally thick' and 'thermally thin' metallic coatings. Results presented in this chapter show that the role of thermal contact resistance (R_{TC}) is strong enough for 'all gold' thickness below 0.5 μm when the thermal diffusivity of a surface material is considered to be 0.1 cm²/s, and the processing laser fluence is maintained at 750 mJ/cm².

4.6 References

[1] S. Zakipur, J. Tidblad, C. Leygraf, 'Atmospheric corrosion effects of SO_2, NO_2, and O_3', *J. Electrochem. Soc.*, **144**, 10, (1997) 3518–3525

[2] A.M. Prokhorov, V.I. Konov, I. Ursu, I.N. Mihailescu, *Laser Heating of Metals*, Adam Higher Series, IOP Publishing, Bristol (1990)

[3] J. Dutta Majumdar, R. Galun, B.L. Mordike, I. Manna, 'Effect of laser surface melting on corrosion and wear resistance of a commercial magnesium alloy', *Materials Science and Engineering A* **361** (2003) 119–129

[4] C.H. Tang, F.T. Cheng, H.C. Man, 'Effect of laser surface melting on the corrosion and cavitation erosion behaviors of a manganese–nickel–aluminium bronze', *Materials Science and Engineering A* 373 (2004) 195–203

[5] W. Khalfaoui, E. Valerio, J-E. Masse, M. Autric, 'Excimer laser treatment of ZE41

magnesium alloy for corrosion resistance and microhardness improvement', *Optics and Lasers in Engineering* **48** (2010) 926–931

[6] F. Viejo, A.E. Coy, F.J. Garcia-Garcia, Z. Liu, P. Skeldon, G.E. Thompson, 'Relationship between microstructure and corrosion performance of AA2050-T8 aluminium alloy after excimer laser surface melting', *Corrosion Science* **52** (2010) 2179–2187

[7] Z. Liu, P.H. Chong, A.N. Butt, P. Skeldon, G.E. Thompson, 'Corrosion mechanism of laser-melted AA 2014 and AA 2024 alloys', *Applied Surface Science* **247** (2005) 294–299

[8] K.G. Watkins, M.A. McMahon, W.M. Steen, 'Microstruture and corrosion properties of laser surface processed aluminium alloys: a review', *Materials Science and Engineering A* **231** (1997) 65–81

[9] Z. Liu, P.H. Chong, P. Skeldon, P.A. Hilton, J.T. Spencer, B. Quayle, 'Fundamental understanding of the corrosion performance of laser-melted metallic alloys', *Surface & Coating Technology* **200** (2006) 5514–5525

[10] N. Zaveri, M. Mahapatra, A. Deceuster, Y. Peng, L. Li, A. Zhou, 'Corrosion resistance of pulsed laser-treated Ti-6Al-4V implant in simulated biofluids', *Electrochimica Acta* **53** (2008) 5022–5032

[11] T.M. Yue, L.J. Yan, C.P. Chan, C.F. Dong, H.C. Man, G.K.H. Pang, 'Excimer laser surface treatment of aluminium alloy AA7075 to improve corrosion resistance', *Surface & Coating Technology* **179** (2004) 158–164

[12] Y.S. Touloukian, D.P. DeWitt, *Thermal Radiative Properties: Metallic elements and alloys*, Vol 7, IFI, Plenum, New York (1972)

[13] H.C. Hottel, A.F. Sarofim, *Radiative Transfer*, McGraw-Hill, New York (1967)

[14] Y.S. Touloukian, T. Makita, *Thermal Properties of Matter*, Plenum, New York, (1970)

[15] Norm Bellcore 'Generic Requirements for Separable Electrical Connectors used in Telecommunications Hardware' GR-1217- CORE, Issue 1, November 1995

[16] C. Georges, H. Sanchez, N. Semmar, C. Boulmer-Leborgne, C. Perrin, D. Simon, 'Laser treatment for corrosion prevention of electrical contact gold coating', *Applied Surface Science*, **186**, 1–4, (2002) 117–123

[17] C. Georges, N. Semmar, C. Boulmer-Leborgne, C. Perrin, D. Simon, 'Laser treatment for corrosion prevention' Proceedings of SPIE, Vol 4760, (2002) 986–993

[18] N. Semmar, C. Boulmer-Leborgne, 'Thermal modelling of surface pulsed laser annealing', *J. Phys. IV*, **120**, (2004) 413–420

[19] N. Semmar, C. Georges, C. Boulmer-Leborgne, 'Thermal behaviour of electric connector coating irradiated by a laser beam', *Microelectronics Journal*, **33** (2002) 705–710

[20] S. Orain, Y. Scudeller, S. Garcia, T. Brousse, 'Use of genetic algorithm for simultaneous estimation of thin films thermal conductivity and contact resistances', *International Journal of Heat and Mass Transfer*, **44** (2001) 3973–3

5
Laser surface modification of nickel–titanium (NiTi) alloy biomaterials to improve biocompatibility and corrosion resistance

K. W. NG and H. C. MAN, Hong Kong Polytechnic University, China

Abstract: This chapter investigates the feasibility of applying a laser surface alloying technique to improve the corrosion resistance and biocompatibility of nickel–titanium (NiTi) in simulated body fluid. It summarizes the result of laser surface modification of NiTi with Mo using a CO_2 laser into three sections. The microstructure, chemical composition, surface morphology, hardness, corrosion resistance, nickel release rate, wettability, bone-like apatite formation and cell adhesion behavior of the surface alloyed layer were analyzed using scanning electron microscopy (SEM), energy dispersive analysis by X-rays (EDAX), X-ray diffractometry (XRD), Vicker's microhardness, polarization tests, atomic absorption spectrometry, sessile drop technique, immersion test and cell adhesion analysis.

Key words: nickel–titanium (NiTi), molybdenum (Mo), laser surface alloying, biocompatibility, cell adhesion analysis.

5.1 Introduction

Biomaterials are widely used in medicine, dentistry and biotechnology. Artificial medical devices have been developed to sustain, support, or significantly improve the life of patients. The population with implants grows at over 10% per year. The market size of the biomaterials industry is estimated to be over $9 billion per year in the United States [1]. The need for organ and tissue replacement is expected to be enormous [1]. However, the history of the development of modern biomaterials is still relatively short. Inadequate knowledge of the long-term behavior of these materials *in vivo* may result in various types of problems such as loss of implant function and release of toxic chemicals into the human body. Research to improve the safe use of implant materials for long-term implantation has increased dramatically recently.

5.1.1 Nickel–titanium (NiTi) as a biomedical implant material

Nickel–titanium shape memory alloy (NiTi SMA) has been increasingly used in the medical devices industry. The NiTi alloy used in biomedical applications is a near-equiatomic intermetallic which is a solid solution with 55 wt% of Ni and 45 wt% of Ti [2]. The shape memory effect (SME), superelasticity (SE) and good biocompatibility of this alloy are the driving forces for its early applications in commercial medical implants [3–7]. In addition to shape memory properties, NiTi alloy exhibits excellent corrosion resistance, wear resistance and mechanical damping capacity. Numerous successful medical applications of NiTi alloy have been reported recently. The alloy can be applied for the manufacturing of implants in orthopedics and dentistry, such as compression staples, bone plates and stents [8, 9].

5.1.2 Need of surface treatment

NiTi alloy exhibits desirable and superior properties for medical applications. However, the existence of a high concentration of nickel in the alloy has caused concern for its use in long-term implantation. It has been reported that the high Ni content in the alloy may be released in an aggressive physiological *in vivo* environment and the release of nickel may induce toxic and allergic response to body tissues [10–13]. It is a challenge to develop safe biomaterials for long-term implantation. As the excellent properties of NiTi are related to its high Ni content, the reduction of Ni content of the bulk material is therefore non-negotiable. However, if the surface of the NiTi can be modified such that the biocompatible properties of the surface can be improved without sacrificing the bulk composition of the alloy and hence its characteristic properties, it will be a most desirable alloy.

5.1.3 Importance of laser surface treatment

Among many different types of surface treatment, laser surface treatment has been shown to be a viable means for modifying the surface properties of materials [14–20]. Lasers have been employed for surface modification because of their flexibility, high precision and intense beam power [21]. The electromagnetic radiation of a laser beam is absorbed within the surface layer for metals. The laser energy can be deposited precisely at the point where it is needed. The bulk properties of NiTi alloy will not be affected after laser surface treatment [22]. The treated surface layer has a strong metallurgical bond with the substrate. The controlled thermal penetration and non-contact feature of laser minimize the need for the post-machining processes. Laser surface treatment is a clean and fast process for surface engineering. The technique can be applied to bone fixation plates or implants with relatively large surface area by overlapping the melt tracks.

5.2 Fundamental characteristics of nickel–titanium (NiTi)

NiTi SMA can exist in two different temperature-dependent crystal structures (phases) called austenite (high-temperature or parent phase) and martensite (low-temperature or daughter phase). The austenite phase is a cubic structure (i.e. face-centered cubic, fcc structure), and the martensite phase has a lower symmetry structure (i.e body-centered tetragonal, bct structure). Since the austenite crystal structure consists of an ordered row–column matrix of atoms as shown in Fig. 5.1, it is strong and rigid. On the other hand, the twinned martensite crystal structure is made up of a lattice of 'zig zagged' atoms (Fig. 5.2); it is soft and ductile and can be easily deformed.

In relation to the thermodynamic aspects of martensite transformation, there are four critical transformation temperatures as shown in Fig. 5.2. When austenitic NiTi is cooled, it will eventually change to the martensite phase. The temperature at which this transformation (martensitic transformation) starts is the martensite start temperature (M_s). The temperature at which martensitic transformation is completed is the martensite finish temperature (M_f). When martensitic NiTi is heated, it will eventually change to austenite.

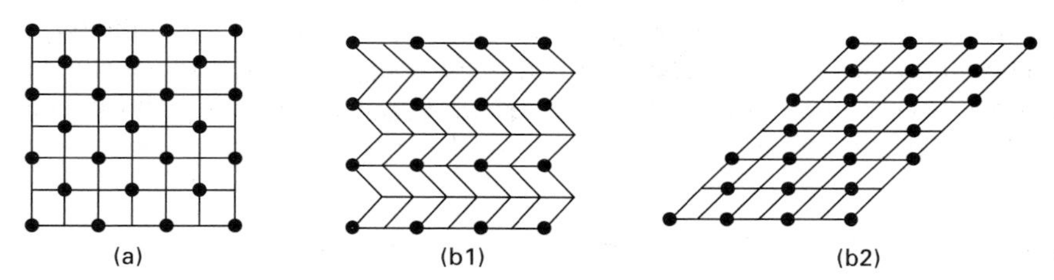

5.1 Different lattice elements in the (a) austenite phase and (b) martensite phase including (b1) twinned martensite phase and (b2) detwinned martensite phase.

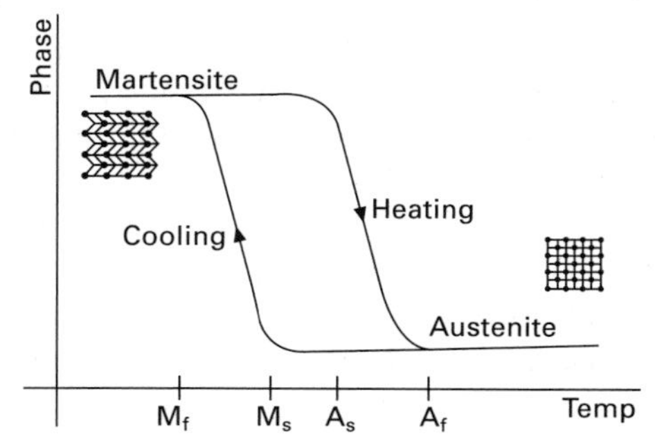

5.2 Martensitic transformation upon cooling and heating. A_s = austenite start, A_f = austenite finish, M_s = martensite start, M_f = martensite finish.

The transformation commences at the austenite start temperature (A_s) and is completed at the austenite finish temperature (A_f) [23]. The austenite–martensite transformation described above results in two main characteristics of NiTi: SME and SE.

5.2.1 Mechanical properties of NiTi

NiTi can change shape, stiffness and other mechanical characteristics in response to temperature [24, 25]. The basic mechanism governing the properties of NiTi is the austenite–martensite transformation. The crystal structure changes in response to the heating cycle and reverts to the original structure during cooling. The low-temperature martensite phase is easily deformed to several percent of strain at quite a low stress, whereas the high-temperature austenite phase has much higher stiffness and yield strength. Full austenitic NiTi at body temperature generally has suitable properties for surgical implantation.

NiTi responds to applied stress by simply changing the orientation of its crystal structure through the movement of twin boundaries while other materials deform by slip or dislocation movement. Thus NiTi exhibits up to 8% strain deformation and is fully recoverable by heating. NiTi has a high damping capacity below A_s. This particular property is beneficial in orthopedic applications to cushion the stress existing between the bone and the implant. The low elastic modulus of NiTi is much closer to the bone elastic modulus than that of any other implant metal, which can reduce the risk of osteoporosis. The unique high fatigue and ductile properties are also favorable for orthopedic implants. The good corrosion and wear resistance are comparable to other implant materials such as CoCrMo alloy and stainless steel [26].

5.2.2 Nickel aspect of NiTi

Nickel is a constituent part of all human organs. The normal nickel concentrations in human tissues are (microgram/kg of dry weight): 173 in lung, 62 in kidney, 54 in heart, 50 in liver, 44 in brain, 37 in spleen and 34 in pancreas [27]. Lower nickel intake reduces growth and skin changes, including altered skin pigmentation, parakeratosis and uneven hair development [28, 29]. Nickel deficiency may reduce iron absorption and result in anemia. Although nickel is nutritionally essential, it is well known that nickel is capable of eliciting toxic and allergic responses. The high nickel content of NiTi (55% by weight) may cause biocompatibility problems if nickel dissolves in deleterious amounts in the human body. The toxicity, carcinogenicity and allergic hazards caused by the release of nickel are of special concern [30].

The toxicity of nickel had been studied *in vitro* and *in vivo* using nickel salts, solid nickel or particulate forms of nickel [31, 32]. Putters *et al.* found that nickel had toxic effects with cellular damage in cell cultures at high concentrations [32]. In *in-vivo* tests, high-dose nickel salts were injected into mice. Accumulation and some deleterious effects were seen in the liver, kidney and spleen [33]. Nickel can cross the placental barrier and influence prenatal development on the embryo. Fetal death and malformations have been reported for various species of nickel compounds injected in experimental animals [34].

Pure nickel implant has been found to cause local tissue irritation and to have a high carcinogenic effect [35]. Cancer has been observed in a variety of organs of experimental animals following the injection of nickel compound. Significantly higher risk for cancer of lungs and nasal cavity has been reported from nickel refinery workers. Workers who were exposed to various nickel compounds were found to inhale higher nickel content than the non-occupationally exposed population [34].

Apart from the toxicity and carcinogenicity of nickel, nickel was also found to be the major cause of allergic problems [36]. Previous studies had indicated that the release of nickel ions from NiTi causes allergic problems to the body system [37, 38]. An initial nickel release rate of $0.5\,mg/cm^2/$ week has been reported [39]. NiTi was found to release more nickel in the cell culture medium than stainless steel [40]. The major risk associated with NiTi is the breakdown of the passive film and the deleterious effects of Ni ion release through damaged passive film.

Nickel ion release from surgical implants probably results primarily from corrosion [41]. The corrosion of metals may then lead to release of higher metal ions and consequently to higher cytotoxicity [42]. Systemic toxicity may be caused by the accumulation and subsequent reaction of the host to corrosion products [43–45]. There are reports about the potential negative impacts of nickel release from NiTi and other alloys on biological systems. Es-Souni *et al.* [42] studied the properties of NiTi and stated that Ni ions release would lead to lower biocompatibility. Granchi *et al.* showed in a joint prosthesis study that the type of cell death is dependent on the concentration of metal ions released from the prosthesis [46]. Wever *et al.* stated that the decrease in the cell activities *in vitro* may be attributed to a release of Ni ions [47]. The high cell death rate in NiTi samples was partly ascribed to the dissolved nickel from the samples in biological environments [48, 49]. Nickel ion dissolution has also been demonstrated to adversely affect the healing of bone [50].

In order to reduce the initial release of nickel out of NiTi, modification to the surface of NiTi seems to be feasible, to further enhance the corrosion resistance of NiTi. Laser surface treatment can produce a rapidly resolidified surface layer, in which both the microstructure and the distribution of the surface alloying elements can be greatly modified. Previous studies [17, 51] have demonstrated that laser surface treatment is an effective way to reduce the elemental Ni level at the surface.

5.2.3 Corrosion behavior of NiTi

The biological aqueous environment of the human body is reported as being one of the factors facilitating the corrosion of metallic implants inside the human body [52]. The concentration of the aqueous media containing ions of blood plasma is shown in Table 5.1 [53]. The physiological fluids and tissues can cause drastic changes in the chemistry surrounding the implant, including a drop in pH and release of strong oxidizing agents. Metal ion release from surgical implants may be toxic and dissolve in body fluids, and adversely affect biocompatibility by corrosion. Characterization and prediction of corrosion are important to determine the adverse biological reaction in the body and to forecast the functionality of the metallic implant under long-term implantation inside the human body.

The corrosion resistance of NiTi in simulated physiological environment is comparable to or even higher than that of Co–Cr–Mo alloy and 316L stainless steel [23, 54]. However, NiTi contains more than 50% of Ni by weight. Immersion tests of NiTi in simulated body fluids (SBF) at a temperature close to human body conditions and pH between 7.3 and 7.5 were found to have satisfactory corrosion resistance (<0.01 mm/year) [55]. Oshida and Miyazaki [56] compared the corrosion of the fabricated orthodontic wire and reported a leaching of nickel from the inhomogeneous structure. The corrosion resistance of NiTi is mainly based on passivity of the surface [57]. A comparison between NiTi and stainless steel reported a better corrosion resistance in Hanks' solution because of the TiO_2 passive film on the NiTi surface [47]. The passive film reduced the nickel release rate significantly from the initial $15.5 \times 10^{-7}\,\mu g\,cm^{-2}\,s^{-1}$ to an undetectable level by atomic absorption spectrometer after 10 days. Contaminants of Ca and P were generally seen in the oxide layer [58]. The biocompatibility of NiTi alloys has been inferred from their high corrosion resistance [47, 59, 60].

Table 5.1 Ionic concentration and pH of blood plasma

Blood plasma ion	Ion concentration (mmol/L)
Na^+	142.0
K^+	5.0
Mg^+	1.5
Ca^{2+}	2.5
Cl^-	103.8
HCO_3^-	27
HPO_4^{2-}	1.0
SO_4^{2-}	0.5
pH	7.40

Various structural properties of metallic implants such as their surface finish quality, porosity, roughness, lattice defects, impurities, contaminants and the degree of homogeneity of microstructures may affect the corrosion reaction. The corrosion resistance can be improved by a variety of surface treatments [20, 56, 61, 62]. However, no long-term *in-vivo* analysis data are available.

5.3 Laser surface alloying of nickel–titanium (NiTi) with molybdenum (Mo)

This section investigates the effects of laser surface alloying of NiTi with Mo on the long-term biocompatibility of NiTi. Mo was chosen to be the alloying element because of its good biocompatibility and non-toxicity feature. Molybdenum is one of the most widely used metals for strengthening alloys through solid solution hardening effects [63]. Mo significantly improves the hardenability and toughness of steels, while it prevents aging and increases resistance to tempering, wear and corrosion [64]. Good oxidation behavior was observed on the AISI 316L containing 2% Mo. Mo played a protective role to hinder iron diffusion and led to a lower growth rate and better scale adherence [65]. Molybdenum surface alloying improved the hardness and reduced the wear rate of austempered ductile iron [66]. With an increase in Mo contents, a microstructural change was responsible for the improvement in the mechanical properties such as hardness and wear resistance of the Mo-modified Stellite 6 hardfacing alloy [67]. The pitting potential of 304 type stainless steel films became more noble and more reproducible with increasing Mo content [68, 69]. Neither the International Agency for Research on Cancer nor the US Toxicology Program lists molybdenum as a carcinogen [70]. Mo is thus considered as one of the most desirable alloying materials for achieving high surface properties.

The optimum laser alloying parameters for Mo-alloyed NiTi were 800 W, 900 mm/min and 0.4 mm in powder paste thickness. This set of optimum parameters minimized the dilution ratio and the weight % of Ni of the alloyed layer.

5.3.1 Microstructural analysis

The cross-section microstructures of Mo-alloyed layer are shown in Fig. 5.3. It can be seen that a laser alloyed surface layer of Ni–Ti–Mo alloy is about 500 μm thick. The interface is of metallurgical nature and no cracking or porosity was found in either layer. The alloyed layer achieved a remarkable microstructural refinement and macroscopic compositional homogeneity following laser surface alloying. The alloyed element, Mo, appeared to have formed a solid solution with the NiTi matrix. The difference in contrast between the top, middle and bottom regions in Fig. 5.3 reflects the difference in thermal gradient, solidification rate and Mo concentration in the regions concerned.

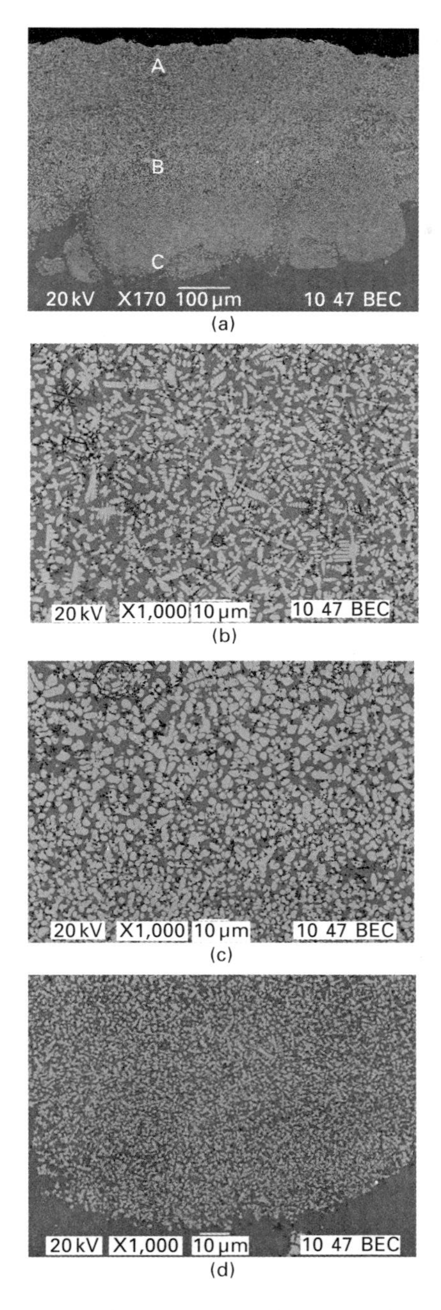

5.3 (a) Cross-section of the Mo-alloyed surface (800 W, 900 mm/min), (b) SEM image of the top region (area A of (a)), (c) SEM image of the middle region (area B of (a)), (d) SEM image at the interface (area C of (a)).

According to Fig. 5.3, different solidified microstructures are observed on the top region A (Fig. 5.3(b)), the middle region B (Fig. 5.3(c)) and the interface C (Fig. 3d) of the laser alloyed NiTi with Mo. Three different phases, grey solid solution, white dendrite and black dendrite, were mainly formed in the alloyed layer. Figure 5.4 shows the results of compositional analysis on a particular area of the alloying zone. It is found that the grey solid solution, the white dendrite and the black dendrite are the NiTi substrate, the Mo-rich phase and Ti-rich phase, respectively.

As shown in Fig. 5.3(b), the top region of the alloyed zone mainly consists of an Mo-rich dendrite phase with a little Ti-rich dendrite phase. Cellular Mo-rich phases are observed in the middle layer of the alloyed zone (Fig. 5.3(c)). At the interface of the alloyed area (Fig. 5.3(d)), relatively small dendrites in the Mo-rich phase are identified. The different observation of the dendritic formation with the distance from the surface of the alloyed layer could be attributed to the different thermal characteristics.

A laser is an ideal tool for producing rapid solidification. The local melting and resolidification of a small portion of the surface of a material with a laser beam presents a unique opportunity for controlled crystal growth under high temperature gradient and high cooling rate.

When the laser beam irradiates the surface, the solidification begins at the interface as shown in Fig. 5.3(d). The large thermal gradient and the small solidification rate results in an infinite G/R ratio as the solidification begins. Near the interface, numerous nuclei developed rapidly. The nuclei grow into the cellular structure in the crystal growing direction. The structure is cellular although the shape of the cross-section of the cells is seen to be influenced by the crystallographic effects. Further from the interface, the cross-section

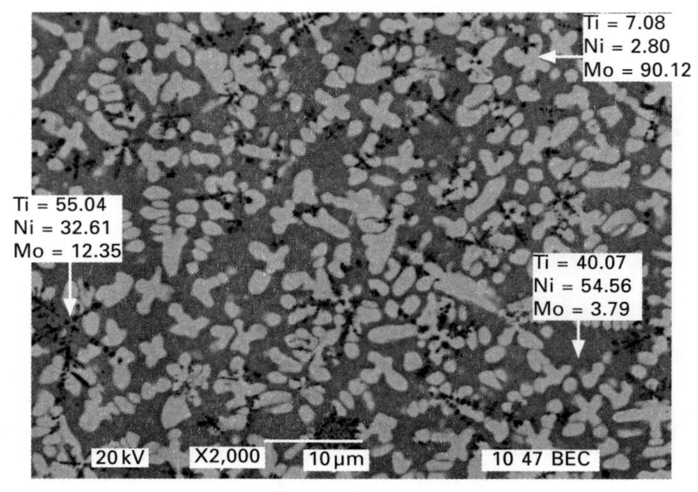

5.4 Composition analysis of different phases of the Mo-alloyed layer.

of the cells became flanged and dendritic structures become visible (Fig. 5.3c). Because the decreasing thermal gradient and the increasing growth rate result in a small *G*/*R* ratio, solidification is no longer with a plane front but dendrites form. Crystallographic effects begin to exert an influence and the cell-growth direction deviates toward the preferred crystallographic growth direction. Simultaneously, the cross-section of the cell generally begins to deviate from its previously circular geometry owing to crystallography effects. Also, the alloyed layer solidifies as a nearly uniform mush with equiaxed dendritic structure. When the alloy of wide freezing range and high conductivity is solidified, only very small temperature differences exist across the thickness and along the length of the alloyed layer. The small temperature differences result in very small differences across the thickness and length of the plate and in an equiaxed structure. As the growth rate increases still further (further from the interface), the cross-structure first became more apparent and then serrations appeared in the flanges of the cross; that is, secondary dendrite arms become discernible.

In Fig. 5.3(b), many such primary and secondary dendrite arms are seen. The central portion of the structure which is growing in approximately the heat-flow direction is termed a primary dendrite arm; the rod-like protrusions perpendicular to the primary dendrite arm are secondary dendrite arms. All the primary dendrite arms in this figure have grown from the same nucleus and have nearly the same crystallographic orientation. They are thus a part of the same dendrite. The size of the dendritic structure near the top region was relatively large. It is because the liquid became richer in solute during subsequent cooling and solidification, and so the solid that formed was of higher solute content at later stages of solidification. It was found that room temperature mechanical properties are generally increased with microstructural refinement because of the rapid solidification.

5.3.2 Compositional characterization

Figure 5.5 shows the variation of composition profile in the cross-section of the alloyed zone as a function of the depth from the surface. Earlier, a similar compositional variation has already been evidenced in the microstructure of the cross-section in Fig. 5.3. The composition profile indicates that Mo can successfully reduce the amount of Ni content of the surface of NiTi. There is an 80% reduction of Ni content on the surface of the Mo-alloyed layer. The composition profile is consistent with the laser-treated thickness. Beyond the depth of 500 μm, it is the substrate that mainly consists of Ni and Ti. A higher Mo content in the near surface region is beneficial for enhancing the surface properties. Thus, Fig. 5.3 indicates that the alloyed zone developed by the optimized laser surface alloying conditions is likely to offer an improved resistance to corrosion.

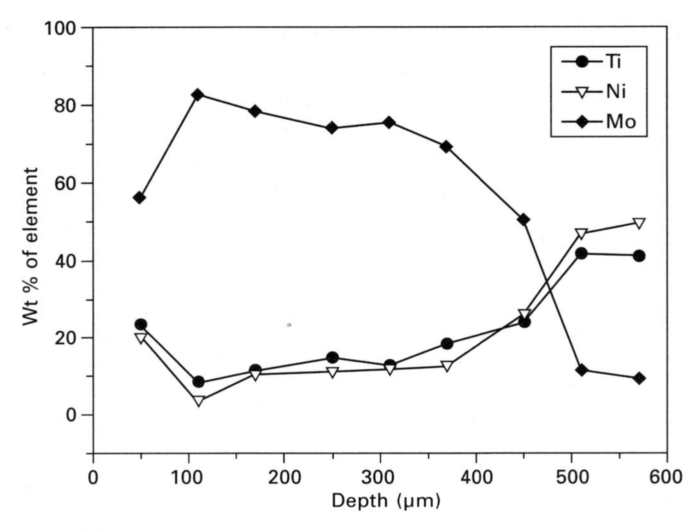

5.5 EDS composition profile of Mo-alloyed specimen.

Phases present

Figure 5.6 presents typical X-ray diffraction (XRD) patterns of the Mo-alloyed NiTi. According to the XRD patterns, several Mo peaks are present. Mo did not form any intermetallic compound with Ni and Ti. When Mo was alloyed at the surface of NiTi, the alloyed layer mainly consists of the austenitic phase NiTi and the Mo is present in solid solution. Figure 5.7 shows the XRD patterns of the as-received NiTi, Mo powder and Mo-alloyed specimen. There is no phase shift after the laser alloying process. The majority of the peaks of the XRD patterns of NiTi and Mo powder are retained for the Mo-alloyed sample. The previous microstructural investigation on the top and cross-section have revealed that the optimum processing conditions resulted in complete melting and mixing of the Mo to form an alloyed zone at the surface of NiTi.

Hardness profile

Figure 5.8 presents the hardness profile as a function of depth measured on the cross-section of the Mo-alloyed NiTi. The maximum hardness value of the Mo-alloyed layer is 690 HV while the hardness for the NiTi substrate is only 220 HV. Thus laser surface alloying with Mo has resulted in an increase in hardness in the alloyed zone by a factor of 3. Higher Mo content was found in the alloyed zone. It is plausible that this relatively higher hardness is related to the higher Mo content [69]. The significant improvement in surface hardness could be due to the synergistic effect in grain refinement

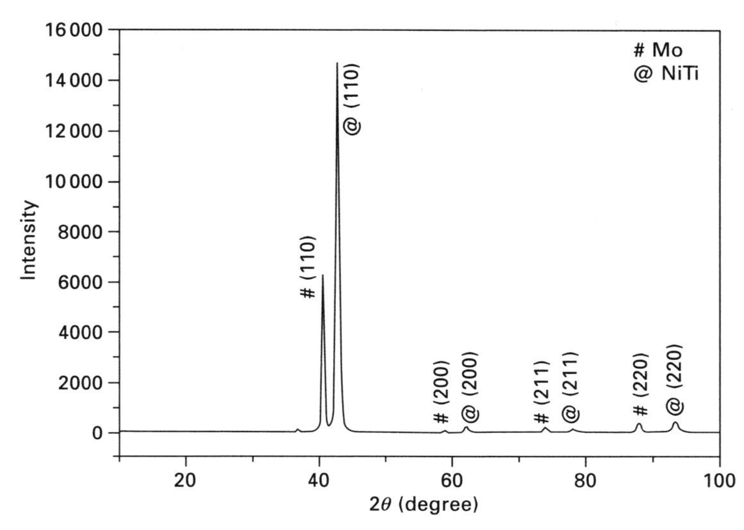

5.6 X-ray diffraction patterns of Mo-alloyed specimen.

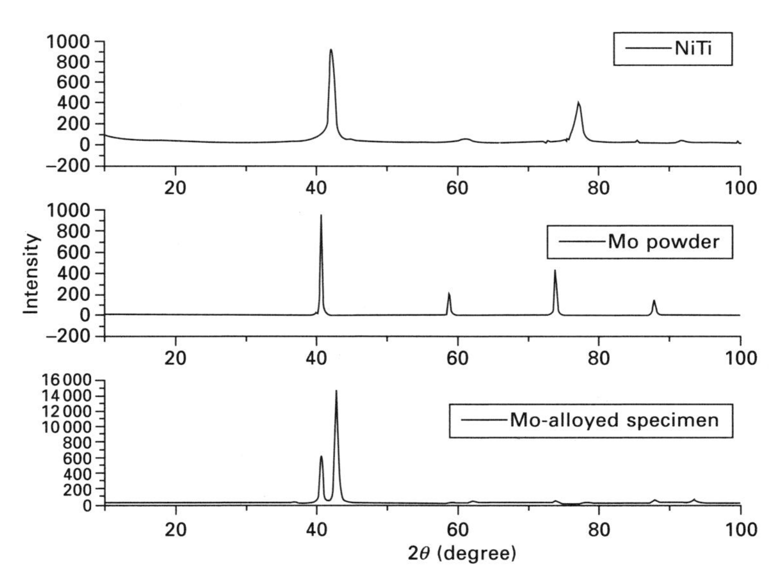

5.7 X-ray diffraction patterns: comparison between Mo powder, untreated NiTi and Mo-alloyed NiTi.

and Mo solution hardening in the laser surface alloyed zone. It may be noted that the hardness, which is a function of the processing parameters, remains uniform throughout the alloyed zone. This uniformity in hardness suggests a macroscopic homogeneity in the alloyed zone.

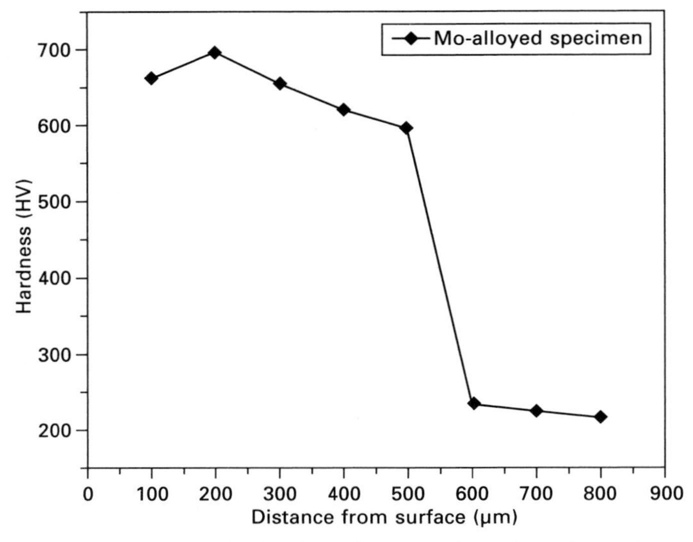

5.8 The hardness distribution along the Mo-alloyed specimen.

Corrosion resistance characteristics

The polarization behavior of the as-received and Mo-alloyed NiTi in Hanks' solution is depicted by the polarization curves in Fig. 5.9. Values for the corrosion potential E_{corr}, pitting potential E_{pit} and corrosion current density I_{corr}, extracted from the curves, are shown in Table 5.2. As shown in Fig. 5.9, a significant improvement in the corrosion resistance is evidenced by a large increase of the breakdown potential (370 mV) of the as-received NiTi. The passive film of Mo-alloyed specimen did not break down even at a potential as high as 2000 mV, indicating that no pitting corrosion occurred. The potentiodynamic corrosion properties reveal unambiguously a higher corrosion resistance of the Mo-alloyed NiTi. In particular, the passive current density for the Mo-alloyed specimens is around 10^{-11} A/cm^2, an order of magnitude lower than that of the as-received NiTi. Clearly, the Mo-alloyed specimen presents higher corrosion resistance than the substrate.

The improved corrosion resistance could be attributed to the refined microstructure, smaller grain size, the absence of inclusions and the existence of elemental Mo in the Mo-alloyed layer. The structural and chemical homogeneities are very important characteristics in resisting localized corrosion [68]. An abrupt reduction of grain size occurs at critical undercooling [71]. Optical micrographs demonstrate that the addition of Mo decreases the grain size considerably. The smaller grain size thus reduces the relative surface homogeneity and makes the alloy less susceptible to corrosion. The microstructural refinement is as crucial as the composition in the alloyed layer to impart the desired resistance to localized corrosion.

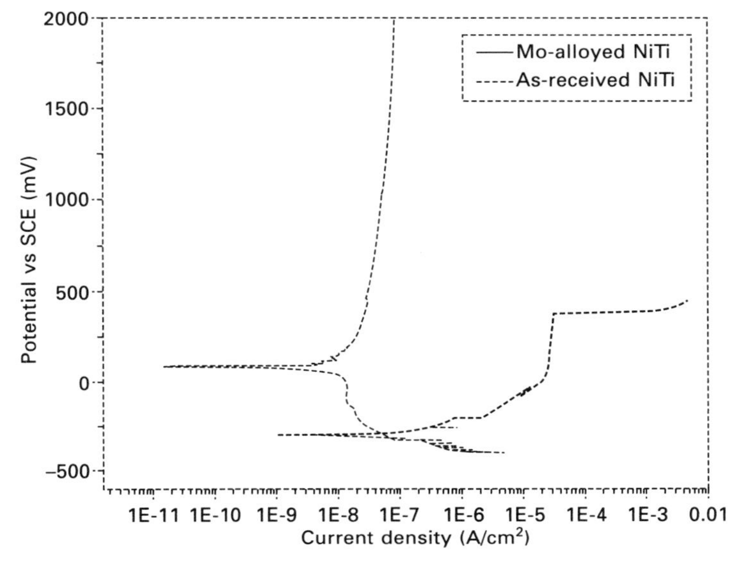

5.9 Polarization curves of Mo-alloyed specimen and as-received NiTi in Hanks' solution.

Table 5.2 Corrosion parameters of Mo-alloyed and as-received NiTi in Hank's solution

	E_{corr} (mV)	I_{corr} (A/cm^2)	E_{pit} (mV)
As-received NiTi	–305	1.34×10^{-9}	370
Mo-alloyed NiTi	89	2.42×10^{-11}	–

The influence of Mo on the corrosion properties of stainless steels has been studied [68]. Mo is well established to be corrosion resistant due to the formation of a passive layer on its surface in aqueous environments [72, 73]. Mo addition makes the open-circuit potentials more positive, thus decreasing the oxidation tendency of the Ti-Mo alloys [74]. Mo confers special resistance to chloride attack. Mo added in the NiTi alloy strengthens the passive film and improves pitting resistance [75] because Cl^- ions have greater difficulty in penetrating the oxide layer due to some chemical alteration during pitting nucleation [76].

In titanium alloys, the resistance of the film increases with an increase in the Mo content due to a selective dissolution of α-phase followed by an accumulation of the β-phase enriched with Mo on its surface [77]. Pitting corrosion does not occur on titanium alloys containing Mo and that passive film has good stability [78]. A region of formation and growth of the anodic oxide in the passive region of the Ti–Mo alloy is obtained even in solutions containing Cl^- ions and results in no pitting corrosion until the end of the growth potential [79]. The increased Mo content in NiTi alloys decreases the alloy's susceptibility to corrosion in Hanks' solution. The laser surface alloying technique is capable of imparting an excellent superficial microstructure and resistance to corrosion by the synergistic effect of grain refinement, microstructure homogenization and the Mo alloying effect [18].

Figure 5.10 and 5.11 show the surface morphology of the as-received NiTi and Mo-alloyed NiTi respectively after the polarization test. Many large pits

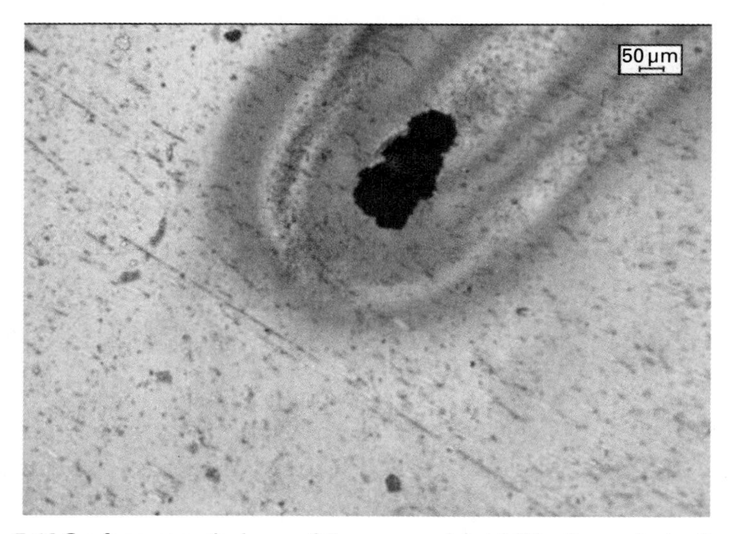

5.10 Surface morphology of the as-received NiTi after polarization test.

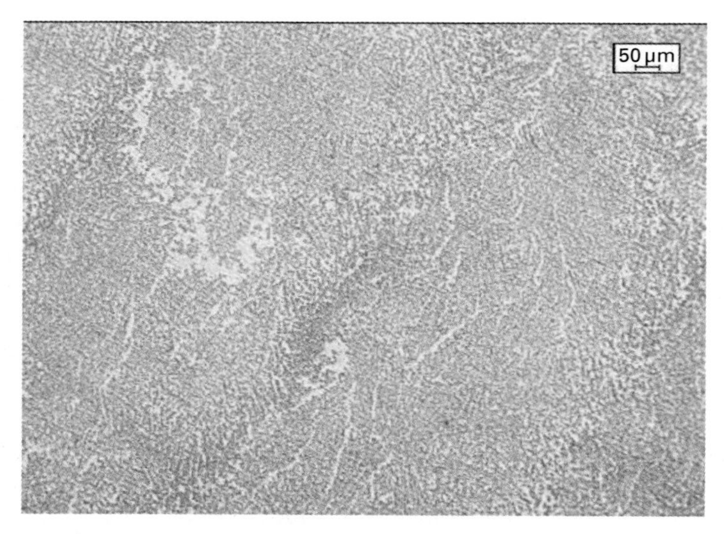

5.11 Surface morphology of the Mo-alloyed NiTi after polarization test.

are found on the surface of as-received NiTi where the largest one is about 150 μm, depicting the corrosion attack on the surface. The surfaces of the Mo-alloyed specimens are comparatively smooth with no pits. The relative poor corrosion resistance of the as-received NiTi is due to the existence of inclusions on the surface. As a result, lots of pits are formed after the polarization test [80]. Additions of Mo give improved pitting corrosion resistance and microstructural stability [75, 81].

Nickel release rate

NiTi SMAs are promising materials in orthopedic applications and for surgical implants because of their unique SME and SE. However, for prolonged use in a human body, deleterious ions released from NiTi to living tissues may become a critical issue [82]. The long-term biocompatibility of NiTi is still under discussion due to its high nickel content [83]. The results of this investigation on laser surface alloying of NiTi with Mo (which acts as a Ni stabilizer) for reducing the elemental Ni at the surface have shown promising effects.

Figure 5.12 shows the amount of Ni ion release into Hanks' solution after the specimens were immersed at different time intervals. The results show that by laser surface alloying NiTi with Mo, the release of Ni ions can be greatly reduced. For as-received NiTi, the highest Ni release occurs within the first day. Then, the nickel release rate decreases with time and tends to

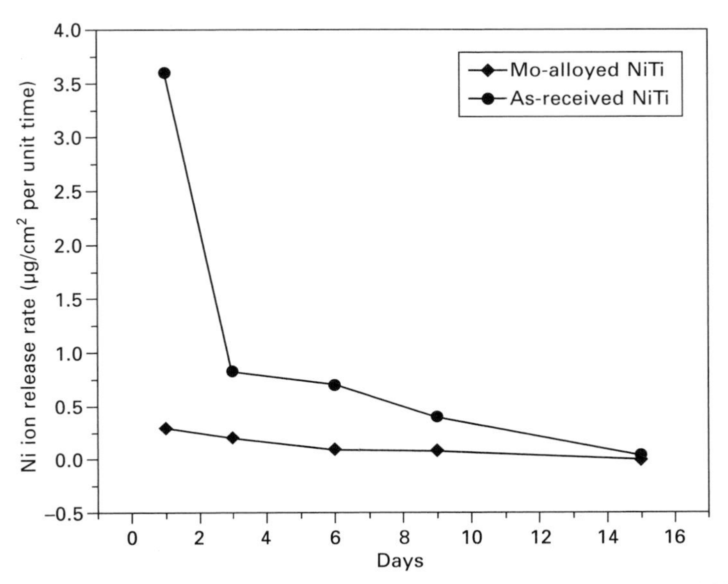

5.12 The nickel release rate of the as-received and Mo-alloyed NiTi in Hanks' solution at 37 °C.

almost zero at the end of the testing period. On the other hand, the amount of Ni ions released from the Mo-alloyed NiTi is very small throughout the test period. The results of Ni release rate are similar to previous studies [15, 18, 47, 84].

The significant reduction in nickel release rate after laser surface alloying could be attributed to the improvement in surface structure [85]. Laser surface alloying creates a new and homogeneous surface. The formation of the uniform and stable passive layer improves the corrosion resistance. The Ni release rate is in agreement with the results of corrosion tests [86]. The value of I_{corr} of Mo-alloyed NiTi is lower and so is its Ni ion release rate.

Although the quantity of Ni ions released is much below the critical concentration (600–2500 μg) to induce allergy [87], this quantity might be enough to induce long-term inflammatory responses or alter cell behavior [88]. Thus, the results obtained for Mo-alloyed NiTi might be of great importance to improve the long-term biocompatibility properties of the material and to reduce sensitization to Ni and allergies.

Bone-like apatite formation

According to the histological examinations in the early implantation period [89], an apatite layer was formed on the implant surface in which the bone matrix integrated into the apatite layer. This apatite layer mainly consists of hydroxyapatite (HA), with composition $Ca_{10}(PO_4)_6(OH)_2$. The features of HA are very similar to the mineral phase in bone; thus osteoblast cells can preferentially proliferate on the apatite layer and differentiate to form an extracellular matrix composed of biological apatite. As a result, the formation of a layer of biologically bone-like apatite on the implant surface is essential for improving the bioactivity of the artificial material to bone in the human body. The biological activity of implant materials is determined by their apatite-forming ability which is the starting point in bone reconstruction [90]. Immersion tests in SBF for a period of a few weeks are commonly used for the *in-vitro* evaluation of apatite-forming ability [53, 91].

For the present work, the samples were soaked in SBF for 14 days. There was no or only a small amount of apatite formation for the untreated NiTi after immersion (see Fig. 5.13). It is possibly because the oxide layer on the sample surface is not bioactive as most metallic surfaces [92]. On the contrary, there was a substantial growth of HA on the Mo-alloyed NiTi as shown in Fig. 5.14. The obvious peaks of the HA in XRD patterns validate its formation on the surface of the Mo-alloyed NiTi, as shown in Fig. 5.15. It can be seen that the surface of the Mo-alloyed NiTi was continuously covered by many spheroidal particles. The easy formation of apatite on laser-treated NiTi in SBF could be attributed to the change of chemistry and wettability of the surface layer.

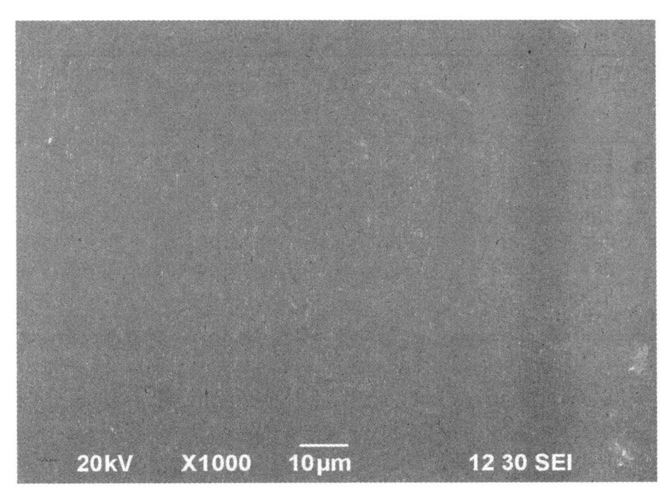

5.13 SEM micrograph of as-received NiTi showing surface after immersion in Hanks' solution for 14 days.

Osteoblast cell behavior

Although cell culture studies cannot directly mimic the cellular response and the environment *in vivo*, the toxic effects of different metals can be well quantified *in vitro*, because the cell culture method is a sensitive method for toxic screening [38]. The first physiological process that occurs within the initial stage of exposure is the adsorption of biomolecules onto the surface which is usually followed by cellular interactions. The quality of this adhesion will influence their morphology and their future capacity for proliferation and differentiation. The biological response to artificial materials is primarily driven by interfacial phenomena at biomaterial surfaces and wetting measurements directly probe these interfacial properties.

Cell adhesion is considered the pre requisite factor prior to subsequent cell proliferation and differentiation before tissue formation. The cells will firstly attach and grow on the surface; the cytoplasm will then web and flatten the cell mass [93]. The cellular interaction is classified by the adsorption process and the cell spreading ability.

Figure 5.16 shows the adhesion of osteoblast cells on the untreated NiTi and laser-treated NiTi. Fewer cells appeared on the surface of untreated NiTi, indicating that the surface is not favorable for cell attachment. When the NiTi was laser alloyed with Mo, more cells attached and grew on the Mo-alloyed NiTi as shown in Fig. 5.17. Laser surface alloying process resulted in a better cell attachment of NiTi surface compared with the untreated NiTi. The difference in cell attachment before and after laser treatment is likely to be a function of wettability and surface chemistry.

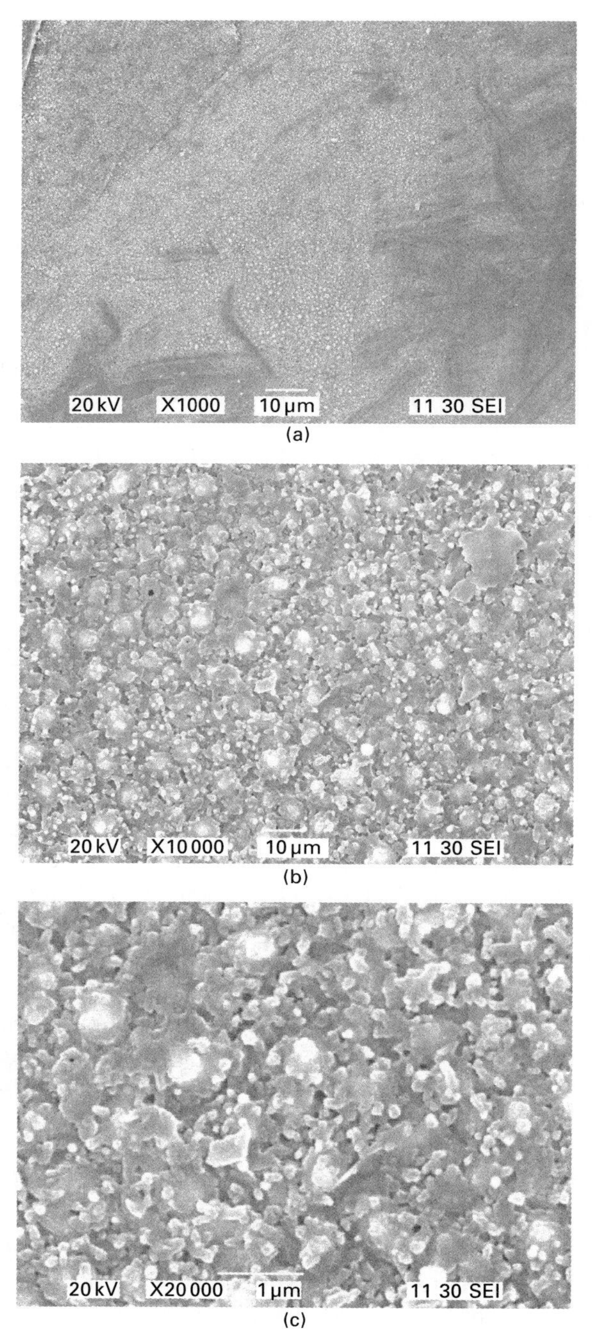

5.14 SEM micrographs of the Mo-alloyed NiTi showing surface after immersion in Hanks' solution for 14 days: (a) 1000× magnification; (b) 10000× magnification; (c) 20000× magnification.

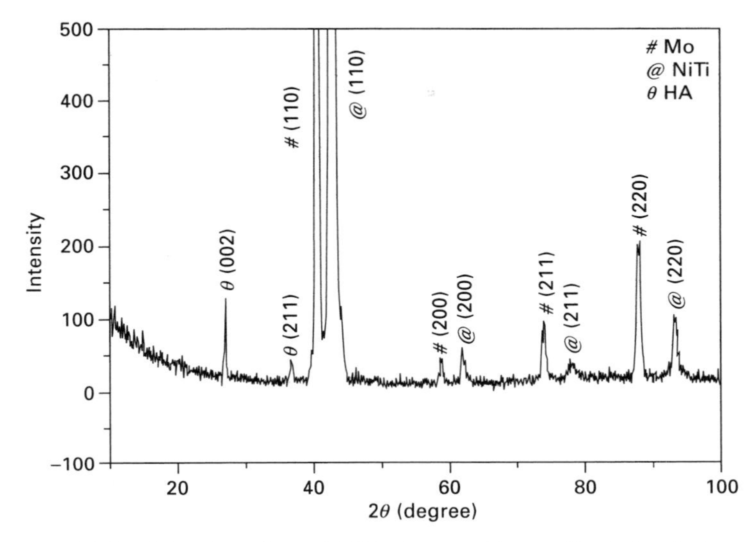

5.15 XRD of the Mo-alloyed NiTi after soaking in Hanks' solution.

The cells align themselves along the groove axis in relation to surface roughness produced during laser treatment as shown in Fig. 5.17. The effect of the groove depth and width on cell alignment has been made by Clark *et al.* [94]. The cells were found to react to the shape of grooved environment. The features explained that the cells accommodate the curvature of the topography and allow cell spreading in the direction of the laser irradiation. The osteoblasts oriented and elongated along the grooves of the Mo-alloyed NiTi sample. Since the dissimilarity of the surface structure resulted in various cell proliferations, topography could be a factor for cell response and biocompatibility. A significant increase in cell proliferation was observed after laser surface alloying process.

5.4 Conclusion

Mo was successfully alloyed on NiTi by the laser surface alloying technique. The following conclusions can be made:

- With proper selection of laser parameters, a good metallurgical bond was observed at the interface of the Mo-alloyed NiTi. The alloyed layers were free of microcracks and pores.
- The hardness of the Mo-alloyed NiTi (690 HV) was about three times that of the NiTi substrate (200 HV). A higher Mo content in the near-surface region is beneficial for enhancing the surface properties.
- Results of corrosion study reveal unambiguously a higher corrosion resistance of the Mo-alloyed NiTi. The passive film of Mo-alloyed NiTi

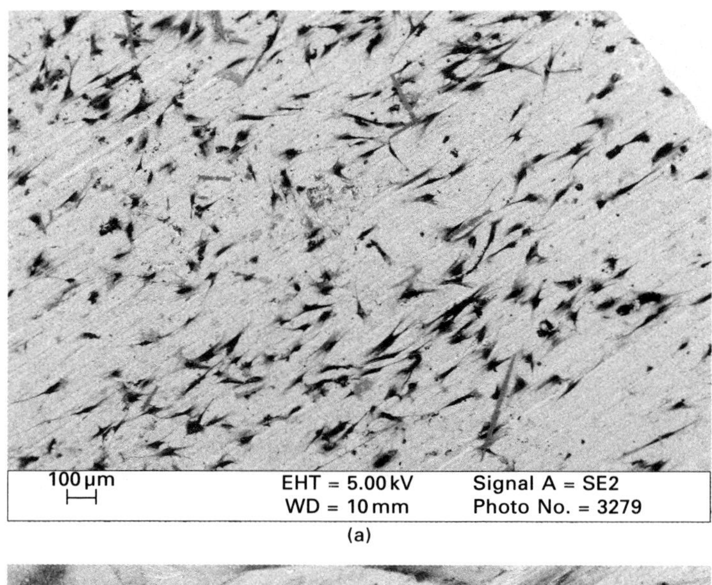

(a)

(b)

5.16 SEM micrographs of osteoblast cells on as-received NiTi: (a) lower magnification; (b) higher magnification.

does not break down even at a potential as high as 2000 mV. Mo is well known to be able to strengthen the passive layer on a metallic surface in a chloride environment.

- The amount of nickel release from the Mo-alloyed NiTi is about eight times lower than that of the as-received NiTi in Hanks' solution, indicating a safer surface for NiTi implants.

- Substantial growth of HA on the Mo-alloyed NiTi was observed. The sample surface was preferable for apatite formation, possibly due to the creation of a more favorable surface chemistry, which is pivotal for bioactivity.

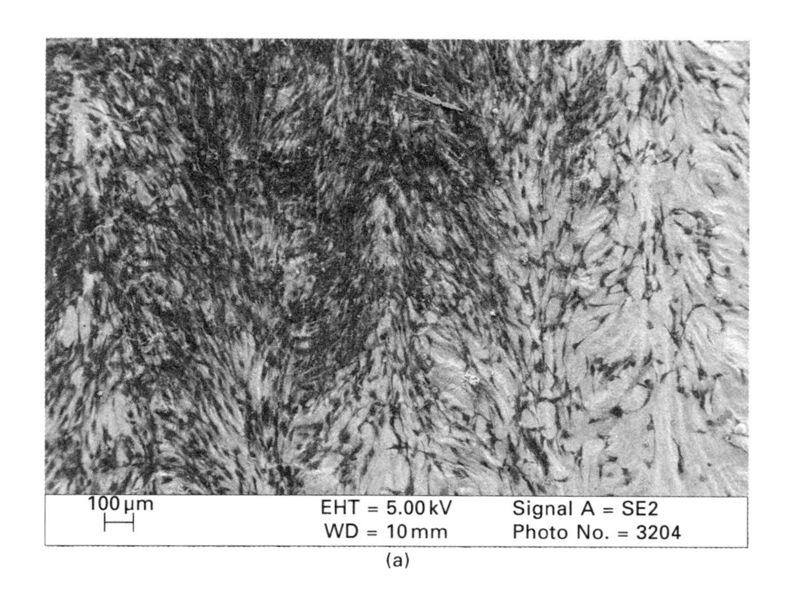

(a)

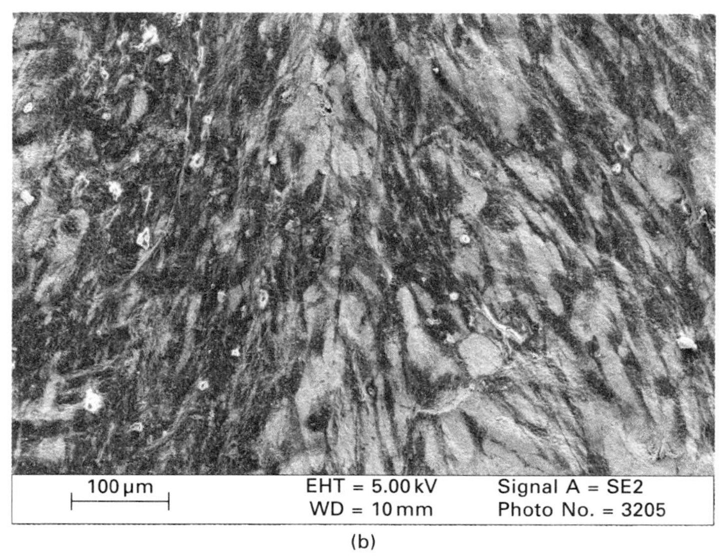

(b)

5.17 SEM micrographs of osteoblast cells on the edge of a Mo-alloyed NiTi sample with increasing magnification from (a) to (c).

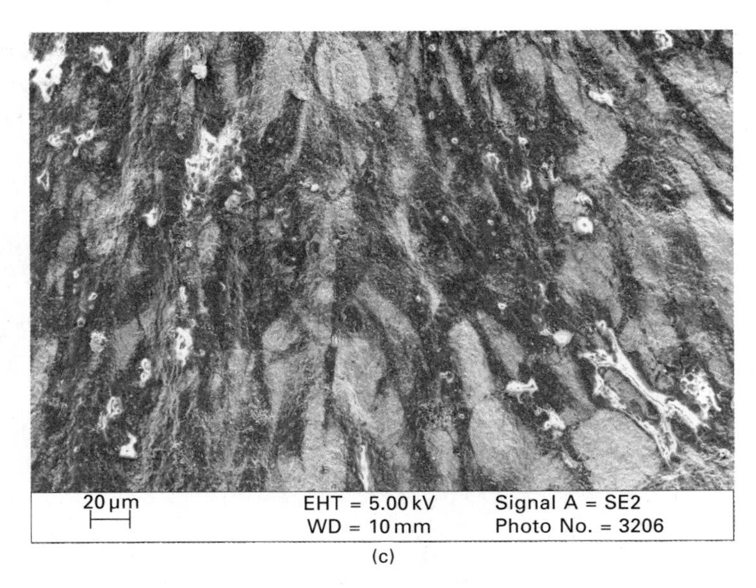

(c)

5.17 Continued

- Very densely packed cells appeared on the Mo-alloyed NiTi. The laser surface alloying process resulted in a better cell attachment of Mo-alloyed NiTi surface compared with the untreated NiTi.

5.5 References

[1] Temenoff, J.S., *Biomaterials: The intersection of biology and materials science*. 2008: Upper Saddle River, NJ: Pearson Prentice Hall. 3.

[2] Brunette, D.M., Titanium in medicine: material science, surface science, engineering, biological responses, and medical applications. *Titanium–nickel shape memory alloys in medical applications*, ed. P. Filip. 2001: Berlin; New York: Springer, p 54.

[3] Kasano, F. and T. Morimitsu, Utilization of nickel–titanium shape memory alloy for stapes prosthesis. *Auris Nasus Larynx*, 1997. **24**(2): p. 137–142.

[4] Shabalovskaya, S.A., On the nature of the biocompatibility and on medical applications of NiTi shape memory and superelastic alloys. *Biomedical Materials and Engineering*, 1996. **6**: p. 267–289.

[5] Prince, M.R., F.J., Schoen A.M., Palestrant and M. Simon, Local intravascular effects of the nitinol wire blood clot filter. *Investigative Radiology*, 1988. **23**: p. 294–300.

[6] Ryhänen, J.M.K., W. Serlo, P. Perämäki, J. Junila, P. Sandvik, E. Niemelä and J. Tuukkanen, Bone healing and mineralization, implant corrosion, and trace metals after nickel–titanium shape memory metal intramedullary fixation *Journal of Biomedical Materials Research*, 1999. **47**: p. 472–480.

[7] Castleman, L.S., F.P. Alicandri, V.L. Bonawit and A.A. Johnson, Biocompatibility of nitinol alloy as an implant material. *Journal of Biomedical Materials Research*, 1976. **10**: p. 695–731.

[8] Kennedy, J.B., *Shape Memory Alloys*. 1987, New York: Gordon and Breach Science Publishers. 226.

[9] Duerig, T., A. Pelton, and D. Stockel, An overview of nitinol medical applications. *Materials Science and Engineering A*, 1999. **273–275**: p. 149–160.

[10] Ponsonnet, L., *et al.*, Effect of surface topography and chemistry on adhesion, orientation and growth of fibroblasts on nickel–titanium substrates. *Materials Science and Engineering*: C, 2002. **21**(1–2): p. 157–165.

[11] Lincks, J., *et al.*, Response of MG63 osteoblast-like cells to titanium and titanium alloy is dependent on surface roughness and composition. *Biomaterials*, 1998. **19**(23): p. 2219–2232.

[12] Deligianni, D.D., *et al.*, Effect of surface roughness of the titanium alloy Ti–6Al–4V on human bone marrow cell response and on protein adsorption. *Biomaterials*, 2001. **22**(11): p. 1241–1251.

[13] Anselme, K., *et al.*, The relative influence of the topography and chemistry of TiAl6V4 surfaces on osteoblastic cell behaviour. *Biomaterials*, 2000. **21**(15): p. 1567–1577.

[14] Cui, Z.D., H.C. Man and X.J. Yang, Characterization of the laser gas nitrided surface of NiTi shape memory alloy. *Applied Surface Science*, 2003. **208–209**: p. 388–393.

[15] Cui, Z.D., H.C. Man and X.J. Yang, The corrosion and nickel release behavior of laser surface-melted NiTi shape memory alloy in Hanks' solution. *Surface and Coatings Technology*, 2005. **192**(2–3): p. 347–353.

[16] Hiraga, H., *et al.*, Fabrication of NiTi intermetallic compound coating made by laser plasma hybrid spraying of mechanically alloyed powders. *Surface and Coatings Technology*, 2001. **139**(1): p. 93–100.

[17] Man, H.C., Z.D. Cui and T.M. Yue, Corrosion properties of laser surface melted NiTi shape memory alloy. *Scripta Materialia*, 2001. **45**(12): p. 1447–1453.

[18] Man, H.C., K.L. Ho and Z.D. Cui, Laser surface alloying of NiTi shape memory alloy with Mo for hardness improvement and reduction of Ni2+ ion release. *Surface and Coatings Technology*, 2006. **200**(14–15): p. 4612–4618.

[19] Man, H.C. and N.Q. Zhao, Phase transformation characteristics of laser gas nitrided NiTi shape memory alloy. *Surface and Coatings Technology*, 2006. **200**(18–19): p. 5598–5605.

[20] Villermaux, F., *et al.*, Excimer laser treatment of NiTi shape memory alloy biomaterials. *Applied Surface Science*, 1997. **109–110**: p. 62–66.

[21] Ion, J.C., *Laser processing of engineering materials: Principles, procedure and industrial application*. 2005: Oxford: Elsevier Butterworth-Heinemann.

[22] Steen, W., *Laser material processing*. 1991, London: Springer-Verlag.

[23] Peng, J.D.K., Plunging method for Nd: YAG laser cladding with wire feeding. *Journal of Materials Processing Technology*, 2000. **104**: p. 284–293

[24] Duerig, T.W. and Stöckel, D., The utility of superelasticity in medicine. *Biomedical Materials and Engineering*, 1996. **6**: p. 255–266.

[25] Brunette, D.M., Titanium in medicine: material science, surface science, engineering, biological responses, and medical applications. *Titanium–nickel shape memory alloys in medical applications*, ed. P. Filip. 2001, New York : Springer. 54–81.

[26] Funakubo, H., *Shape memory alloys*. 1987, New York: Gordon and Breach Science Publishers.
[27] Rezuke, W.N. and F.W.J. Sunderman, Reference values for nickel concentrations in human tissues and bile. *American Journal of Industrial Medicine*, 1987. **11**: p. 419–426.
[28] Szilagyi, M. and Balogh I, Effect of nickel deficiency on biochemical variables in serum, liver, heart and kidneys of goats. *Acta Veterinaria Hungarica*, 1991. **39**: p. 231–238.
[29] Anke, M., H. Kronemann and M. Grun, *Nickel – an essential element*. IARC. Science Publication, 1984. **53**: p. 339–365.
[30] Starosvetsky, D. and I. Gotman, TiN coating improves the corrosion behavior of superelastic NiTi surgical alloy. *Surface and Coatings Technology*, 2001. **148**(2–3): p. 268–276.
[31] Takamura, K.H., N. Ishinishi, T. Yamada and Y. Sugioka, Evaluation of carcinogenicity and chronic toxicity associated with orthopedic implants in mice. *Journal of Biomedical Materials Research*, 1994. **28**(5): p. 583–589.
[32] Putters, J.L., A. Bijma and P.A. Besselink, Comparative cell culture effects of shape memory metal (Nitinol), nickel and titanium: a biocompatibility estimation. *European Surgical Research*, 1992. **24**: p. 378–382.
[33] Pereira, M.C. and J. P. Sousa, Evaluation of nickel toxicity on liver, spleen, and kidney of mice after administration of high-dose metal ion. *Journal of Biomedical Materials Research Part A*, 1998. **40**: p. 40–47.
[34] *Chromium, nickel and welding*. Vol. 49. 1990: Lyon: IARC monographs on the evaluation of carcinogenic risks to humans. 257–445.
[35] Laing, P.G. and E.S. Hodge. Tissue reaction in rabbit muscle exposed to metallic implants. *Journal of Biomedical Materials Research* 1967. **1**: p. 135–149.
[36] Peltonen, L., Nickel sensitivity in the general population. *Contact Dermatitis*, 1979. **5**: p. 27–32.
[37] El Medawar, L., *et al.*, Electrochemical and cytocompatibility assessment of NiTiNOL memory shape alloy for orthodontic use. *Biomolecular Engineering*, 2002. **19**(2–6): p. 153–160.
[38] Ryhänen, J., W. Serlo, E. Niemelä, P. Sandvik, H. Pernu and T. Salo, Biocompatibility of nickel-titanium shape memory metal and its corrosion behavior in human cell cultures. *Journal of Biomedical Materials Research Part A*, 1997. **35**: p. 451–457.
[39] Menne, T., Prevention of nickel allergy by regulation of specific exposures. *Annals of Clinical and Laboratory Science*, 1996. **26**: p. 133–138.
[40] Ryhänen, J., *Biocompatibility evaluation of nickel–titanium shape memory alloy*. 1999, Oulu University.
[41] Davis, J.R., *Handbook of materials for medical devices Corrosion of metallic implants and prosthetic devices*. 2003: Materials Park, OH: ASM International.
[42] Es-Souni, M. and H. Fischer-Brandies, On the properties of two binary NiTi shape memory alloys. Effects of surface finish on the corrosion behaviour and *in vitro* biocompatibility. *Biomaterials*, 2002. **23**(14): p. 2887–2894.
[43] Ishimatsu, S., K. Matsuno and Y. Kodama, Distribution of various nickel compounds in rat organs after oral administration. *Biological Trace Element Research*, 1995. **49**: p. 43–52.
[44] Lugowski, S.J., A.D. McHugh and J.C. Van Loon, Release of metal ions from dental implant materials *in vivo*: determination of Al, Co, Cr, Mo, Ni, V, and Ti in organ tissue. *Journal of Biomedical Materials Research*, 1991. **25**: p. 1443–1458.

[45] Bergman, M. and R. Soremark, Tissue accumulation of nickel released due to electrochemical corrosion of non-precious dental casting alloys. *Journal of Oral Rehabilitation*, 1980. **7**: p. 325–330.

[46] Granchi, D, Ciapetti, G., Savarino, L., Stea, S., Gamberini, S., Gori, A. and Pizzoferrato, A. Cell death induced by metal ions: necrosis or apoptosis? *Journal of Materials Science: Materials in Medicine*, 1998. **9**(1): p. 31–37.

[47] Wever, D.J., *et al.*, Electrochemical and surface characterization of a nickel–titanium alloy. *Biomaterials*, 1998. **19**(7–9): p. 761–769.

[48] Kapanen, A., *et al.*, Behaviour of Nitinol in osteoblast-like ROS-17 cell cultures. *Biomaterials*, 2002. **23**(3): p. 645–650.

[49] Rocher, P., *et al.*, Biocorrosion and cytocompatibility assessment of NiTi shape memory alloys. *Scripta Materialia*, 2004. **50**(2): p. 255–260.

[50] Lee, T.M., E. Chang and C.Y. Yang, Attachment and proliferation of neonatal rat calvarial osteoblasts on Ti6Al4V: effect of surface chemistries of the alloy. *Biomaterials*, 2004. **25**(1): p. 23–32.

[51] Man, H.C. and T. M. Yue, Surface characteristics and corrosion behavior of laser surface nitrided NiTi shape memory alloy for biomedical applications. *Journal of Laser Applications*, 2002. **14**: p. 242–247.

[52] Temenoff, J.S., *Biomaterials : the intersection of biology and materials science*. 2008: Upper Saddle River, NJ: Pearson Prentice Hall. 177–204.

[53] Kokubo, T. and H. Takadama, How useful is SBF in predicting *in vivo* bone bioactivity? *Biomaterials*, 2006. **27**(15): p. 2907–2915.

[54] Speck, K.M., Anodic polarization behavior of Ti–Ni and Ti–6A1–4V in simulated physiological solutions. *Journal of Dental Research*, 1980. **59**: p. 1590–1595.

[55] Yoshida, Y.S.M., Corrosion and biocompatibility of shape memory alloys. *Corrosion Engineering*, 1991. **40**: p. 1009–1025.

[56] Oshida, Y.S.R., and S. Miyazaki, Microanalytical characterization and surface modification of TiNi orthodontic archwires. *Biomedical Materials and Engineering*, 1992. **2**: p. 51–69.

[57] Rubin, L.R., Biomaterials in reconstructive surgery. *Fundamental aspects of the corrosion of metallic implants*, 1983: Mosby, St. Louis. 145–157.

[58] Davies, J.E., The bone-biomaterial interface. *The biomaterial–tissue interface and its analogues in surface science and technology*, ed. Kasemo B. 1991: University of Toronto Press, Toronto. 19–32.

[59] Ryhanen, J., *et al.*, Bone modeling and cell–material interface responses induced by nickel–titanium shape memory alloy after periosteal implantation. *Biomaterials*, 1999. **20**(14): p. 1309–1317.

[60] Wever, D.J., *et al.*, Cytotoxic, allergic and genotoxic activity of a nickel–titanium alloy. *Biomaterials*, 1997. **18**(16): p. 1115–1120.

[61] Grant, D.M., S.M. Green and J.V. Wood, The surface performance of shot peened and ion implanted NiTi shape memory alloy. *Acta Metallurgica et Materialia*, 1995. **43**(3): p. 1045–1051.

[62] Green, S.M., D.M. Grant and J.V. Wood, XPS characterisation of surface modified Ni–Ti shape memory alloy. *Materials Science and Engineering A*, 1997. **224**(1-2): p. 21–26.

[63] Pantelis, D.I., A. Griniari and C. Sarafoglou, Surface alloying of pre-deposited molybdenum-based powder on 304L stainless steel using concentrated solar energy. *Solar Energy Materials and Solar Cells*, 2005. **89**(1): p. 1–11.

[64] Liu, Z. and M. Hua, Wear transitions and mechanisms in lubricated sliding of a molybdenum coating. *Tribology International*, 1999. **32**(9): p. 499–506.

[65] Buscail, H., *et al.*, Characterization of the oxides formed at 1000C on the AISI 316L stainless steel – Role of molybdenum. *Materials Chemistry and Physics*, 2008. **111**(2–3): p. 491–496.

[66] Amirsadeghi, A. and M.H. Sohi, Comparison of the influence of molybdenum and chromium TIG surface alloying on the microstructure, hardness and wear resistance of ADI. *Journal of Materials Processing Technology*, 2008. **201**(1–3): p. 673–677.

[67] Shin, J.-C., *et al.*, Effect of molybdenum on the microstructure and wear resistance of cobalt-base Stellite hardfacing alloys. *Surface and Coatings Technology*, 2003. **166**(2–3): p. 117–126.

[68] Kraack, M., *et al.*, Influence of molybdenum on the corrosion properties of stainless steel films. *Surface and Coatings Technology*, 1994. **68–69**: p. 541–545.

[69] Majumdar, J.D. and I. Manna, Laser surface alloying of AISI 304-stainless steel with molybdenum for improvement in pitting and erosion–corrosion resistance. *Materials Science and Engineering A*, 1999. **267**(1): p. 50–59.

[70] Merian, E., M. Ihnat and M. Stoeppler, Elements and their compounds in the environment: occurrence, analysis and biological relevance. Chapter 18: *Molybdenum*. Vol. 2. 2004: Weinheim: Wiley-VCH. 1028–1029.

[71] Charlmers, B., *Principles of solidification*. 1964, New York: John Wiley & Sons.

[72] Hull, M.N., The anodic oxidation of molybdenum in hydroxide ion solutions. *Journal of Electroanalytical Chemistry*, 1971. **30**(1): p. App1-App3.

[73] Hull, M.N., On the anodic dissolution of molybdenum in acidic and alkaline electrolytes. *Journal of Electroanalytical Chemistry*, 1972. **38**(1): p. 143–157.

[74] Capela, M.V., *et al.*, Repeatability of corrosion parameters for titanium–molybdenum alloys in 0.9% NaCl solution. *Journal of Alloys and Compounds*, 2008. **465**(1-2): p. 479–483.

[75] Jones, D.A., *Principles and prevention of corrosion*. 1996: Upper Saddle River, NJ: Prentice Hall.

[76] Ilevbare, G.O. and G.T. Burstein, The role of alloyed molybdenum in the inhibition of pitting corrosion in stainless steels. *Corrosion Science*, 2001. **43**(3): p. 485–513.

[77] Gonzalez, J.E.G. and J.C. Mirza-Rosca, Study of the corrosion behavior of titanium and some of its alloys for biomedical and dental implant applications. *Journal of Electroanalytical Chemistry*, 1999. **471**(2): p. 109–115.

[78] Gordin, D.M., *et al.*, Synthesis, structure and electrochemical behavior of a beta Ti–12Mo–5Ta alloy as new biomaterial. *Materials Letters*, 2005. **59**(23): p. 2936–2941.

[79] Oliveira, N.T.C., *et al.*, Development of Ti–Mo alloys for biomedical applications: Microstructure and electrochemical characterization. *Materials Science and Engineering : A*, 2007. **452–453**: p. 727–731.

[80] Ng, K.W., H.C. Man and T.M. Yue, Corrosion and wear properties of laser surface modified NiTi with Mo and ZrO_2. *Applied Surface Science*, 2008. **254**(21): p. 6725–6730.

[81] Zhang, X., *et al.*, The pitting behavior of Al-3103 implanted with molybdenum. *Corrosion Science*, 2001. **43**(1): p. 85–97.

[82] Yeung, K.W.K., X.Y. Liu, J.P.Y. Ho, C.Y. Chung, P.K. Chu, W.W. Lu, D. Chan and K.M.C. Cheung, Corrosion resistance, surface mechanical properties, and

cytocompatibility of plasma immersion ion implantation-treated nickel-titanium shape memory alloys. *Journal of Biomedical Materials Research*, 2005. **75**(2): p. 256–267.

[83] Bogdanski, D., *et al.*, Easy assessment of the biocompatibility of Ni–Ti alloys by *in vitro* cell culture experiments on a functionally graded Ni–NiTi–Ti material. *Biomaterials*, 2002. **23**(23): p. 4549–4555.

[84] Thierry, B., C. Trepanier, O. Savadogo and L'H. Yahia, Effect of surface treatment and sterilization processes on the corrosion behavior of NiTi shape memory alloy. *Journal of Biomedical Materials Research*, 2000. **51**: p. 685–693.

[85] Miao, W., *et al.*, Effect of surface preparation on corrosion properties and nickel release of a NiTi alloy. *Rare Metals*, 2006. **25**(6, Supplement 2): p. 243–245.

[86] Esenwein, S.A., *et al.*, Influence of nickel ion release on leukocyte activation: a study with coated and non-coated NiTi shape memory alloys. *Materials Science and Engineering: A*, 2008. **481–482**: p. 612–615.

[87] Huang, H.-H., *et al.*, Ion release from NiTi orthodontic wires in artificial saliva with various acidities. *Biomaterials*, 2003. **24**(20): p. 3585–3592.

[88] J. C. Wataha, M. Marek and M. Ghazi, Ability of Ni-containing biomedical alloys to activate monocytes and endothelial cells *in vitro*. *Journal of Biomedical Materials Research*, 1999. **45**(3): p. 251–257.

[89] Kokubo, T., H.-M. Kim and M. Kawashita, Novel bioactive materials with different mechanical properties. *Biomaterials*, 2003. **24**(13): p. 2161–2175.

[90] Rey, C., Orthopedic biomaterials, bioactivity, biodegradation; A physical-chemical approach. *Journal of Biomechanics*, 1998. **31**(Supplement 1): p. 182–182.

[91] Regina Filgueiras, M. and L.L. Hench, Solution effects on the surface reactions of a bioactive glass. *Journal of Biomedical Materials Research*, 1993. **27**: p. 445–453.

[92] Wong, M.H., F.T. Cheng and H.C. Man, Characteristics, apatite-forming ability and corrosion resistance of NiTi surface modified by AC anodization. *Applied Surface Science*, 2007. **253**(18): p. 7527–7534.

[93] Lee, J.H., *et al.*, Cell behaviour on polymer surfaces with different functional groups. *Biomaterials*, 1994. **15**(9): p. 705–711.

[94] Clark, P., A.S. Curtis, J.A. Dow and C.D. Wilkinson Topographical control of cell behaviour: II. Multiple grooved substrata. *Development*, 1990. **108**: p. 635–644.

Part II

Improving erosion–corrosion resistance

6

Laser surface modification of metals for liquid impingement erosion resistance

M. DURAISELVAM, National Institute of Technology, India

Abstract: This chapter discusses laser cladding and laser surface alloying techniques for surface modification of commercial turbine blades, AISI 420 martensitic stainless steel and Ti–6Al–4V, to combat liquid impact erosion. The materials are laser surface modified using nickel and titanium-based intermetallic composites strengthened by ceramic reinforcements. Among the different coatings tested, the grain boundary strengthened, WC-reinforced nickel aluminide matrix showed highest erosion resistance up to 1408 min/mm^3 compared with the base material erosion resistance of 79.6 min/mm^3. The experiments revealed the dominance of microstructural features and the importance of different mechanical properties in improving the erosion resistance. The intermetallic-based composite could be a better surface coating for turbine blades for improved tribological properties.

Key words: laser cladding, laser surface alloying, turbine blade, intermetallic matrix composites, liquid impact erosion.

6.1 Introduction

Erosion of solid surfaces can occur through the impact of liquid through damage induced by the formation and collapse of cavities or from the high-velocity impacts of liquid droplets. It has been defined as the 'progressive loss of material from a solid surface due to continued exposure to impacts by liquid drops or jets' (Heymann, 1992). When a liquid droplet impacts the material at high velocity, the surface of the material is subjected to high pressure caused by the compressed liquid. Subsequently, the liquid/solid contact periphery grows faster when the droplet further approaches the solid surface. At this stage, the liquid inside the droplet escapes in the form of a pressurized high-velocity jet when the shock wave propagates ahead of the contact periphery. This micro-jetting action is primarily responsible for material erosion (Heymann, 1969; Haller *et al.*, 2003).

The erosion in turbine blades, pipes in power plants and nuclear industry, missiles, heat exchangers and aircraft traveling at high speed through rain are typical examples of liquid impact erosion (Brunton and Rochester, 1979; Preece, 1979; Lesser and Field, 1983; Jilbert and Field, 2000). In the present investigation, steam and hydro turbine blades are considered since they are critical components in the power-generation industry. In steam turbines, the condensed water droplets at the low-pressure stage are deposited on the concave side of the stator blade. The deposited droplets erode the convex side of the moving blades as they are accelerated by the steam flow. In hydro turbines, the stream of liquid that impinges on the blade surface itself can cause material removal by plastic deformation and crater formation.

This kind of erosion can be minimized to some extent through design changes but this will become tedious and not feasible due to the complications and cost involved. However, substituting an erosion-resistant material proves uneconomical. Surface modification techniques are a viable and economical solution to improving the service life. Many surface-modification techniques, such as high-velocity oxy-fuel spraying (Mann and Arya, 2003), plasma spraying (Mann *et al.*, 1985) and ion nitriding (Chen *et al.*, 2002), have been found to be more promising in modifying the surface of turbine blade for better erosion resistance. Apart from these techniques, the use of laser surface engineering in turbine industries is gaining much attention owing to its localized heat input, reduced distortion and high performance, which can produce distinctive surface and microstructural characteristics without affecting the bulk properties (Mordike, 1997; Kathuria, 2000; Shepeleva *et al.*, 2000). Laser cladding (Steen, 1991) is one such technique which involves the development of a metallurgically bonded layer over the surface of the material by depositing the intended alloy powder. Laser surface alloying is another technique that involves melting and solidifying a pre-placed alloy layer or externally supplied alloy powder using laser. Both processes have equal industrial preference with the latter being restricted in some applications where dilution of base material is a constraint.

Many investigations have been done in the past in order to improve the erosion resistance of materials used in different industries. NiCrFeSiBC–Co-based coatings were developed on mild steel using laser surface alloying (Tomlinson *et al.*, 1987). The vibratory cavitation test shows improved erosion resistance of the laser alloyed surface. A direct relationship was found to exist between hardness and erosion resistance. Laser surface alloying was performed on UNS S31603 austenitic stainless steel using NiCrSiB powder (Kwok *et al.*, 1998). Hardness was identified as the crucial parameter deciding the erosion resistance. Nb, Fe–Cr and SiC was laser alloyed on constructional steel materials such as low alloy steel (15GA), carbon steel (15HN45) and chromium steel (2H13) (Boleslaw and Szkodo, 2001). At high-intensity rotary cavitation, there is no correlation with hardness and erosion resistance. The erosion resistance improvement through the formation of borides and boro-carbides was demonstrated through laser surface alloying of NiCrSiB on AISI 1050 mild steel (Kwok *et al.*, 2001). The interfacial bonding between the matrix or binder and the ceramic phase in laser surface alloying of WC, Cr_3C_2, SiC, TiC, CrB_2 and Cr_2O_3 on UNS S31603 stainless steel was found to be important in vibratory cavitation erosion (Cheng *et al.*, 2001). The erosion resistance of the coating exhibited no correlation with hardness. The initiation of erosion preferably happens at interfacial boundaries. Co–WC and

Ni–WC laser clad S31603 stainless steel exhibited 10 to 45 times improved erosion resistance (Cheng *et al.*, 2002). The reinforcing ceramic particles acted as obstacles to the merging of craters during erosion. Superplastic NiTi was laser alloyed on AISI 316L stainless steel in order to improve the erosion resistance (Chiu *et al.*, 2005). The surface hardness and elasticity of the coating were regarded as contributing factors to erosion resistance improvement. Zhang and Zhang (2005) produced Ni–Cr_3C_2 and Ni–WC coatings with overlapping clad tracks on a martensitic stainless steel with 0.2 wt% carbon and improved the erosive–corrosive wear performance.

Intermetallics are the best substitute for commercial hard coatings which generally offer better creep, oxidation and erosion resistance. Nickel, ferrous and titanium aluminides, silicides such as Ni_3Si, and $MoSi_2$ and NiTi are the widely investigated intermetallics for structural applications. Buehler and Wang (1968) first reported the high cavitation erosion resistance of NiTi in their pioneering work on the development of NiTi. Cheng *et al.* (2004) melted NiTi wire to obtain a clad track on AISI 316 stainless steel substrate. The formation of intermetallics offered increased oxidation, corrosion and erosion resistance in comparison with super-alloys. Apart from these positive aspects, the intrinsic grain boundary brittleness is the prime limiting factor in practical applications of these intermetallics. Particulate reinforcement is one of the options to circumvent this limitation which can reduce the tendency to brittleness at room temperature and incorporate additional strength (Xu *et al.*, 2003; Zhong *et al.*, 2004).

In general, Ni and Co-based alloys were predominantly laser clad/alloyed to improve the erosion resistance. Hardness was found to dominate the erosion resistant characteristics in most of the cases with due consideration to other mechanical properties in some investigations. However, the use of intermetallics for turbine components and their erosion characteristics have been rarely reported. In the present study, laser cladding and laser surface alloying were carried out separately on two different commercial turbine blade materials and their liquid impact erosion characteristics were studied. The laser cladding experiment was carried out on as-received martensitic stainless steel AISI 420 (AR-420) with Ni/Al–TiC and Ni/Al–WC powders and the laser surface alloying was performed on Ti–6Al–4V with Ni/Al–VC and Al–VC precursor powder mixtures.

6.2 Experimental procedures

6.2.1 Laser cladding and laser surface alloying: experimental set-up and material/alloy preparation

The specimen for laser cladding was prepared by cutting the AR-420 into dimensions of $100 \times 60 \times 10\,mm^3$. The chemical composition of AR-420 is shown in Table 6.1.

Table 6.1 Chemical composition of AISI 420 martensitic stainless steel

Element (wt%)	Fe	C	Si	Mn	P	S	Cr	Ni
AR-420	bal.	0.17	0.39	0.34	0.017	0.014	13.4	0.18

A pre-alloyed Ni/Al powder supplied by Sulzer Metco was used to clad nickel aluminide-based intermetallic composite (IC) coating. The powder mixture for intermetallic matrix composites (IMC) was prepared with 80 wt% Ni/Al–20 wt% TiC and 80 wt% Ni/Al–20 wt% WC through mechanical mixing.

Laser surface alloying were performed on Ti6Al4V, a light commercial turbine blade material, using Ni, Al and VC precursor powder mixture. During exposure to the laser, it will form intermetallics based on titanium alloyed from the substrate. In one set of specimens, 50 wt% Ni–Al and 50 wt% VC was used as a precursor powder. In the other set of experiments, pure aluminum powder of 60 wt% was added to 40 wt% VC in an attempt to form titanium aluminide-based intermetallic.

A 4 kW CW Nd:YAG (yttrium aluminum garnet) laser with a lateral powder feeding unit was used for laser surface modification of the substrate material. The angle of the powder feeding nozzle is about 50° to the specimen surface and the stand-off distance was varied between 12 and 15 mm, which should be precisely adjusted to get the maximum catchment of the powder. The alloy powder mixture was fed into the molten pool in a controlled feed rate by a rotary-disk powder feeding system as shown in Fig. 6.1. An argon gas shroud was used to provide a protective atmosphere against oxidation during laser processing. The specimen was kept at a defocused position to avoid excessive melting of the surface. The process parameters used for laser cladding and laser surface alloying are given in Table 6.2. The individual laser tracks were overlapped at 50% overlap ratio to cover the larger surface area. The laser scanning speed and the overlapping ratio are controlled by a CNC program. The laser power was varied between 1.5 and 1.75 kW and the scanning speed was kept constant at 100 mm/min for laser cladding. A higher scanning speed of 150 and 300 mm/min and lower powder feeding rate was set for laser surface alloying experiment. This will enable effective dilution of titanium from the substrate and form titanium-based intermetallics.

6.2.2 Liquid impact erosion test

The liquid impact erosion experiments were carried out using the set-up shown in Fig. 6.2 as per ASTM Standard G73-98 (1998). The specimen was held stationary instead of mounting on a rotary attachment which permits simulation under accelerated erosion conditions.

The water jet was directed at high velocity (~150 m/s) towards the

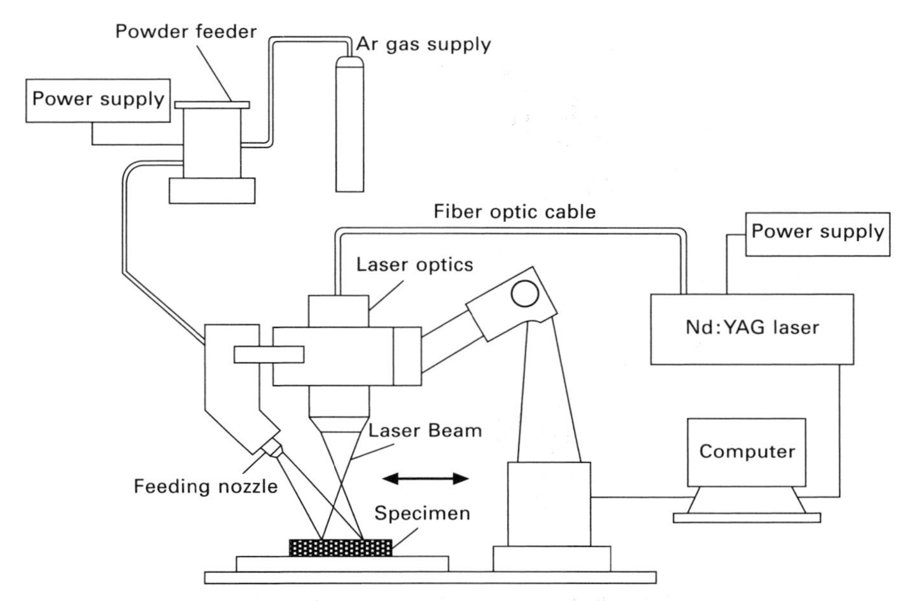

6.1 Set-up for laser surface treatment experiment.

Table 6.2 Laser process parameters

Specimens	Substrate	Laser power (kW)	Scanning speed (mm/min)	Powder feeding rate (g/min)	Coating density (g/cm^3)
Ni–Al	AR-420	1.5	100	8.5	6.29
Ni–Al–TiC	AR-420	1.5	100	6.5	5.70
Ni–Al–WC	AR-420	1.5	100	5.25	7.04
Ni–VC	Ti–6Al–4V	1.75	150	1.0	5.15
Al–VC	Ti–6Al–4V	1.75	300	1.0	4.62

specimen surface through a nozzle at an angle of 90°. The stand-off distance was set at 100 mm to create a 5 mm diameter impact area on the specimen. The specimens were weighed at 10 min intervals for 60 min test duration. The volume loss was computed by dividing the weight loss with density of respective specimen. The rate of volume loss (RV_l) and erosion resistance (E_R) were calculated by the following equations:

$$RV_l (mm^3/min) = \frac{\Delta W}{\rho_s \Delta t} \quad [6.1]$$

$$E_R \ (min/mm^3) = \frac{1}{RV_l} \quad [6.2]$$

where ΔW is the weight loss in grams, ρ_s is the density of the substrate in g/mm^3 and Δt is the time interval in minutes. The density of clad and alloyed

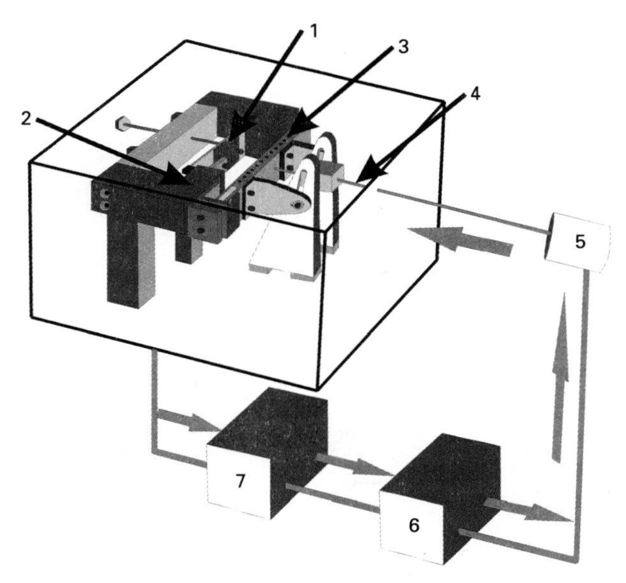

6.2 Liquid impact erosion experimentation set-up (1 specimen, 2 holder, 3 shutter, 4 water inlet, 5 high pressure pump, 6 filter, 7 collection tank).

layers was estimated as follows. A thin layer of the coating was cut from the specimen and weighed in air and methylbenzol. The density of the IMC coatings was computed using the weight difference as shown in Eq. (6.3).

$$\text{Coating density } (\rho) = \rho_l \times \frac{m_a}{m_a - m_l} \qquad [6.3]$$

ρ_l is the density of methylbenzol (0.864–0.868 g/cm^3 at 20 °C), m_a is the weight of laser clad/alloyed layer in air (g) and m_l is the weight of laser clad/alloyed layer in liquid (methylbenzol) (g).

6.2.3 Metallographic and mechanical characterization

The metallographic characterization was performed using optical microscopy (OM), scanning electron microscopy (SEM), energy dispersive spectrography (EDS) and X-ray diffraction (XRD) analysis. In EDS, quantitative compositional analysis, line scan and point analysis were performed. The quantitative compositional analysis was performed in a 100 × 100 μm^2 area to identify the atomic percentage of various elements present in the laser surface treated region. The line scan was performed over an area where carbide, matrix or grain boundaries were crossed. The elemental composition at a particular point of interest was ascertained using point analysis. The XRD analysis was performed using a D-500 Siemens X-ray diffractometer.

The CuK$_\alpha$ radiation was used at a voltage of 40 kV and a current of 40 mA. The specimens were scanned from 20° to 100° at a scan speed of 0.5°/min. The diffraction peaks were indexed using the Joint Committee on Powder Diffraction Standards (JCPDS) software.

The microhardness test and depth-sensing indentation experiment were performed to determine primary mechanical properties of the coating such as hardness, elasto-plastic properties, elastic modulus, yield strength and strain hardening exponent. The cross-sectional microhardness was measured at 1.96 N load for 20 s. The hardness measurements were also made in the cross-section close to the eroded surface to identify work hardenability of the coating. Zwick ZHU 2.5 with a Vickers indenter tip was used for depth-sensing indentation experiments as per EN ISO 14577-1 standards (2003). A 5 N load with loading and unloading speeds of 1 mm/min and 0.2 mm/min was used in the present study. The elastic modulus was determined using software attached to the system. Image analysis software was used to measure the appropriate areas under the loading and unloading cycles to compute irreversible indentation work (W_{plast}) and reversible indentation work (W_{elast}). The strain hardening exponent (n) and yield strength were computed using procedures explained elsewhere (Giannakopoulos and Suresh, 1999; Alcala *et al.*, 2000).

6.3 Coating characteristics

6.3.1 Microstructure of clad and alloyed layers

The macro cross-section of the nickel aluminide IC coating on AR-420 with overlapped clad tracks is shown in Fig. 6.3(a). The clad layer was found to be free of cracks and porosity. The thickness of the coating was approximately 2 mm. The slight undulations observed at the top of the coating may be machined before final assembly of the blade. The microstructure (Fig. 6.3(b)) of the Ni–Al IC coating revealed the presence of equiaxed grains and a dendritic structure which were distributed uniformly throughout the matrix. The XRD analysis confirmed the presence of Ni–Al, Ni_3Al and Fe_3Al. The intentional dilution of Fe from the substrate was made in order to reduce the cracking tendency of nickel aluminide (Liu *et al.*, 1997). This resulted in the formation of Fe_3Al.

The specimen Ni–Al–TiC contains unmelted TiC particles predominantly distributed in the Ni–Al matrix with a few re-precipitated and partially melted TiC particles as shown in Fig. 6.4(a). An increase in laser power or a reduction in scanning speed might result in complete melting. But it might lead to excessive dilution of Fe from the substrate. The Ni–Al–WC (Fig. 6.4(b)) exhibited a microstructure with reprecipitated WC particles distributed along the grain boundary. This kind of grain boundary strengthening effect could lead to improved tribological properties.

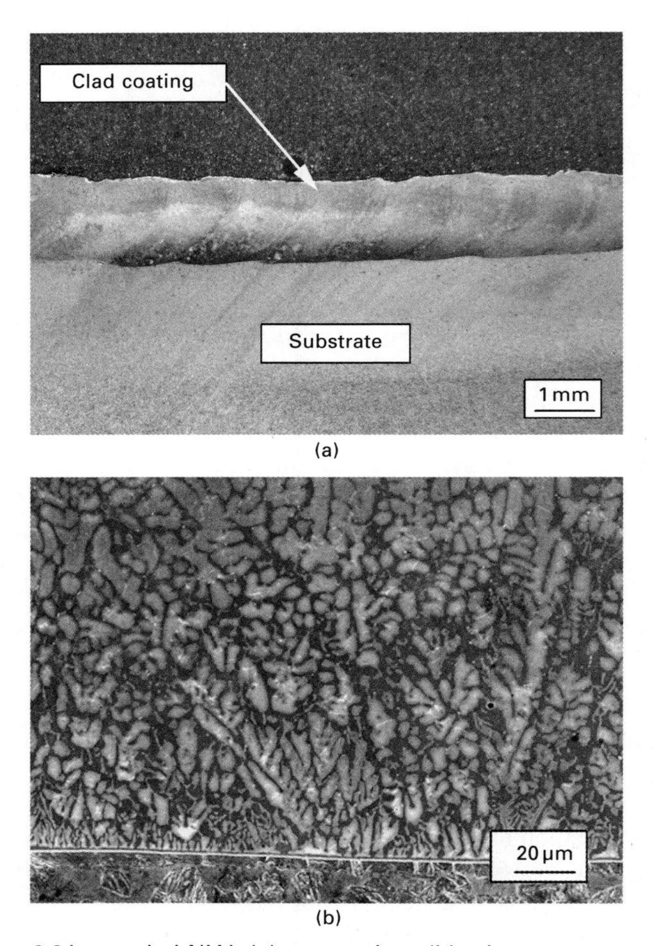

6.3 Laser clad NiAl: (a) macro-view; (b) microstructure.

The laser surface alloyed specimens are shown in Fig. 6.5(a, b). The powder feeding rate was kept at lower levels to promote dilution of titanium from the substrate material which will react with the supplied alloy powder to form the desired intermetallic phases. The cross-sectional micrograph of Ni/Al–VC alloyed layer is shown in Fig. 6.5(a) and exhibited TiC dendrites distributed in titanium aluminide matrix phase. Apart from these major phases, the presence of NiTi and Ti_2Ni was also evident from XRD analysis. The specimen Al–VC was alloyed with Al and VC precursor powder in order to form a single phase IMC coating. The microstructure shown in Fig. 6.5(b) exhibited a Ti_3Al intermetallic matrix with TiC reinforcement. However, the specimen exhibited few macro-cracks which require further optimization. In both specimens, the VC was preferred to other carbides because of its positive influence on the coating characteristics. During laser processing, the

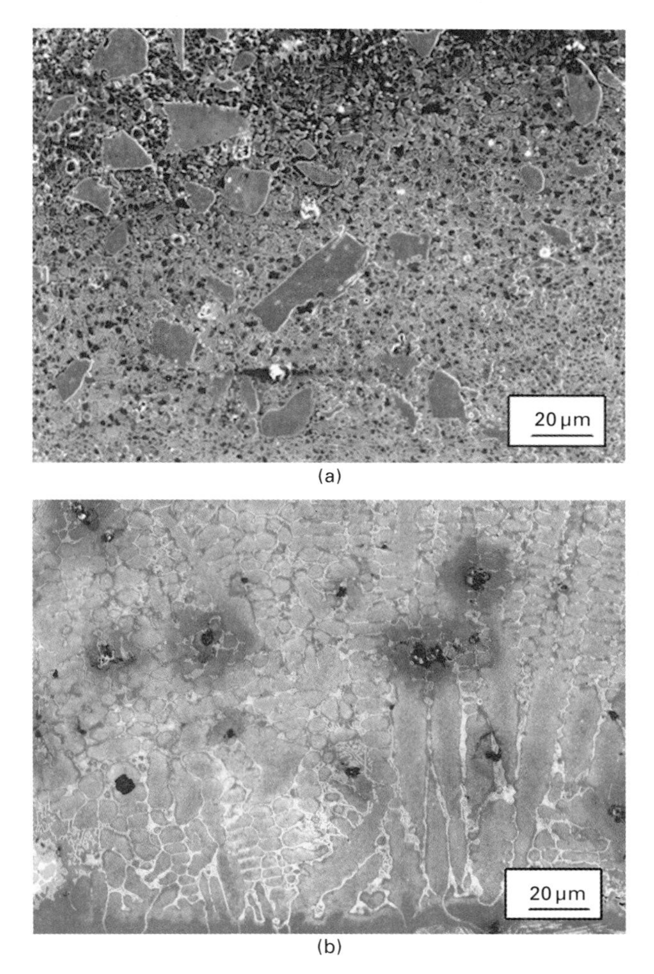

6.4 Microstructure of laser-clad IMC coatings: (a) Ni–Al–TiC; (b) Ni–Al–WC.

VC melts and forms TiC by reacting with the Ti from the substrate. Due to higher free enthalpy of formation, the TiC reprecipitated as a primary phase during solidification (Rapp and Zheng, 1991). The disassociated vanadium dissolved in the matrix will effectively lower the α/β transformation temperature since it is a β stabilizer (Lütjering and Williams, 2003).

6.3.2 XRD: Phase identification of different clad and alloyed coatings

The XRD analysis of Ni–Al IC coating on AR-420 (Fig. 6.6) revealed the presence of Ni_3Al, Ni–Al and Fe_3Al. The NiAl–Ni_3Al dual phase matrix

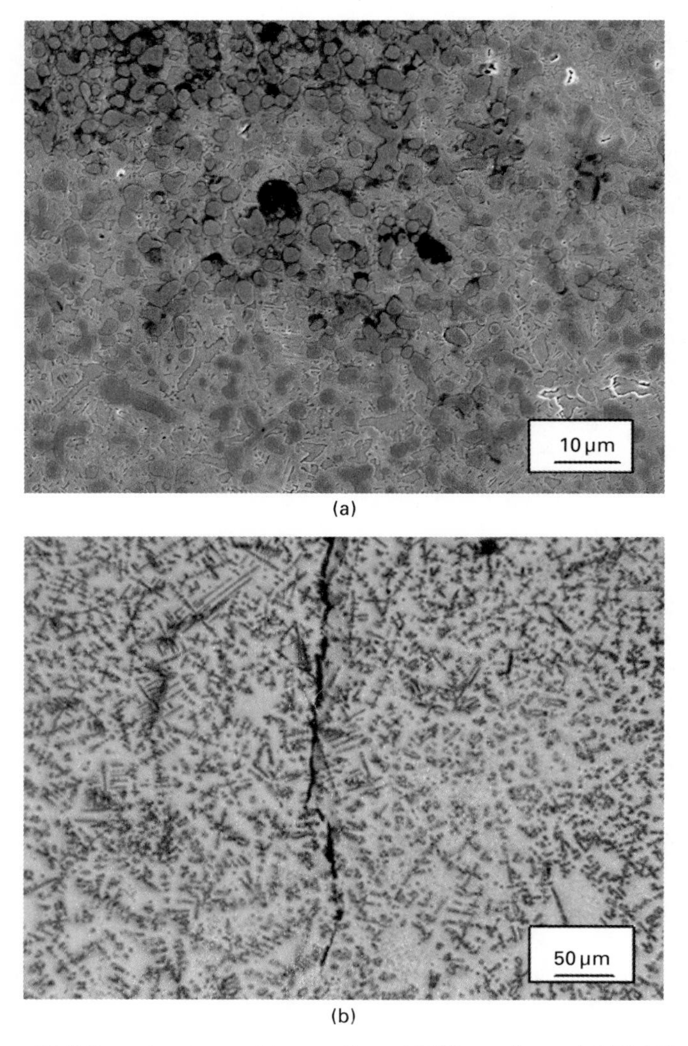

6.5 Microstructure of laser alloyed IMC coatings: (a) Ni–VC; (b) Al–VC.

probably exists as a solid solution with Fe_3Al. The XRD of specimen Ni–Al–TiC is similar to Ni–Al except for the presence of additional TiC peaks. The Ni–Al–WC exhibited M_6C (Fe_3W_3C, Fe_4W_2C) in addition to NiAl, Ni_3Al and WC. The decomposition of WC and Fe dilution from the substrate might have resulted in M_6C formation.

In the laser alloyed Ti6Al4V, the decomposition of VC to form TiC is confirmed from the XRD spectrum as shown in Fig. 6.7. The specimen Ni–VC exhibited multiphase intermetallics of NiTi, Ti_2Ni and Ti_3Al. In

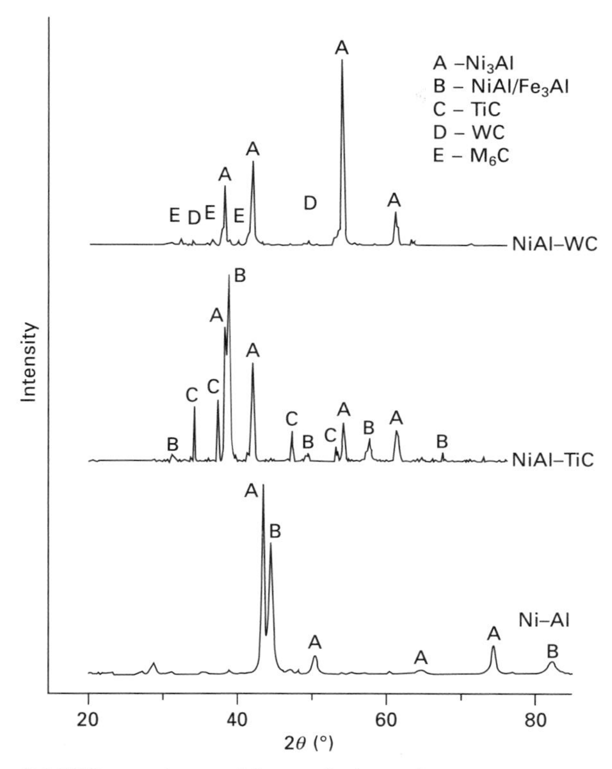

6.6 XRD spectrum of laser clad specimens.

general, Ti_2Ni is highly brittle compared with other intermetallics and may lead to crack formation in the specimen. However, the very weak peaks and the crack-free microstructure indicate the secondary nature of the presence of this intermetallic, which could improve the hardness. The XRD spectrum of Al–VC indicates that Ti_3Al is primarily intermetallic with TiC.

6.3.3 Microhardness: cross-sectional hardness variation of laser-treated specimens

The cross-sectional microhardness of the clad and alloyed layers measured as a function of depth is shown in Fig. 6.8. The average hardness of Ni–Al IC coating was 328 $HV_{0.2}$ which is approximately 100 $HV_{0.2}$ higher than the base material hardness. The IMC layer with TiC reinforcement exhibited hardness values in the range of 580–650 $HV_{0.2}$ compared with those with WC reinforcement which were in the range of 400–430 $HV_{0.2}$. The retention of unmelted TiC particles and the low wettability of TiC compared with WC with Fe, Ni and Al might have led to the improved hardness of the clad

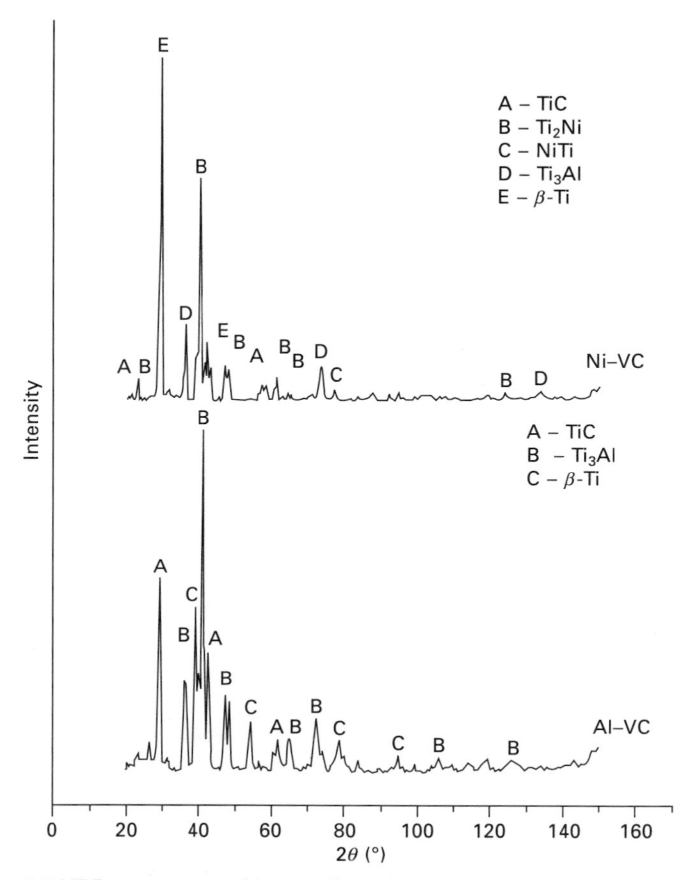

6.7 XRD spectrum of laser alloyed specimens.

layers with TiC reinforcement (Lalitha *et al.*, 2000). However, the uniform dispersion and precipitation of WC in the Ni–Al–WC specimen led to less variation in hardness over the cross-section. The alloyed layers exhibited still higher hardness values compared with clad layers, which were above 800 $HV_{0.2}$. The Ti_2Ni precipitation in Ni–VC and Ti_3Al formation in Al–VC might have contributed to this improvement in hardness.

6.3.4 Depth sensing indentation: pseudo-elastic behavior of clad and alloyed layers

The loading and unloading cycle of different specimens measured by the depth-sensing indentation technique is shown in Fig. 6.9. The maximum plastic deformation (W_{plast}) was observed in AR-420 with little elastic recovery (W_{elast}). This clearly indicates that the material has poor resistance

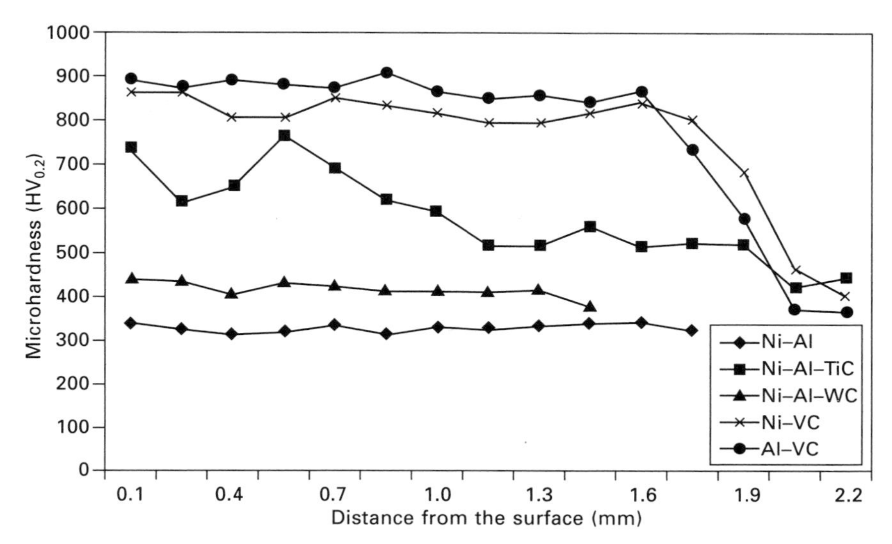

6.8 Microhardness of laser clad and alloyed specimens.

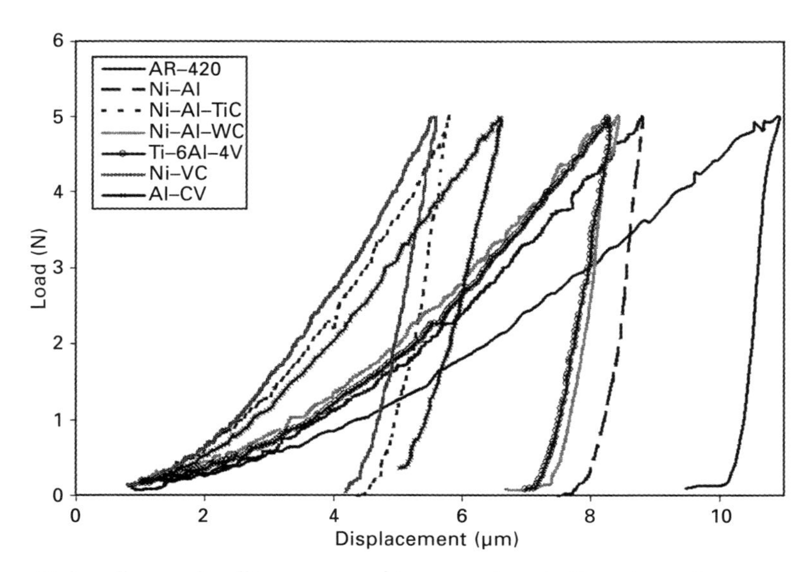

6.9 Loading–unloading curves of laser clad and alloyed specimens.

to the impact pressure generated by the liquid droplets and readily deforms to absorb the stresses. The tendency for plastic deformation was considerably reduced in Ni–Al. This was further improved in the case of WC due to the additional grain boundary strengthening effect. The elasto-plastic property of Ti–6Al–4V was in comparable with the specimen with WC reinforcement due to the intrinsic mechanical properties of this alloy. The plastic deformation of Ni–VC was the lowest of all tested specimens due to its high hardness. In Ni–Al–TiC, the presence of unmelted TiC reduced the intensity of plastic deformation. The plastic deformation in Al–VC was slightly higher than in Ni–VC and Ni–Al–TiC.

6.4 Liquid impact erosion characteristics

The cumulative volume loss of the laser clad and alloyed specimens eroded for 60 mins is shown in Fig. 6.10. The highest volume loss was observed for the specimen AR-420 and lowest for the specimen Ni–Al–WC. Among the laser clad and alloyed specimens, the volume loss of Al–VC was higher. The volume loss of Ti–6Al–4V was comparable to Ni–Al. The Ni–Al–TiC and Ni–VC showed slightly lower volume loss than Al–VC. Figure 6.11 shows that the specimen AR-420 exhibited lowest erosion resistance of 79.6 min/mm^3. The highest erosion resistance achieved with Ni–Al–WC which was 1408 min/mm^3. In the laser clad specimens, the influence of carbide reinforcement can be clearly seen as Ni–Al–TiC and Ni–Al–WC showed improved resistance to erosion compared with Ni–Al. In the laser alloyed specimens, the Al–VC failed to provide enhanced erosion resistance compared with untreated Ti–6Al–4V. The few pre-existing micro cracks might have influenced the erosion resistance of Al–VC.

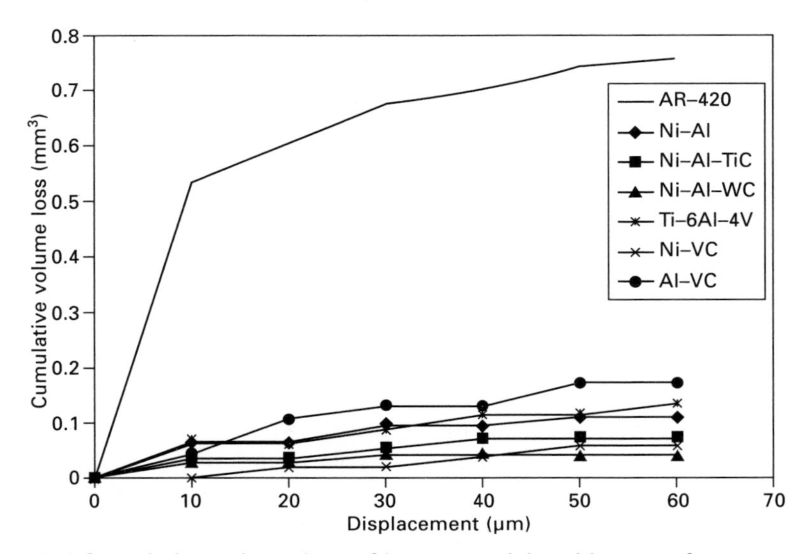

6.10 Cumulative volume loss of base material and laser surface treated specimens.

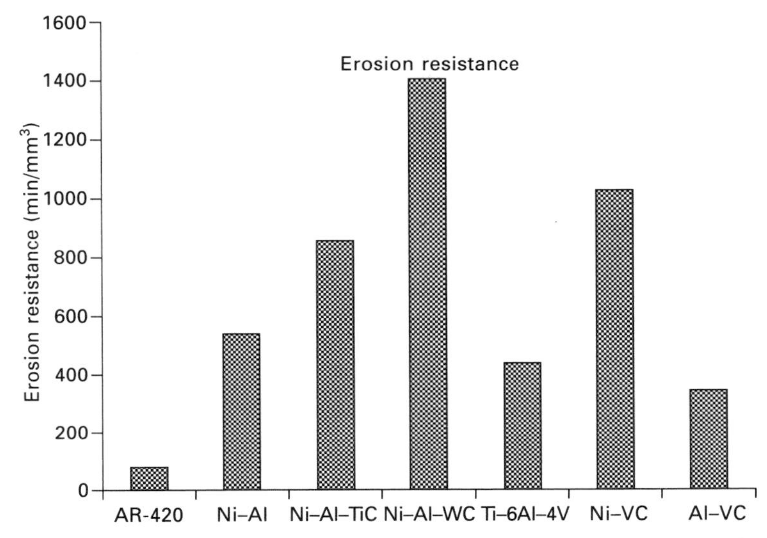

6.11 Erosion resistance of base material and laser surface treated specimens.

6.5 Eroded surface morphology

The surface morphology of all tested specimens is shown in Figs 6.12 and 6.13. AR-420 (Fig. 6.12(a)) exhibited deep craters with undulations and cracks. The high-pressure pulses generated by the collapsing cavities plastically deformed the material and craters was formed. In addition, the lateral dispersion of micro-jets either tends to the initiation of cracks or enlargement of the craters. The piled up material around the crater periphery might be removed as the erosion progress, leading to accelerated erosion rates.

There was no evidence of deep crater formation in the Ni–Al specimen shown in Fig. 6.12(b). The pseudo-elastic property and the work hardening ability of the Ni–Al matrix prevented the surface from severe plastic deformation. However, a few deep straight grooves were formed over the surface with build-up of material around the lips of the groove. This indicates the least resistance of the matrix to lateral micro-jets. The breadth of the grooves is wider at one end, diminishes uniformly and becomes shorter at the other end. The collapse of the droplet might have happened at the wider region and the propagation of microjet penetrates the material in the lateral direction to form grooves. As the jet travels further, the velocity decreases and the pressure drops which leads to shallow and narrower groove profile. In TiC and WC reinforced specimens (Fig. 6.12(c, d)) no such groove formation is evident, except for a few shallow craters. The absence of grooves in these specimens indicates the resistance of reinforced layer to lateral micro-jets. The

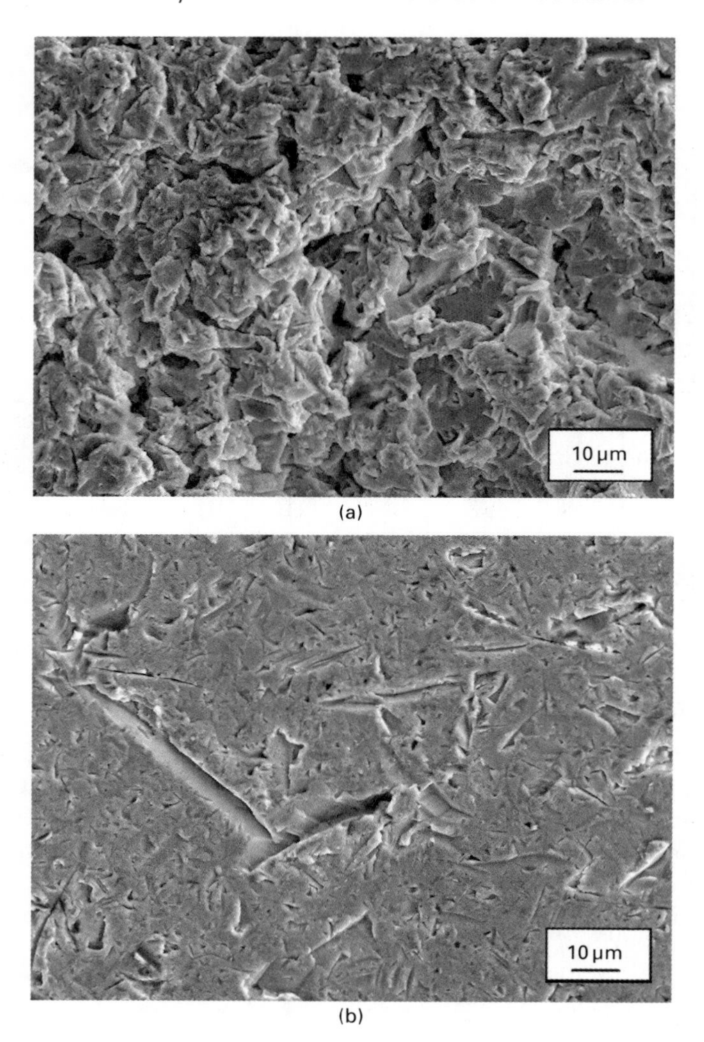

6.12 SEM micrographs of eroded specimens: (a) AR-420; (b) Ni–Al; (c) Ni–Al–TiC; (d) Ni-Al-WC.

selective removal of unmelted TiC particles might have reduced the erosion resistance even though it exhibits a very good combination of mechanical properties compared to Ni–Al–WC. This clearly proves the dominance of microstructural features rather than the mechanical properties in deciding the erosion resistance characteristics.

In the case of laser alloyed specimens, Ti–6Al–4V (Fig. 6.13(a)) exhibits surface topography similar to Ni–Al with groove formation. In addition to that, the lips of the grooves were sharper with severe undulations, suggesting a brittle mode of material removal. In Ni–VC (Fig. 6.13(b)), sparsely distributed

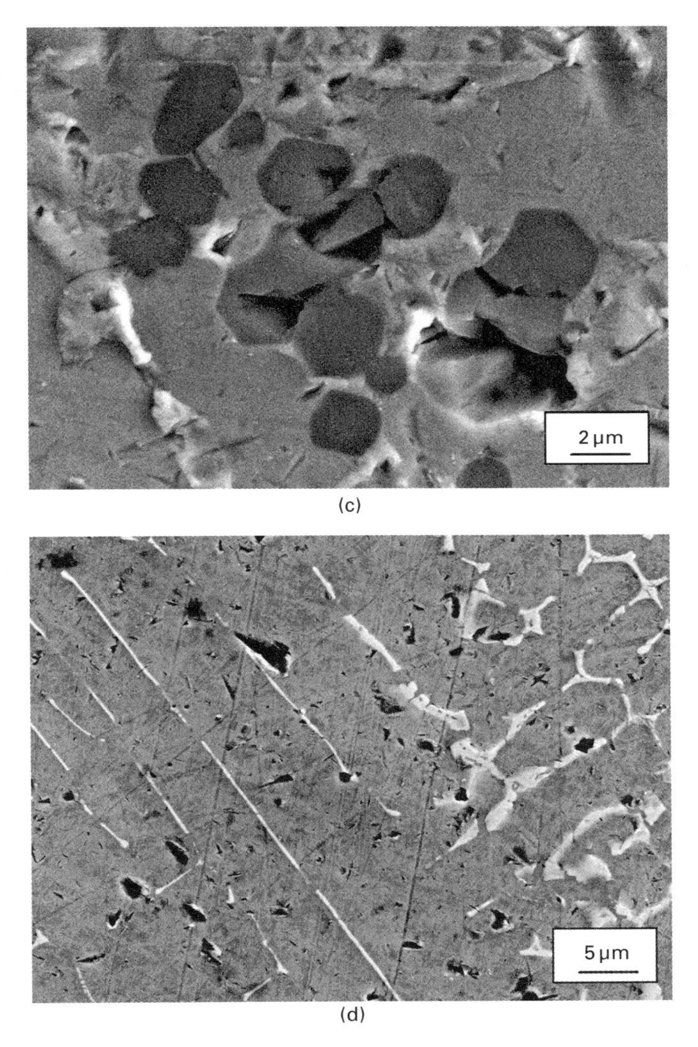

6.12 Continued

pits can be seen, which clearly indicates the significance of NiTi–Ti_3Al solid solution strengthened by TiC in improving the erosion resistance. On the other hand, the Al–VC shown in Fig. 6.13(c) exhibited a few deep craters and fragmentation effects similar to brittle fracture.

6.6 Correlation between mechanical properties and erosion resistance

The mechanical properties measured using depth sensing indentation (DSI) instrument are listed in Table 6.3. The high hardness, strain hardening

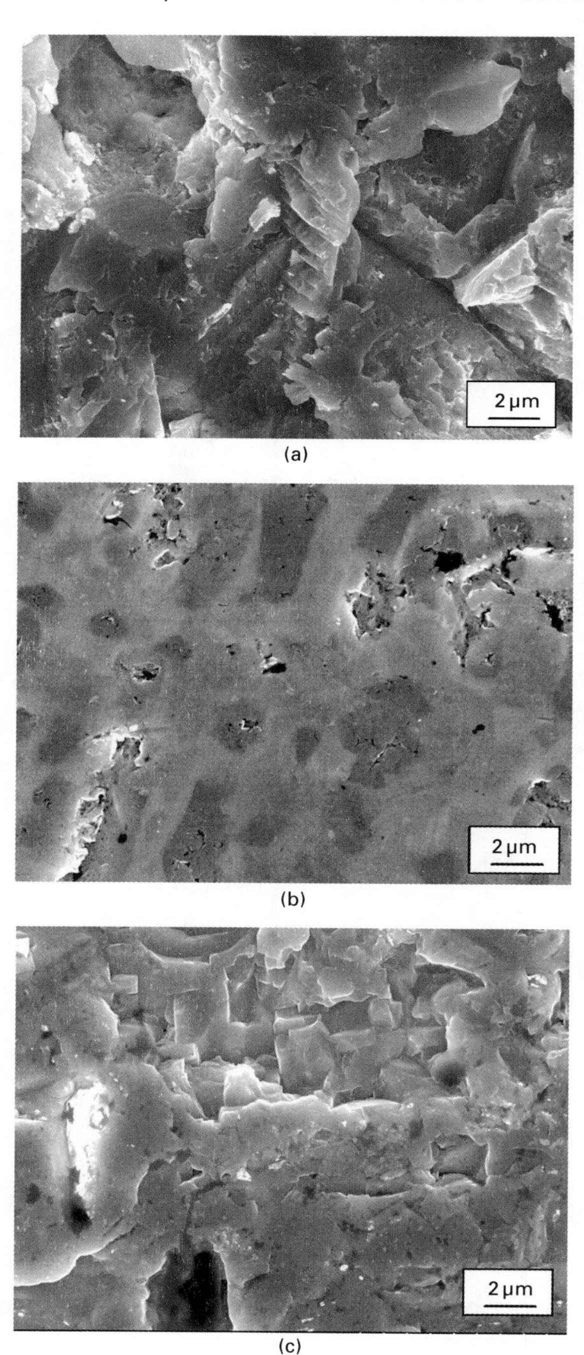

6.13 SEM micrographs of eroded specimens: (a) Ti–6Al–4V; (b) Ni–VC; (c) Al–VC.

Table 6.3 Correlation of erosion resistance and mechanical properties

Specimen	Hardness (GPa)	Elastic modulus (GPa)	W_{plast} (J) (10^{-6})	W_{elast} (J) (10^{-6})	n	σ_y (GPa)	Erosion resistance (min/mm^3)
AR-420	1.775	200.2	19.59	1.78	0.280	0.663	79.6
Ti6Al4V	3.207	128.5	11.94	1.71	0.129	1.458	441.9
Ni–Al	2.863	125.9	13.13	2.08	0.187	1.282	539.14
Ni–Al–TiC	6.096	199.4	8.12	2.51	0.102	3.383	855
Ni–Al–WC	3.447	162.4	12.67	2.14	0.116	1.406	1408
Ni–VC	7.855	191.1	7.05	2.26	0.455	5.212	1030.8
Al–VC	6.345	130.9	8.60	2.38	0.234	2.974	346.6

exponent and yield strength and the lower W_{plast} of Ni–VC and Ni–Al–TiC primarily contributed to the improved erosion resistance. The specimen Ni–Al–WC did not possess this combination of mechanical properties but still provides excellent erosion resistance. This clearly indicates the dominant influence of the grain boundary strengthening effect. The specimen Al–VC includes a better set of mechanical properties than Ni–VC. However, few pre-existing surface defects might have caused the favorable coating characteristics to deteriorate. The close similarity in the erosion performance of Ni–Al and Ti–6Al–4V can be justified by the similar mechanical properties of the specimen. In AR-420, except the strain hardening exponent, all other mechanical properties are inferior to other specimens which results in low erosion resistance.

6.7 Conclusions

The commercial turbine blade material such as AISI 420 martensitic stainless steel and Ti–6Al–4V was laser clad and alloyed using nickel and titanium aluminide-based intermetallics to improve the liquid impact erosion resistance. Based on the experimental results, the following conclusions can be drawn:

- The hardness of laser clad Ni–Al IC on AISI 420 was improved to 328 $HV_{0.2}$. It was further increased to 650 $HV_{0.2}$ and 430 $HV_{0.2}$ for Ni–Al–TiC and Ni–Al–WC IMC, respectively. The laser surface alloyed specimens (Ni–VC and Al–VC) exhibited higher hardness in the range of 800 $HV_{0.2}$.
- The high work-hardening ability and pseudo-elastic property of laser clad Ni–Al IC improved the erosion resistance up to 539.14 min/mm^3 higher than that of AR-420 (79.6 min/mm^3).
- The erosion resistance was further enhanced to 855 and 1408 min/mm^3 for Ni–Al–TiC and Ni–Al–WC, respectively. This was primarily attributed to the strength and toughness provided by the reinforcements. The grain

boundary strengthening of WC shows phenomenal improvement in erosion resistance compared with TiC reinforced matrix.

- The combination of mechanical properties of Ni–Al–TiC did not yield improved erosion resistance compared with Ni–Al–WC due to the preferential removal of unmelted TiC particles from the matrix. This clearly indicates that the microstructural features are dominant over the mechanical properties in deciding the erosion resistance.
- The highest strain hardening exponent of 0.28 was observed in AR-420 compared with laser clad and alloyed specimens. However, it possesses the lowest erosion resistance. This suggests that an improvement in a single mechanical property alone may not be sufficient to combat the complex liquid impact erosion.
- The dual phase NiTi–Ti_3Al IMC with TiC reinforcement exhibited excellent combination of mechanical properties which resulted in higher erosion resistance of 1030.8 min/mm^3.
- The specimen Al–VC exhibited inferior erosion resistance compared with Ti–6Al–4V. Preferential material removal around the pre-existing cracks might have deteriorated the favorable coating characteristics.

6.8 Acknowledgments

I would like to acknowledge ISAF, TU Clausthal, Germany for providing the laser cladding facility, alloy powders and metallographic characterization facilities. I would also like to acknowledge EMPA, Thun, Switzerland for providing the liquid impact erosion test rig.

6.9 References

Alcala J, Baronr A C and Anglada M (2000), 'The influence of plastic hardening on surface deformation modes around Vickers and spherical indents', *Acta Mater.*, **48**, 3451–3464.

ASTM (1998) Standard practice for liquid impingement erosion testing, Annual Book of ASTM standards, ASTM Standard G73–98, 03.02, 285–304.

Boleslaw G G and Szkodo M (2001), 'Methodology of cavitation erosion assessment of laser processed materials', *Adv. Mater. Sci.*, **1**, 28–36.

Brunton J H and Rochester M C (1979), 'Erosion of solid surfaces by the impact of liquid drops', in Preece C M, *Treatise on Materials Science and Technology*, Vol. 16, New York, Academic Press.

Buehler W J and Wang F E (1968), 'A summary of recent research on the Nitinol alloys and their potential application in ocean engineering', *Ocean Eng.*, **1**, 105–120.

Chen K C, He J L, Huang W H and Yeh T T (2002), 'Study on the solid–liquid erosion resistance of ion-nitrided metal', *Wear*, **252**, 580–585.

Cheng F T, Kwok C T and Man H C (2001), 'Laser surfacing of S31603 stainless steel with engineering ceramics for cavitation erosion resistance', *Surf. Coat. Technol.*, **139**, 14–24.

Cheng F T, Kwok C T and Man H C (2002), 'Cavitation erosion resistance of stainless steel laser-clad with WC-reinforced MMC', *Mater. Lett.*, **57**, 969–974.
Cheng F T, Lo K H and Man H C (2004), 'A preliminary study of laser cladding of AISI 316 stainless steel using preplaced NiTi wire', *Mat. Sci. Eng. A*, **380**, 20–29.
Chiu K Y, Cheng F T and Man H C (2005), 'Cavitation erosion resistance of AISI 316L stainless steel laser surface-modified with NiTi', *Mat. Sci. Eng. A*, **392**, 348–358.
EN ISO 14577–1 (2003), Instrumented indentation test for hardness and materials parameters 467–496.
Giannakopoulos A E and Suresh S (1999), 'Determination of elastoplastic properties by instrumented sharp indentation', *Scripta Mater.*, **40**, 1191–1198.
Haller K K, Ventikos Y and Poulikakos D (2003), 'Wave structure in the contact region during high speed droplet impact on a surface: solution of the Reimann problem for the stiffened gas equation state', *J. Appl. Phys.*, **93**, 3090–3097.
Heymann F J (1969), 'High-speed impact between a liquid drop and a solid surface', *J. Appl. Phys.*, **40**, 5113–5122.
Heymann F J (1992), *Liquid Impingement Erosion*, ASM Hand Book, 18, USA, ASM International.
Jilbert G H and Field J E (2000), 'Synergistic effects of rain and sand erosion', *Wear*, **243**, 6–17.
Kathuria Y P (2000), 'Some aspects of laser surface cladding in the turbine industry', *Surf. Coat. Technol.*, **132**, 262–269.
Kwok C T, Cheng F T and Man H C (1998), 'Laser surface modification of UNS S31603 stainless steel using NiCrSiB alloy for enhancing cavitation erosion resistance', *Surf. Coat. Technol.*, **107**, 31–40.
Kwok C T, Man H C and Cheng F T (2001), 'Cavitation erosion–corrosion behaviour of laser surface alloyed AISI 1050 mild steel using NiCrSiB', *Mat. Sci. Eng. A*, **303**, 250–261.
Lalitha R K, Agarwal A and Dahotre N B (2000), 'Laser surface engineered TiC coating on 6061 Al alloy: microstructure and wear', *Appl. Surf. Sci.*, **153**, 65–78.
Lesser M B and Field J E (1983), 'The impact of compressible liquids', *Ann. Rev. Fluid Mech.*, **15**, 97–122.
Liu Z G, Guo J T, Zhou L Z, Hu Z Q and Umemoto M (1997), 'Mechanical alloying synthesis and structural characterization of ternary Ni–Al–Fe alloys', *J. Mater. Sci.*, **32**, 4857–4864.
Lütjering G and Williams J C (2003), *Titanium*, Berlin, Springer-Verlag.
Mann B S and Arya V (2003), 'HVOF coating and surface treatment for enhancing droplet erosion resistance of steam turbine blades', *Wear*, **254**, 652–667.
Mann B S, Krishnamoorthy P R and Vivekananda P (1985), 'Cavitation erosion characteristics of nickel-based alloy–composite coatings obtained by plasma spraying', *Wear*, **103**, 43–55.
Mordike B L (1997), 'Lasers in materials processing', *Prog. Mater. Sci.*, **42**, 357–372.
Preece C M (1979), 'Cavitation erosion', in Preece C M, *Treatise on Materials Science and Technology*, Vol. 16, New York, Academic Press.
Rapp R A and Zheng X-J (1991), 'Thermodynamic consideration of grain refinement of aluminium alloys by titanium and carbon', *Metall. Trans. A*, **22**, 3071–3075.
Shepeleva L, Medres B, Kaplan W D, Bamberger M and Weisheit A (2000), 'Laser cladding of turbine blades', *Surf. Coat. Technol.*, **125**, 45–48.
Steen W M (1991), *Laser Materials Processing*, London, Springer-Verlag.

Tomlinson W J, Moule R T, Megaw J H P C and Bransden A S (1987), 'Cavitation wear of untreated and laser processed hardfaced coatings', *Wear*, **117**, 103–107.

Xu X Y, Liu W J, Zhong M L and Sun H Q (2003), 'Synthesis and fabrication of WC particulate reinforced Ni_3Al intermetallic matrix composite coating by laser powder deposition', *J. Mater. Sci. Lett.*, **22**, 1369–1372.

Zhang D-W and Zhang X-P (2005), 'Laser cladding of stainless steel with Ni–Cr_3C_2 and Ni–WC for improving erosive–corrosive wear performance', *Surf. Coat. Technol.*, **190**, 212– 217.

Zhong M L, Xu X Y, Liu W J and Sun H Q (2004), 'Laser synthesizing NiAl intermetallic and TiC reinforced NiAl intermetallic matrix composite', *J. Laser Appl.*, **16**, 160–166.

7 Laser surface modification of steel for slurry erosion resistance in power plants

R. C. SHIVAMURTHY, M. KAMARAJ and R. NAGARAJAN, Indian Institute of Technology Madras, India and S. M. SHARIFF and G. PADMANABHAM, International Advanced Research Centre for Powder Metallurgy and Newer Materials (ARCI), India

Abstract: This chapter discusses the microstructure and slurry erosion characteristics of laser surface alloyed coatings that are used to mitigate the erosion issues of 13Cr–4Ni martensitic stainless steel. The chapter first describes the present scenario of surface modification methods, such as conventional welding, surface hardening/nitriding, external coatings, etc., which are being utilized to minimize the silt erosion problems of components in hydropower plants. The chapter also explains newer coating methods, such as laser surface alloying. It then describes the usefulness towards erosion mitigation in hydropower plant applications of laser surface alloying of 13Cr–4Ni steels with commercial coating powders such as Colmonoy 88 and Stellite 6. Erosion performances of coatings and substrates are evaluated using a slurry jet erosion test-rig with various parameters, using commercial silica sand and river sand. An evaluation of the effect of boron carbide addition on the erosion performances of laser surface alloyed coatings is given. An attempt has also been made to correlate the quantitative results (erosion performances) with qualitative results (erosion mechanisms) by extensive studies of eroded samples using scanning electron microscopy. By the end of the chapter, a correlation between measured and predicted erosion rates has been established in power-law formulation.

Key words: hydroturbine, silt/slurry erosion, 13Cr–4Ni martensitic stainless steel, laser surface alloying, Stellite 6 and Colmonoy 88 coatings.

7.1 Introduction

7.1.1 Background

The different sources of electrical energy include hydro, thermal, nuclear and non-conventional energy resources such as wind, solar and biomass. Among these different sources of energy, hydropower is recognized as a renewable source of energy, which is economical, non-polluting and environmentally friendly. Water in India and Nepal is one of the major energy resources, formed by the snowcapped mountains, glaciers and regular monsoons. Himalayan topography and continuous sources of water have favoured the country with a huge hydropower potential; however, only 17% of the hydro potential has been utilized so far (Padhy and Saini, 2008). Management of the small hydropower plants for achieving higher efficiency of hydroturbines with time is an important factor, but the turbines show a decline in performance after a few years of operation as they become severely damaged for various reasons such as slurry/silt erosion, cavitation and corrosion (Bajracharya *et al.*, 2008; Padhy and Saini, 2008).

Especially during the monsoon season, most of the Himalayan rivers contain very high sediment concentrations. Major components of this sediment are hard abrasive sand and silts (ranging between 5000 ppm to as high as 20 000 ppm in monsoon), which have a rich quartz content. Singal and Singh (2006) reported that Himalayan silt contains nearly 90% quartz, which is 7 on Mohs hardness scale compared with 10 for diamond; any particle having Mohs hardness of more than 5 is extremely harmful for turbine components. Hydropower plants are facing two main problems concerning sediment aspect. The first is efficient operation of hydropower plants to meet the electricity demand by storing energy in reservoirs that will become filled with sediment over a period of time. This problem has been taken care of by designing sediment-settling systems in power plants. However, huge amounts of unsettled sediment pass through the turbines, and turbine parts operating in such water are exposed to severe erosion. This represents the second challenge. The impact of sediments in the water passing the turbines at high velocity causes severe erosion of turbine components. To combat the impact of sediments on turbine components, special head work, design of civil structures and of mechanical components (such as sediment-settling basins and sediment-flushing systems) has been done, but at increased cost to the hydropower projects. Hence, they are designed to remove only coarser sediment particles. In principle, smaller particles are allowed to pass through the turbine.

Many parts of the hydroturbine components, namely blades, guide vanes, bearing bodies, etc., experience silt erosion problems. Mann and Arya (2002) pointed out that even with many improvements in materials and in hydroturbine design, erosion remains an unsolved problem in hydro power plants. Apart from hydroturbine erosion, erosion phenomena are observed in various other components, viz. (i) propellers, hubs and rudders in case of ships, (ii) high-speed pumps of all types, (iii) regulators, valves and gate valves, (iv) flow-measuring equipment such as orifices, ventures, and (v) sudden enlargements and bends, etc.

The loss of material from turbine components over a period of time implies a geometrical change in the turbine parts. This is caused by erosion, cavitation pitting and corrosion. The eroding particles present in the moving fluid possess high kinetic energy. When these particles strike the surface, wear of material takes place by cutting or deformation of the surfaces.

Erosion depends on several factors such as (i) properties of erodent including average size, shape, hardness, material type; (ii) properties of substrate materials including chemistry, elastic properties, hardness, surface morphology; and (iii) operating environment such as velocity, impingement angle, flux rate or concentration, medium of flow, temperature, etc.

7.1.2 Earlier work

Mann (1998) reported that mainly two types of martensitic stainless steels, forged 12Cr steels and cast 13Cr–4Ni steel, are generally used in water pumps and hydroturbines because of their excellent mechanical properties and adequate corrosion resistance. However, these materials are considerably less resistant to erosive wear and get damaged by excessive solid content entrained with the water.

In order to minimize erosion of turbine components, the study of material characteristics and failure mechanisms helps in understanding the causes of material failure, and this in turn helps to minimize damage to the material. Various surface modification processes or surface coatings are used to protect the surface of hydroturbine components to combat erosive wear. For some applications, however, bulk wear-resistant materials may not be suitable, perhaps for reasons of cost, overall weight and difficulty of fabrication or due to particular mechanical properties. Surface engineering methods can then be used to apply a coating of a wear-resistant material to a substrate with lower wear resistance but with the desired bulk properties.

Attempts are being made to reduce the damage caused by silt erosion either by reducing the particle velocity, by controlling their size and concentration, or by using high velocity oxy fuel (HVOF) spraying of cermet coatings and surface hardening by plasma nitriding. HVOF cermet coatings consist of hard tungsten carbide (WC) embedded in ductile matrix; typically coating systems consisting of WC–Co, WC–Co–Cr, WC–Ni–Cr and FeCrAlY–Cr_3C_2 are being used in different industries. Several coatings have been studied and the most satisfactory results have been obtained with the powder having carbide content of greater than 80% (Stein *et al.*, 1999). Other coatings, such as oxides applied by an atmospheric plasma spraying technique, cladding by laser process, spraying by detonation, physical vapour deposition (PVD) and boronising, are in use, and some of these have been investigated in abrasive and erosive wear using silica sand at different slurry velocities (Wood *et al.*, 1997). Mann and Arya (2001) reported that detonation, as well as HVOF sprayed coatings and boronizing, provides remarkable improvements compared with plasma nitriding at certain test velocities. However, contradictory results regarding plasma nitriding have been reported and depending upon their degree of success, plasma nitriding as well as HVOF coatings are being exploited commercially to overcome the loss of turbine efficiency arising due to excessive erosion of hydroturbines. Even though these HVOF coatings are economical, because of their mechanical bonded nature and defects, reports indicate that these coatings are not advisable over the long term.

7.2 Surface engineering of hydroturbine steels

Though as-cast/tempered 13Cr–4Ni, 16Cr–5Ni and forged 12Cr martensitic stainless steels have excellent mechanical properties and adequate corrosion resistance, they are considerably less resistant to erosive wear. In general, there are three ways to reduce silt erosion damages in turbines (Zhang *et al.*, 1996; Singal and Singh, 2006). One involves optimizing the hydraulic design of equipment based on silt presence, the second involves effective de-silting arrangements, and the third involves developing coatings for the turbines, which can prolong the overhaul interval of hydraulic components. The latter is reasonably economical and is considered in the present investigation. There are various methods to combat erosive wear which are based on either suitably modifying the surface with hard coatings or hardening the surface by plasma nitriding, laser hardening, etc.

The majority of the engineering components fail due to surface degradation process such as corrosion, oxidation, friction, fatigue, wear and abrasion at surfaces because the free surface is more prone to environmental degradation and the intensity of the externally applied load is often highest at the surface. The engineering solution to minimize or eliminate such surface initiated failure lies in tailoring the surface composition and/or microstructure of the near-surface region of a component without affecting the bulk.

7.2.1 Surface modification by welding

Earlier, welding technology was utilized to repair the eroded runners and other components (instead of importing high-cost runners) due to cheap labour cost for welding and grinding. Matsumura and Chen (2002) have reported on the use of bead welding to protect the flow passage of hydroturbines from erosion, and paving welding to attach anti-erosion plates to the surface by spot or braze welding. Sharma (1996) presented the linear relationship between cumulative erosion and time of exposure for welded layers by hardfacing deposition of 13Cr4Ni0.5Mo (welding electrode D and H 444L). He also suggested preheating up to 300 °C to get better mechanical properties and microstructure of 13Cr—4Ni steel, and, in turn, better erosion resistance in weld layer. Goel and Sharma (1996) have collected the weld repair data including type of electrodes used for different type of stainless steels. Most of the electrodes were AWS equivalent E309-15, E309-16, E308-16, E410-15, etc. Some of these electrodes also had a small percentage of Mo and Cb.

However, welding techniques are not recommended for highly stressed components of turbines. Moreover, weldability of martensitic 13Cr–4Ni steel is very poor due to its high strength and hardness compared with other steels. They are less ductile even with heat treatment. The chromium content added to improve cavitation erosion and corrosion properties of 13Cr–4Ni steel also acts as ferrite former; hence, higher a percentage of carbon content is required to ensure formation of martensite. But with higher carbon contents, cracking at the heat-affected zone (HAZ) becomes a more serious problem (Thapa, 2004).

7.2.2 Boronizing and hard chrome plating

From earlier research work, Mann (1997) reported that boronization of steel has effectively improved its performance in adhesive, sliding and abrasive wear. In fact, pack-borided 12% chromium steel for steam turbine nozzles has performed exceptionally well and surpassed all other coatings, including plasma and detonation. It has been reported that borided 13Cr–4Ni steel produced by a pack cementation technique has improved resistance to abrasion but showed poor resistance to cavitation–erosion due to extremely brittle borided layer, and loss of elongation and strain energy. He also pointed out that tempered borided 13Cr–4Ni steel shows better cavitation erosion than borided steel; however, it showed less resistance to cavitation than as-received 13Cr–4Ni steel. Mann (1998, 2000) found poor erosion resistance of borided 13Cr–4Ni steel compared with borided 12Cr and T400 steels due to its low hardness and presence of extended surface cracks up to the base. Graham and Ball (1989) showed a better erosion resistance of boronized AISI 440 steel (18Cr–1C) compared to Stellite 6, AISI 440C steel and WC–Co cermets (8 and 11 wt% Co binder), but poor resistance compared to WC–Ni cermets (9 wt% Ni binder).

Mann (2000) reported a slightly better erosion resistance of hard chrome plated 13Cr–4Ni steel compared with as-received 13Cr–4Ni steel, plasma nitrided and D-gun coated (Cr_3C_2–25NiCr) 13Cr–4Ni steel, but poor resistance compared with borided 13Cr–4Ni steel under high particle impact energy.

7.2.3 Plasma nitriding

Plasma nitriding is a modern technique for surface hardening of metallic components to improve their service life. Basically, plasma nitriding is a glow discharge process in a mixture of nitrogen and hydrogen gases. The main advantages of plasma nitriding over conventional nitriding processes are: reduced cycle time, controlled growth of the surface layer, elimination of white layer, reduced distortion, no need of finishing, pore-free surfaces and mechanical masks instead of copper plating. The nitrided layers consist of FeN, $Fe_{2\text{-}3}N$, Fe_4N and Fe_2N_3 diffused layers. The diffused layers range from tens to hundreds of microns and these are ideal for improving wear resistance. By optimizing the nitrogen to hydrogen ratio, it is possible either to eliminate some of the brittle/white layers or to improve the erosion/corrosion properties (Mann and Arya, 2002).

Podgornik and Vizintin (2001) carried out pulse plasma nitriding for AISI 4140 and AISI A355 structural steels and found that no significant improvement is obtained in terms of hardness and wear behaviour. It has been reported that the plasma nitriding of 12Cr and 13Cr–4Ni martensitic steels resulted in improvement in resistance to abrasive and erosive wear. It is also reported that plasma nitrided 12Cr steels performed better than 13Cr–4Ni steel (Mann and Arya, 2002). However, Mann (2000) reported negligible erosion improvement of plasma nitrided 13Cr–4Ni steel compared with as-received 13Cr–4Ni steel, but better erosion resistance compared with D-gun (Cr_3C_2–25NiCr). Mann and Arya (2001) have shown a poor abrasion and erosion resistance of plasma nitrided 13Cr–4Ni steel compared with plasma nitrided 12Cr steels and HVOF sprayed WC–10Co–5Cr due to low hardness and inability to absorb more nitrogen. Corengia *et al.* (2005) have investigated the DC pulsed plasma nitrided 4140 steels and found that the process improved wear resistance. Manisekaran (2005) has reported 2-fold increases in erosion resistances for plasma nitrided 13Cr–4Ni steel compared with as-received 13Cr–4Ni steel but poor performance compared with laser hardened 13Cr–4Ni steel. Wei *et al.* (2006) have produced thick nitrides (ZrN, CrN and TiN) and carbonitrides (ZrSiCN and TiSiCN) using a plasma-enhanced magnetron sputtering (PEMS) technique for sand erosion protection. They reported that carbonitrides of TiSiCN showed at least 25 times less erosion loss than uncoated steels and 5–10 times less than other nitrided steels. High-temperature nitrided AISI 410S and AISI 410 martensitic stainless steels showed better slurry wear resistance than conventional AISI 420 martensitic stainless steel, when tested in substitute ocean water containing quartz particles (Mesa *et al.*, 2003).

7.2.4 Thermal spray techniques

Thermal spraying involves a different process of applying coating to improve wear resistance and corrosion properties. The molten raw materials (powder or wire) are accelerated in the thermal spray system and strike the substrate. Due to heat and impact, the particles flatten in the form of thin platelets and bind together with the substrate. These layers of flattened particles act as a protective coating. The quality of thermal sprayed coating depends on both coating material property and spray process parameters. Hotea *et al.* (2008) have discussed elaborately modern technological applications of various thermal sprayed coatings.

Santa *et al.* (2007) have reported better erosion resistance of two coatings obtained by oxy-fuel powder (OFP) and wire arc spraying (WAS) processes (with WC/Co in Fe–Cr–Ni matrix) on AISI 304 steels, compared with AISI 431 and CA6NM steels. Better resistance of above thermal sprayed coatings is attributed to the presence of hard and wear- resistant particles and a ductile metallic matrix.

Detonation (D-gun) sprayings

Mann (2000) extensively studied the use of D-gun coatings for hydroturbine applications, especially under high-energy particle impact. D-gun coated WC–12Co on 13Cr–4Ni steel showed excellent erosion resistance compared with D-gun sprayed Cr_3C_2–25NiCr on 13Cr–4Ni steel, as-grounded and plasma nitrided 13Cr–4Ni steel. The reasons behind its excellent erosion resistance are high density and hardness of 13Cr–4Ni steel D-gun sprayed WC–12Co. However, he pointed out that the use of D-gun coatings at high-energy particle impacts is limited due to its defects (voids and oxides).

HVOF coatings

In recent years, HVOF spraying has been considered an asset to the family of thermal spray processes, especially for material with a melting point below 3000 K. It has proven successful, its advantages in density and bond strength making it attractive for improved erosive wear and corrosion resistance in turbine applications (Arya *et al.*, 2003). Tungsten carbide powders are widely used in the HVOF spraying systems. These are used to produce dense, hard and excellent wear-resistant coatings generally to combat erosion and corrosion occurring in hydropower plants and pumps. WC cermets with and without nickel or cobalt are utilized to overcome the problems of abrasive and dry or slurry erosive wear of components (Karimi and Verdon, 1993; Karimi *et al.*, 1995; Hawthorne *et al.*, 1999; Mann and Arya, 2003; Barber *et al.*, 2005).

Under identical test conditions of erosion and abrasion, HVOF coatings with WC–10Co–5Cr on 13Cr–4Ni steel have shown much better performance than plasma nitrided 12Cr and 13Cr–4Ni steel (Mann and Arya, 2001). The presence of defects in these coatings leads to deterioration under high-energy particle impacts. Thapa (2004) has reported that HVOF sprayed duplex steel SAF 2304 (23Cr–5Ni) with 86WC–6Co–8Cr shows lower erosion rates than other HVOF sprayed coatings (86WC–10Co–4Cr and 75 Cr_3C_2–25(80Ni20Cr)).

7.2.5 Erosion protection by non-metallic coatings/other protection methods

Hard metallic coating materials were initially thought to be a good choice to resist abrasion. However, an alternative approach employs non-metallic materials (Iwai and Nambu, 1997; Lathabai *et al.*, 1998), benefiting from their ease of application, cost-effectiveness and, in some cases, high deformation capability or high hardness. In order to counteract the detrimental effects of cavitation erosion and silt abrasion, the use of these coatings has been investigated by a number of researchers, both in laboratory tests and in service experiences. Zhang *et al.* (1996) have reported from earlier research that polyurethane elastomer had more than 10 times the erosion resistance of flake graphite cast iron in slurry, and the same elastomer showed good cavitation resistance too. From their work, they concluded that epoxy resin reinforced by synthetic corundum particles and castable polyether-based polyurethane rubber were the most resistant coatings to abrasion and the combined abrasion–cavitation damage, respectively. Chopra and Arya (1996) reported use of ceramic tiles (aluminium-talc-clay) in the Kaplan turbine (Chilla Hydropower Project, India) blade with epoxy adhesive to prevent erosion, but these tiles were washed away very soon due to high silt.

Juliet *et al.* (1996) have found better erosion resistance for the cases of tungsten–carbon-based multilayer coatings obtained by reactive magnetron sputtering (PVD). Wood (1999) has reported significant improvement in erosion protection from boron carbide coated on tungsten carbide by chemical vapour deposition (CVD).

7.2.6 Laser surface engineering

There are several drawbacks to conventional hardening methods (viz. surface hardening, induction hardening and flame hardening) such as length of time/material/energy consumption, requirement of complex heat treatments and poor precision and flexibility). Thermal sprayed coating exhibits defects such as voids and surface cracks. These problems can be overcome by effective utilization of laser (light amplification by stimulated emission of radiation) technology for the surface problems of engineering components (Steen and Watkins, 1993; Majumdar and Manna, 2003).

Laser hardening or laser transformation hardening (LTH)

In laser transformation hardening (LTH), the enhancement of the surface properties resulted mainly from a solid-state transformation to form a harder phase. The depth of hardening and temperature profile can be determined by energy density and spot speed. It has the advantage of local treatment with minimum HAZ on the bulk volume. LTH has become a well-established technique for heat treating steel surfaces, competing against the more widely used flame and induction hardening methods. The process presents considerable advantages over the conventional methods; the most significant are the high degree of controllability and automation, low distortion and capability of hardening very selective areas and precise treatment (no post treatment required).

The LTH process uses the heat generated by the impingement of a laser beam on the surface of the steel to austenize it. The austenized layers are subsequently transformed to martensite due to very rapid conduction of heat into the cold interior of the workpiece. This effect is known as self-quenching, and the cooling rates obtained are usually high enough to allow martensite formation, even in low carbon steels, which have very low hardenability. In addition, the hardened microstructure obtained is usually finer than that from conventional hardening techniques, leading to increased hardness and consequently to increased wear and fatigue resistance.

Lo (2003) found that the LTH of AISI 440C high carbon martensitic stainless steels resulted in significant improvement in resistance to cavitation erosion. Tinamin (2003) found that LTH enabled 2Cr martensitic stainless steels to obtain superior surface properties in terms of oxidation resistance, hardness and impact wear resistance. Mann and Arya (2003) found that the LTH of 12Cr forged hydroturbine steels resulted in excellent performance in jet impingement erosion for the applications of steam turbine. Manisekaran (2005) reported better erosion resistance of laser hardened 13Cr–4Ni steel (due to tempered martensite and retained austenite) over plasma nitriding and laser nitriding.

Laser surface melting (LSM)

In laser surface melting (LSM), enhancement in surface properties results from the refinement and homogenization of microstructure. The LSM requires higher power densities than laser hardening (Srinivasan, 2008). The LSM of martensitic stainless steel UNS S42000 using a 3.5 kW continuous wave CO_2 laser has been reported (Kwok *et al.*, 2000). The cavitation erosion resistance of laser melted specimens using a power of 1.7 kW and a scanning speed of 25 mm/s was reported to be 70 times that of the as-received (annealed) steels, and 1.8 times that of conventionally heat-treated steel. The excellent cavitation erosion resistance was due to the combined effect of a high volume fraction of retained austenite (89%) and moderate hardness (450 HV). By using different processing parameters, it was found that the cavitation erosion resistance of the laser-melted specimens increased with increase in volume fraction of retained austenite, a result attributable to the high martensitic transformability of the austenite in UNS S42000 steel. However, cavitation erosion resistance increased with the increase in hardness to a maximum value and then dropped with further increase in hardness. This indicated that martensitic transformability played a more important role than hardness in cavitation erosion. Owing to pitting potential variation, the pits formed in the laser melted specimens were shallower than those formed in as-received and hardened UNS S42000 steel. The improvement in pitting corrosion resistance resulted from the dissolution or refinement of carbide particles and the presence of retained austenite, as evidenced by the fact that the pitting potential increased linearly with the amount of retained austenite.

LSM for improving cavitation erosion resistance by means of a continuous Nd : YAG (yttrium aluminium garnet) laser has been attempted on austenitic stainless steels UNS S31603 and UNS S30400, and super duplex stainless steel UNS S32760. It was reported that the cavitation erosion resistance of the LSM stainless steels was highly dependent on the microstructural changes and the residual stress in the laser-melted layer (Kwok *et al.*, 1998). Rao *et al.* (1993) investigated the laser surface melted 0.4% C low-alloy steel and conducted erosion tests with SiC as erodent for two different velocities (46 and 96 $m\,s^{-1}$) at three impact angles (30°, 60° and 90°). They found that LTH did not improve the erosion resistance of 0.4% C low alloy steel due to reduction in ductility. Lo (2003) found that the LSM of AISI 440C high carbon martensitic stainless steels resulted in better corrosion resistance by carbide refining.

Laser nitriding

Nitriding of steels is well known to improve hardness, wear and corrosion resistance of surfaces. Recently, it has been found that the simple irradiation of steel with the pulses of an excimer laser in air or in nitrogen atmosphere leads to the large dissolution of nitrogen into the irradiated surface. It is expected that there will be an increase in the hardness and wear resistance with the number of pulses. Pressure and plasma-enhanced dissolution and diffusion of nitrogen in combination with the macroscopic material transport (piston effect, convection, fall-out) are other important effects. In comparison to the conventional nitriding methods, laser nitriding results in a small HAZ, in depth and lateral dimensions, and accurate spatial control of the surface treatment without any undesired heating of the substrate (Schaff, 1995).

Carpene (2002) investigated the efficiency of the laser nitriding process for Fe, Al, Ti, Si and austenitic stainless steels with different types of pulsed lasers. It was demonstrated for the first time that the irradiation of iron samples with excimer laser pulses in nitrogen atmosphere and in air led to a significant nitriding effect with very interesting nitriding results for various carbon steels such as Ck 15, Ck 45, C60, C80, C105 (Schaff, 1995). Manisekaran (2005) has attempted to enhance erosion resistance of 13Cr–4Ni steel by laser nitriding; however, he did not succeed as the interaction time for laser nitriding was low (hence insufficient nitrogen diffusion). But the reported slurry jet erosion performance was slightly better than as-received 13Cr–4Ni steel when tested at two particle sizes.

Laser cladding (LC)

Laser cladding (LC) is capable of producing a wide range of surface alloys and composites of required properties. The application of laser beam cladding in surface engineering enables sound surface cladding (free from porosity and cracks) containing uniformly distributed hard particles in the softer and tougher matrices. During the process, the material is fed into the substrate surface under the laser beam by a number of routes including powder blowing, wire feeding, pre-placed powder coating, etc. The desired properties from LC can be achieved by appropriate selection and control of processing parameters such as laser beam power density, laser beam travel speed, and laser beam diameter at the workpiece surface. The main advantages of LC over plasma spraying and arc welding include reduction in dilution, reduction in thermal distortion (since in LC substrate will absorb very little energy compared to other alternatives), deposit porosity and also its capability of producing near net shape of the components (Steen, 1991).

The use of LC for the slurry erosion resistance purposes was reported by Tucker *et al.* (1984). They incorporated TiC, WC and $MoSi_2$ into Co-based Stellite 6 alloy by using a 5 kW transverse-flow CO_2 laser source and found that WC reinforced Stellite 6 claddings show higher erosion resistance. Zhang and Zhang (2005) have reported the applications of LC (cladded with Ni–Cr_3C_2 and Ni–WC on 13Cr martensitic stainless steels) for erosive–corrosive (EC) environments. They have reported that the EC rates of laser-clad Ni–Cr_3C_2 and Ni–WC composite coatings decreased by approximately 60% and 30% compared with that of substrate, respectively. The increase in EC resistance is closely related to structure state, kinds and amount of carbide, microhardness and toughening ability of the clad layer. Slurry erosion resistances of nickel–tungsten (40, 50 and 60%) functionally graded materials produced (on AISI 4140 steels) by LC were extensively discussed by Yarrapareddy *et al.* (2006). Jiang and Kovacevic (2004) have reported a better slurry erosion resistance of laser clad Fe–Cr–B–Si coatings on AISI 4140 steel compared with as-received AISI 4140 steels and claddings of Cr_2C_3–Ni20Cr and W_2C/WC–12%Co. Better erosion properties of Fe–Cr–B–Si claddings have been attributed to higher ductility and toughness compared with other carbide claddings.

7.3 Materials and processes

7.3.1 Turbine materials

Important properties such as high fatigue, erosion, cavitation and corrosion resistance must be considered while selecting materials for turbine components. Basically, various steels are used for high head turbines, whereas cast iron and bronze are used for some low-head small turbines. Earlier studies (Goel and Sharma, 1996) confirmed the use of 18Cr–8Ni austenitic stainless steels with less than 0.2% Mo to fabricate runners in Pathri (Uttar Pradesh) and Mettur (Tamilnadu) hydropower plants in India. However, these steels showed poor fatigue behaviour. Also, there is a self-hardening effect caused by mechanical abrasion; hence, this steel was not recommended for runners, guide vanes and other moving parts of turbine. At later stages, forged martensitic stainless steels 12Cr and 13Cr–1Ni have been developed to substitute 18Cr–8Ni steel, but these steels also showed poor erosion, cavitation and corrosion resistance because of low hardness, toughness and also reduced Ni content. The weldabilty of this steel was also poor. Owing to problems with restrictions in usage of patented Swedish quality steel, such as 13Cr–6Ni, various other steels such as 13Cr–4Ni, 16Cr–5Ni and 17Cr–4Ni have been developed, among which 16Cr–5Ni and 13Cr–4Ni steel are widely used to fabricate various parts of turbines. Both 13Cr–4Ni and 16Cr–5Ni steels have excellent cavitation erosion and corrosion resistance.

7.3.2 Coating alloys

Coating alloys fall into three basic categories: (a) cobalt-base and nickel-base alloys, (b) Iron-base alloys and (c) steel or alloys containing tungsten carbide. Most of the hardfacing alloys consist of hard particles within the matrix, and, for many, it is the hard constituent that provides resistance to wear. However, the cobalt-base alloys are an exception in this regard as they exhibit wear resistance largely by virtue of the deformation and fracture characteristics of the cobalt-rich matrix. According to the microstructure, coating alloys can be divided into five types: carbide-type, boride-type, silicide-type, intermetallic-type and solid solution-type alloys (Gurumoorthy, 2006).

Selection of coating powders

The selection of the coating alloy depends to a great degree on the nature of wear the component will encounter. The primary inputs for selecting an alloy system to combat wear includes the type of wear, method of application and cost. Some coating powders used according to latest research and experiments are CrC_2–NiCr–WC–CoCr, WC–Cr_3C_2–Ni–alumina–titania

(97 : 3) and (87 : 13), chromium oxide (99%), chromium oxide–silica–titania (95 : 3 : 2), Triballoy-400, T410 (Singal and Singh, 2006).

The substrate material used in this work was 13Cr–4Ni steel, which is widely used to fabricate various components of a hydroturbine. These steels are generally used in quenched and tempered conditions and with a hardness of 240–250 HV. Two commercially available coating materials, Co-based Stellite 6 (Delero Stellite Co., UK) and Ni-based Colmonoy 88 (Wall-Colmonoy Co., USA) were selected for laser surface alloying on 13Cr–4Ni steel. The nominal compositions of 13Cr–4Ni steel and selected coating powders are shown in Table 7.1.

Commercial boron carbide (B_4C) powder (average size of 40 μm) is added into Stellite 6 and Colmonoy 88 prior to laser surface alloying. Properties of selected coating powders are listed in Table 7.2. The SEM morphologies of Stellite 6, Colmonoy 88 and B_4C coating powders are shown in Fig. 7.1.

7.3.3 Process

Laser surface alloying (LSA)

Laser surface alloying (LSA) is a process wherein thin layer of the metal is melted by a laser beam with the simultaneous addition of the desired alloying elements to the melt pool, thereby changing the surface chemical compositions of the metal. The LSA process can be carried out in a single stage or in two stages (Vilar, 1999). In the single-stage process, alloying elements (in the form of wire, rod and powder) are directly fed to melt pool. The two-stage process consists of first pre-depositing the desired alloying elements by various methods (electroplating, evaporation, sputtering, etc.) and then melting the pre-deposited layers with laser to form protective coatings. The details of the LSA process are given by Draper and Ewing (1984).

Table 7.1 Nominal chemical composition (wt%) of used materials

Material	C	B	Si	Cr	W	Mo	Mn	Co	Ni	Fe
13Cr–4Ni steel	0.058	–	1.00	12.86	–	0.50	1.00	–	3.85	Bal.
Stellite 6	1.17	–	1.33	30.80	4.30	0.10	0.30	Bal.	2.59	2.25
Colmonoy 88	0.82	3.08	3.93	14.87	17.40	–	–	0.18	Bal.	3.90

Table 7.2 Properties of selected coating powders

Properties	Density (g/cm^3)	Melting point (°C)	Particle size (μm)	Hardness (HRC)
Stellite 6	8.44	1285–1410	30–120	36–45
Colmonoy 88	9.89	1100–1180	20–90	59–64

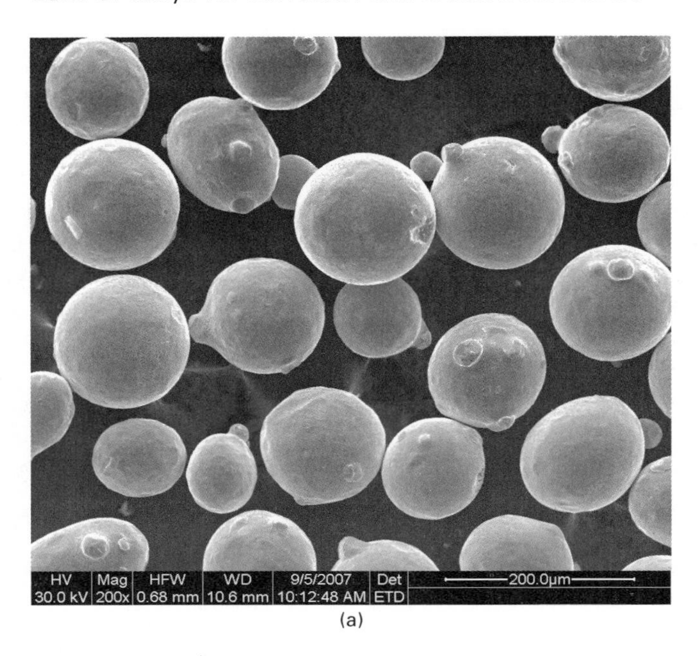

(a)

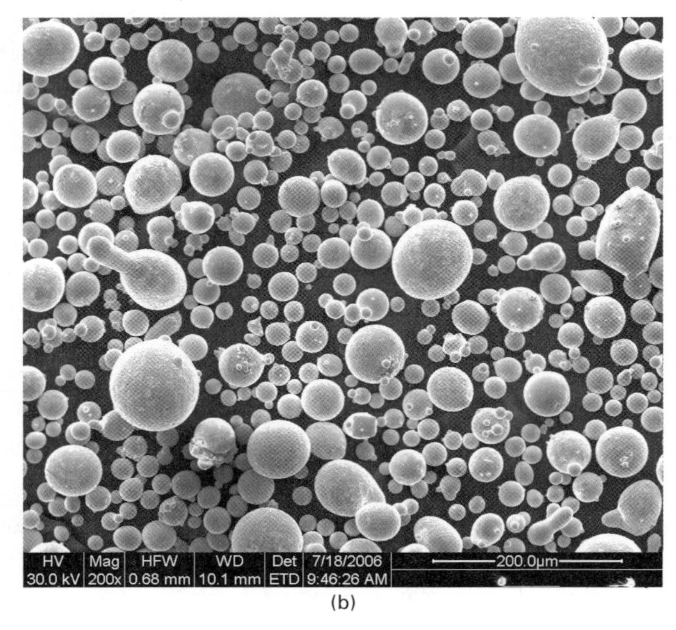

(b)

7.1 Scanning electron microscopy (SEM) morphology of various coating powders: (a) Stellite 6; (b) Colmonoy 88; and (c) B_4C.

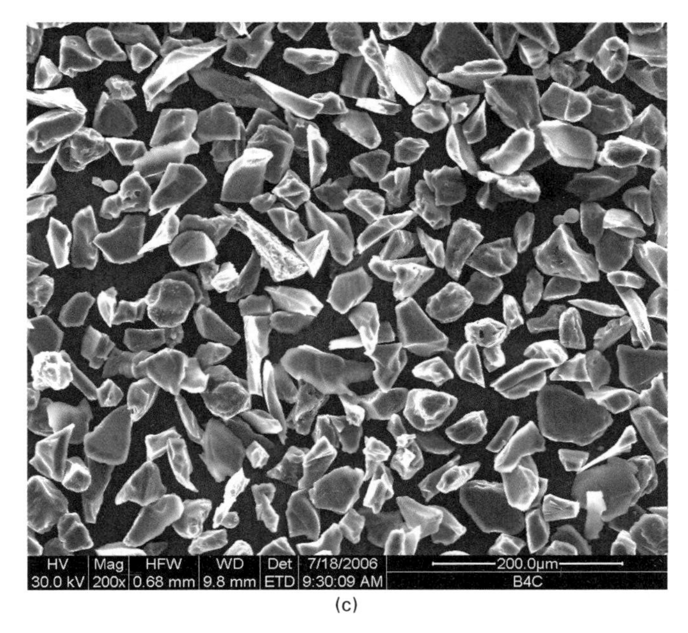

(c)

7.1 Continued

Some of the advantages of LSA over other coating processes are as follows:

- Cost effective: the process substitutes for the massive purchase of expensive or hard to use materials.
- Metallurgy: it offers new solutions by making alloys that are unique in composition and microstructures.
- Geometry: low distortion, high precision and flexible motion.
- No risk of coating delamination.
- Requires a very small amount of modifier alloy.
- Requires little or no surface preparation for certain applications.

Details of surface modification process: two-stage process of LSA

Newer coatings on 13Cr–4Ni steel were applied by using two-stage process of LSA with commercially available Co-based Stellite 6 and Ni-based Colmonoy 88 hardfacing powders. In order to maximize the erosion resistance of the basic LSA coatings, an attempt has been made to reinforce light-weight and high-hardness B_4C into Stellite 6 and Colmonoy 88 at two different weight fractions (5% and 10%) prior to LSA. The LSA work was carried out at Centre for Laser Processing Materials (CLPM), International Advanced Centre for Powder Metallurgy and Newer Materials (ARCI), Hyderabad, India using

9 kW transverse-flow CO_2 laser system (Model No. ML-108, MLI, Israel). The output laser is a continuous wave with wavelength of 10.6 μm. The output beam in the form of ring (70 mm outer diameter and 35 mm inner diameter) in doughnut mode (M2 = 4) is tailored into a rectangular beam of size 18 mm × 1 mm by using 10 mm faceted and cylindrical integrating (zooming) oxygen free high conductivity (OFHC) mirrors. The distance between integrating mirror and faceted mirror has been kept at a calibrated vernier position of 18.5 and the effective working distance has been kept at 143 mm to obtain uniform rectangular beam. The intensity distribution across the beam spot has been assessed by testing on perplex burn print. The desired coating height, depth (250 to 300 μm of coating thickness) and width were achieved by a single pass of the laser beam (no overlapping) with laser power 3 kW, laser scan speed 3 mm/s, laser scan width of 18 mm, focusing distance 130 mm along with N_2 gas at 0.034 MPa.

Prior to the application of laser beam, the 13Cr–4Ni steel plates of dimension 70 × 140 × 8 mm^3 were pasted with coating powders to a thickness of 500 to 600 μm by using polyvinyl acetate (PVA) solution. The procedure adapted to pre-pasting the coating powder to the 13Cr–4Ni steel is as given below.

The PVA of 2–4% powder has to be mixed in boiling water and properly stirred. The substrate, on which coating/alloying is to be done, must be slightly etched with any acid. This will remove the oxide layer and help in uniform spreading of the PVA solution. After etching, acetone cleaning should be done, followed by drying. Now, the PVA solution can be applied to the substrate with a paintbrush such that a uniform layer is achieved. Then, the hardfaced powder can be sprinkled on with a spatula, which must be tapped with a finger to sprinkle a uniform shower of fine powder on the substrate. The thickness achieved by this method depends on thickness of the PVA film on the substrate, and on the powder size. After obtaining a uniform thickness, the substrate can be left in an open atmosphere for 6 hours so that the powder adheres properly after the PVA dries out. One or two more layers of coating powders can be obtained by repeating the above procedure. Figure 7.2 illustrates the schematic of pre-pasting of the coating powder on 13Cr–4Ni steel.

7.4 Metallurgical performance of coatings

7.4.1 Microstructural examination

Microstructural studies were carried out on LSA coatings and substrate 13Cr–4Ni steel. Microstructural examinations of the coatings were carried out with Zeiss MC43 optical microscopy and JSM-840A JEOL scanning electron microscopy (SEM). Energy dispersive spectrum (EDS) analysis was done with ISIS Link Microanalyser, Oxford Instruments. Microscopic

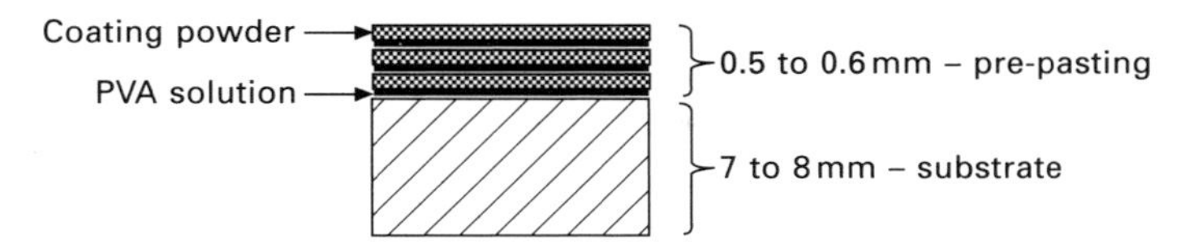

7.2 Schematic of pre-pasted coating powder on steel plates (not to scale).

examination was done using Villela's reagent (mixture of 1 g picric acid, 5 ml HCl and 100 ml methanol) and aqua regia (HNO_3 : HCl in 1 : 3 ratio). Prior to etching, the samples were polished with a series of 220, 320, 400, 600, 800 and 1000 water emery sheets followed by alumina and diamond fine polishing to the range of 0.5–1 μm surface finish.

Measurement of coating characteristics (thickness, width and dilution)

The ideal cross-section of LSA coating characteristics (depth, minimum height, width and thickness of the coating) is given schematically in Fig. 7.3. Generally, desired coating powders are added to the melt pool at the subsurface of the substrate created by laser, thereby reducing the height of the coating. However, there are reports (Damborenea *et al.*, 1994; Lo, 2003; Pang *et al.*, 2005) suggesting that a little height has been maintained during the work of LSA in order to minimize irregular surface finish. This may be effective while doing the two-stage process, especially during mechanical pasting (using PVA or any other binder agent) of the coating powders prior to laser passing. Typical laser-modified coating characteristics, such as coating thickness (height + depth), width and percentage (%) of dilution, were measured from optical micrographs of the cross-sections of Stellite 6 and Colmonoy 88 with and without additions of 5 wt% B_4C coatings and given in Fig. 7.4. The measured characteristics are tabulated in Table 7.3.

For a set of given LSA parameters, all the coatings have large, similar coating thicknesses of about 0.40 mm with dilution in the range from 52 to 68%. For slurry erosion test purposes, a coating thickness of 0.25 to 0.30 mm was considered. The extra coating height given during LSA work has been removed by grinding and then metallurgical polishing performed before conducting slurry erosion tests. Differences in characteristics are attributed to variations in properties such as powder size, alloying elements and melting points of coating powders (Tables 7.1 and 7.2).

A typical acceptable level of dilution in case of laser cladding is approximately 8–10% (Qian *et al.*, 1997). Joshi and Sundararajan (1999) have distinguished the processes based on dilution and, according to them, processes involving less than 10% dilution are referred to as 'cladding'

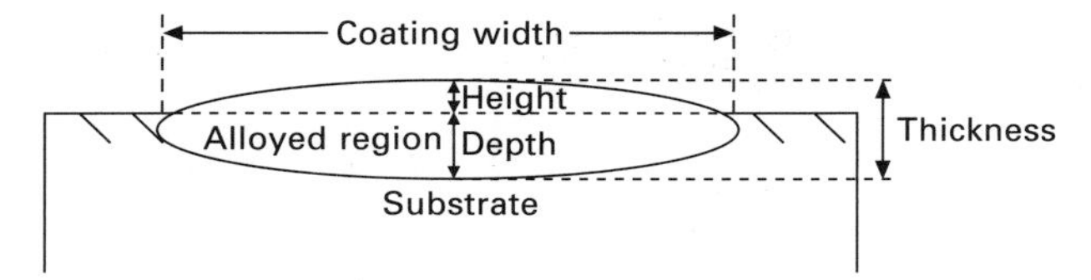

7.3 Schematic of LSA coating characteristics.

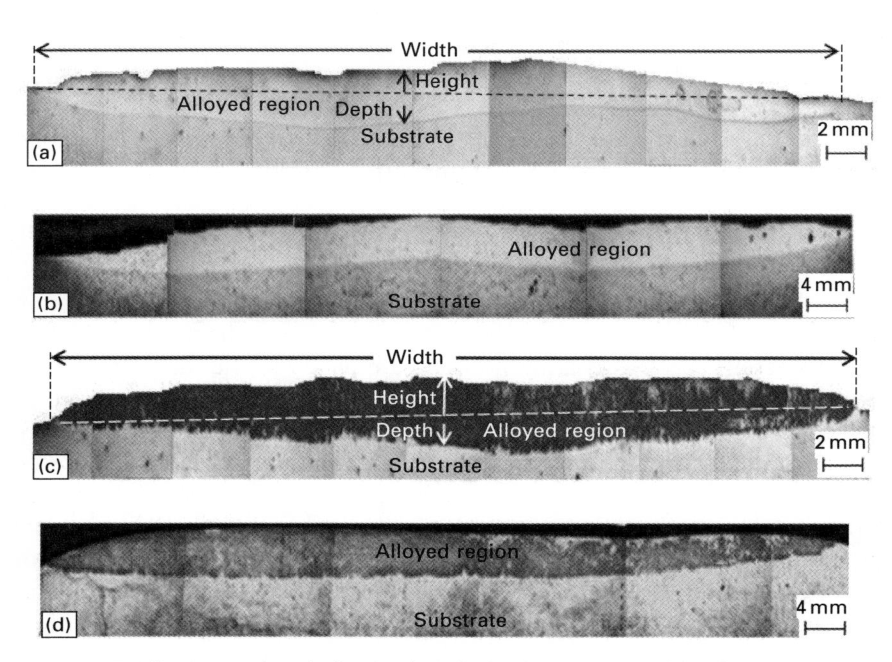

7.4 Cross-sectional views of optical micrographs of (a) Stellite 6, (b) Stellite 6 + 5 wt% B_4C, (c) Colmonoy 88, and (d) Colmonoy 88 + 5 wt% B_4C coatings.

Table 7.3 Typical characteristics of LSA coatings

Coatings	Coating characteristics				
	Height, *a* (mm)*	Depth, *b* (mm)*	Width (mm)	Thickness, (*a* + *b*) (mm)	Dilution (*b*/(*a* + *b*)) × 100 (%)
Stellite 6 (St. 6)	0.15	0.29	17.9	0.44	65.91
St 6 + 5 wt% B_4C	0.13	0.28	17.5	0.41	68.29
Colmonoy 88 (Col. 88)	0.19	0.21	18.2	0.40	52.50
Col. 88 + 5 wt% B_4C	0.18	0.20	17.5	0.38	52.63

* Height and depth of the coating measured at the centre of entire coating width.

while those with dilution exceeding 10% are termed as 'surface alloying'. In the present investigation, LSA coatings showed the dilution was about 50%.

Interface studies

Macroscopically, all the coatings exhibited a good surface appearance with no surface cracks and porosities. In order to understand the interfacial behaviour and bonding between coatings and substrate, backscattered electron images have been taken at low magnifications using high-resolution SEM. Typical cross-sectional features of Stellite 6, Colmonoy 88 and B_4C added coatings are shown in Figs 7.5 and 7.6, respectively. The substrate/coating interfaces were continuous and implies that coating is well adhered to the substrate.

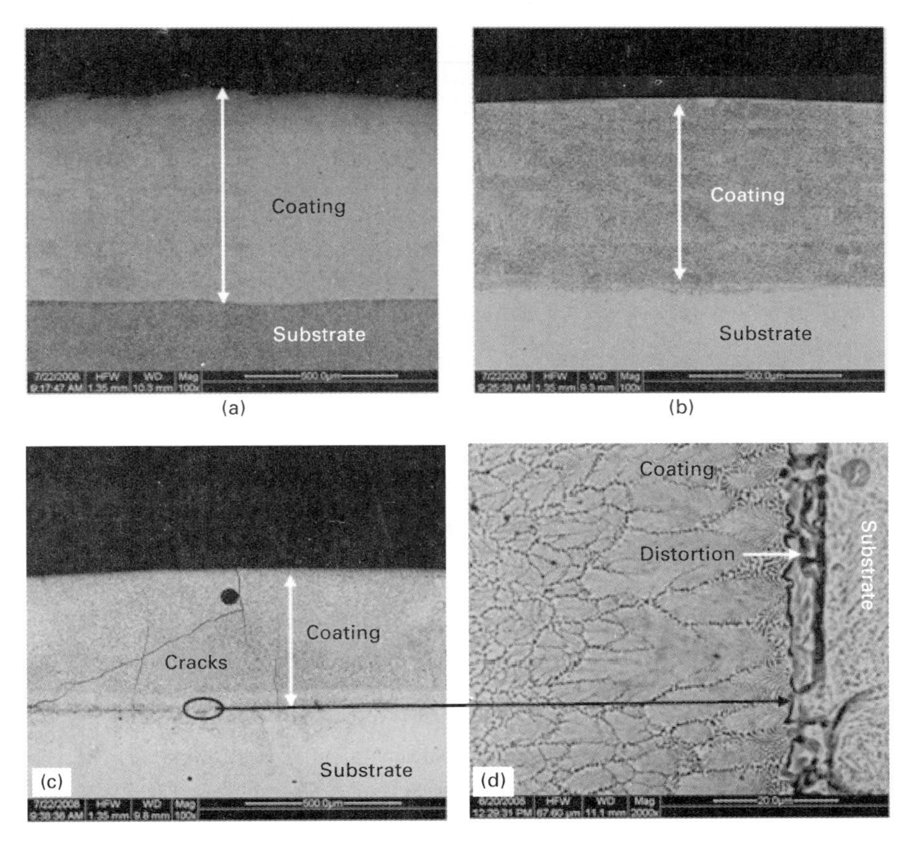

7.5 Cross-sectional BSE images of various Stellite 6 coatings: (a) Stellite 6 (without B_4C); (b) Stellite 6 + 5 wt% B_4C; (c) Stellite 6 + 10 wt% B_4C and (d) closure view at interface of 10 wt% B_4C.

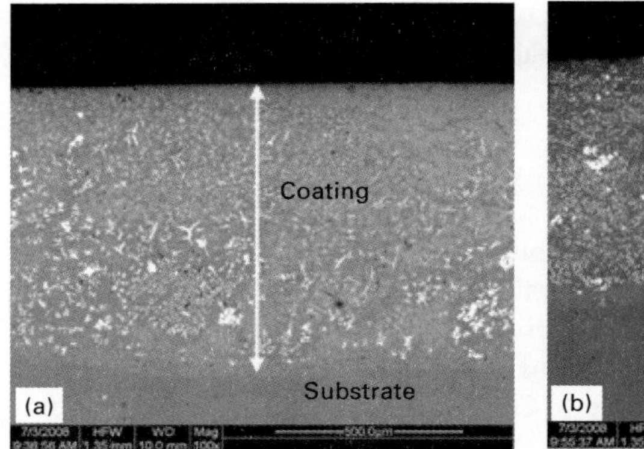

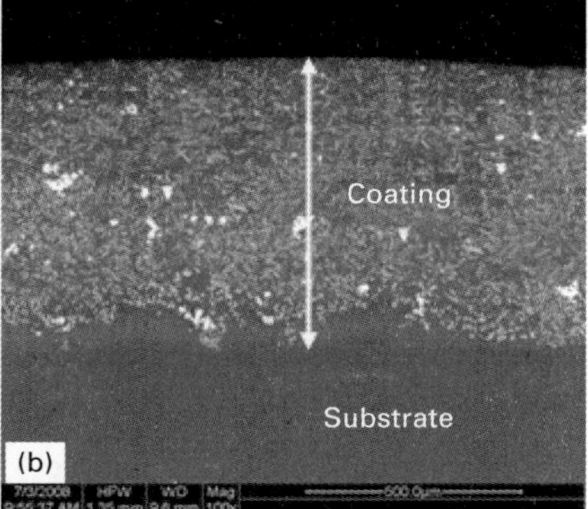

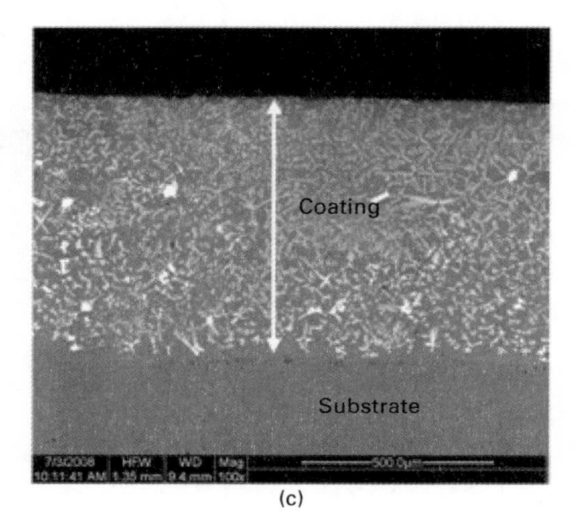

(c)

7.6 Cross-sectional BSE images of various Colmonoy 88 coatings: (a) Colmonoy 88 (without B_4C); (b) Colmonoy 88 + 5 wt% B_4C; and (c) Colmonoy 88 + 10 wt% B_4C.

The alloyed (coating) region is free from defects such as cracks and pores. However, for the given LSA parameters, Stellite 6 + 10 wt% B_4C coatings showed few lateral and vertical cracks, which are observed up to substrate/coating interfaces (Fig. 7.5(c)) and also bonding between substrate and coating was not good and high distortion at interfaces was observed as seen in Fig. 7.5(d). This may be due to thermal mismatch between Stellite 6 and B_4C powders and also may be the selected LSA parameters were not suitable to melt the powders properly. Hence, further studies on Stellite 6 + 10 wt% B_4C coatings were not carried. Figures. 7.5(a)–(c) and 7.6(a)–(c) have revealed the uniform distribution of complex secondary phases along the thickness of LSA coatings.

Optical microstructures

The optical microstructure of 13Cr-4Ni steel taken at two different magnifications is shown in Fig. 7.7. The microstructure consisted of acicular tempered martensite, which is mainly of fine needle shape martensite.

During the LSA process, the pre-placed powder paste was melted together with subsurface of substrate 13Cr–4Ni steel and formed an alloyed (coating) region. Figure 7.8 shows the optical microstructure of near surface regions of Stellite 6 coatings. It consists of primary dendrites (white phases) and interdendritic eutectic regions (dark regions). The eutectic regions in Fig. 7.8 consisted of secondary phases and Co-rich solid solution. Quantitative analysis showed that volume fraction of Co-phase is about 65% and finer carbides were about 35%. It was very difficult to measure the sizes of

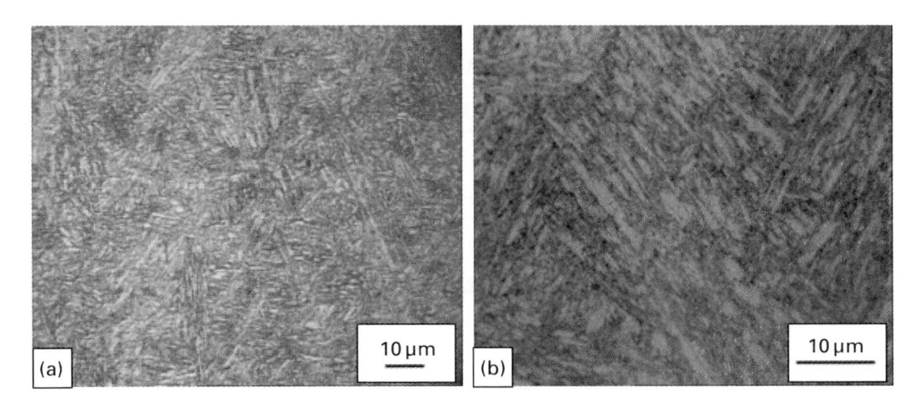

7.7 Optical microstructure of 13Cr–4Ni steel at (a) low and (b) high magnifications.

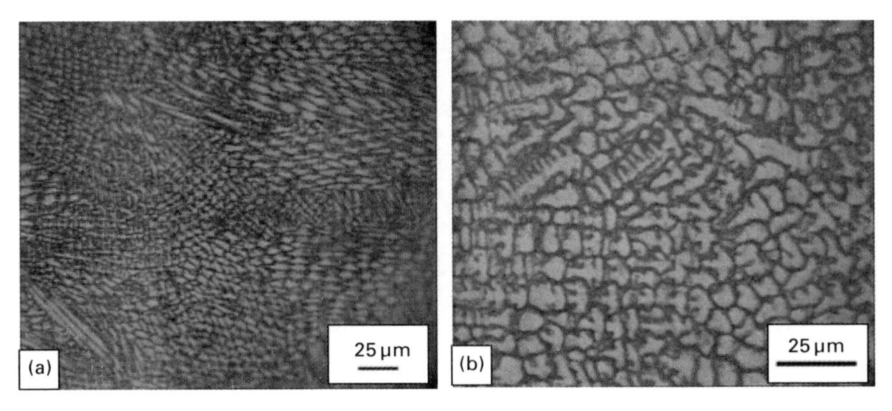

7.8 Optical microstructure of Stellite 6 coatings at (a) low and (b) high magnification.

secondary phases in the case of Stellite 6 coatings as the secondary phases were very fine. However, from previous studies (Tiziani *et al.*, 1987; Xu and Kutsuna, 2006) it can be concluded that secondary phases of (< 1 μm) are obtained during LSA process.

Figure 7.9 shows optical microstructures of Colmonoy 88 coatings obtained at two different magnifications. It is clearly seen from Fig. 7.9 that there is uniform distribution of characteristic secondary phases throughout the coatings. The quantitative analysis showed that about 55% of white phase regions and 45% of secondary phases. Also, secondary phases of 2–10 μm were obtained from LSA.

SEM and EDS analysis of coatings

In order to identify the chemical constituents of eutectics and secondary phases, combined analyses of SEM and EDS have been done on both Stellite 6 and Colmonoy 88 coatings and are presented in Figs 7.10 and 7.11, respectively. As-coated microstructures can be described as a Co-rich matrix with a network of carbides in the interdendritic regions. The microstructure contains fine dendrites as a consequence of the high cooling rates achieved during LSA process. Tiziani *et al.* (1987) have reported that the primary and secondary arm spacings were about 7 and 8 μm, respectively, for Stellite 6 coatings on AISI 316 stainless steels using a CO_2 laser.

It can be seen in Fig. 7.10 that the eutectic regions of Stellite 6 coatings consist of face centred cubic (fcc) Co-, Fe- and also Cr-rich with a network of carbides ($M_{23}C_6$ types, where M = Cr, Fe) in the interdendritic regions. These results agree well with earlier work by Zhong *et al.* (2002), Zielinski *et al.* (2005) and Kashani *et al.* (2007), except for the presence of a higher percentage of Fe in coatings region. This is attributed to the LSA coating

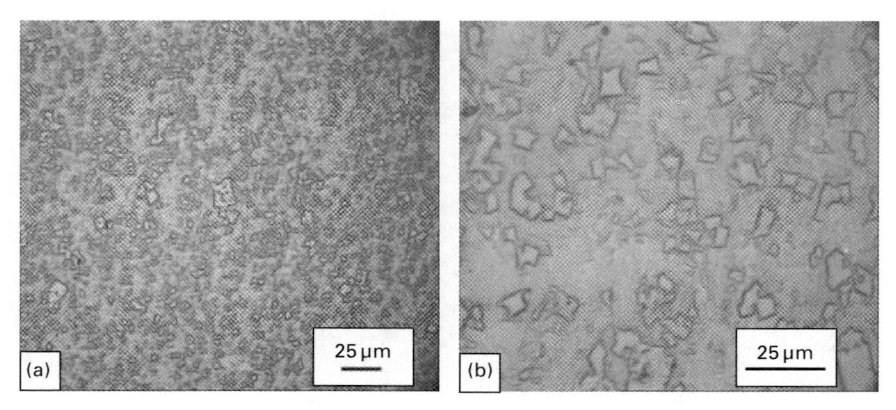

7.9 Optical microstructure of Colmonoy 88 coatings at (a) low and (b) high magnifications.

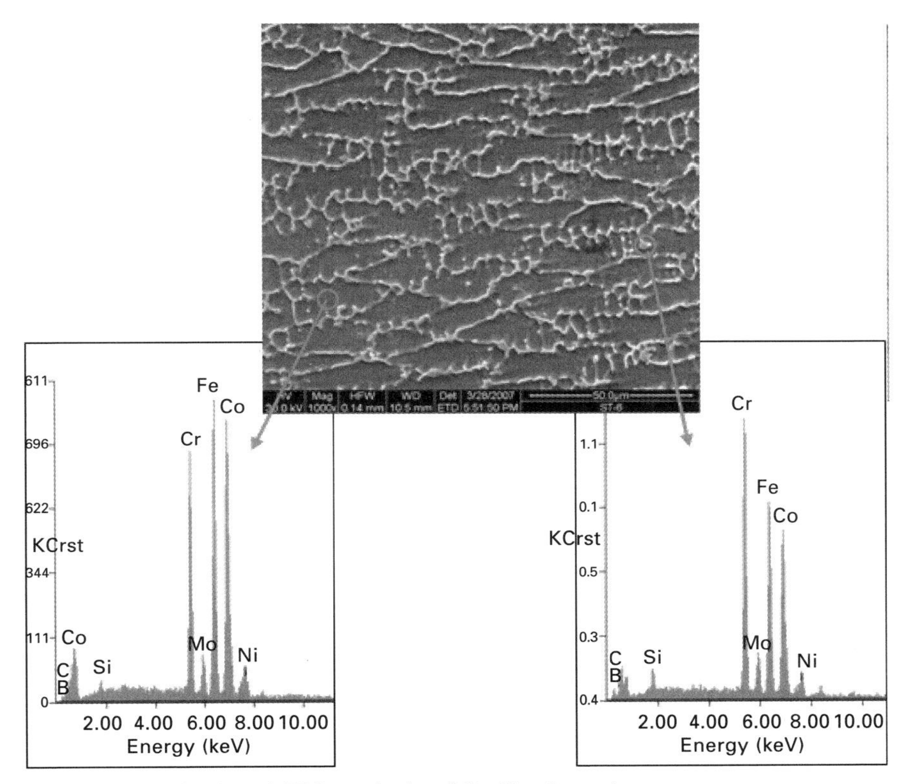

7.10 SEM and EDS analysis of Stellite 6 coatings.

process, wherein coating powder is melted into subsurface of the substrate, instead of developing the layers above the substrate as reported in the literature.

The SEM and EDS analysis of Colmonoy 88 coatings (Fig. 7.11) confirmed the presence of Ni- and Cr-rich and the very low presence of Fe-rich phases of matrix, whereas secondary phases (only rectangular/quadrangular in shape) were richer by tungsten and chromium phases. These results are different from earlier work reported by Qian *et al.* (1997) and Lim *et al.* (1998). They reported two different types of secondary phases (rectangular or quadrangular along with leaf-like morphology), when Colmonoy 88 was laser cladded on AISI 1020 low carbon steel using continuous wave CO_2 laser. There were two major differences between Stellite 6 and Colmonoy 88 coatings: first, the level of Fe-rich regions along with Co/Ni- and Cr-rich matrix, and secondly, size and shape of carbides in coatings. Stellite 6 showed higher levels of Fe at matrix regions along with Co and Cr-rich regions. This is attributed to its highest % of dilution (Table 7.3) among coatings. Stellite 6 coatings showed only very fine (~ < 1 μm) Cr/W-rich carbides, whereas Colmonoy

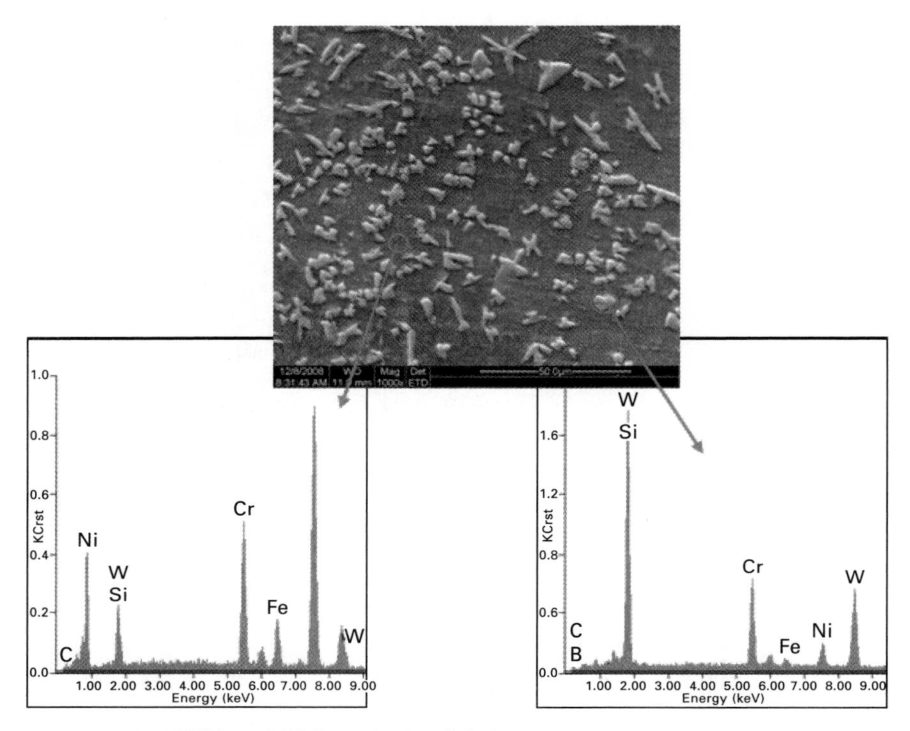

7.11 SEM and EDS analysis of Colmonoy 88 coatings.

88 coatings showed Cr-rich carbides of 2 to 5 μm and W-rich carbides of 6 to 10 μm.

Effect of B_4C reinforcement

A new attempt has been made to increase the hardness (in turn, slurry erosion resistance) of basic LSA coatings by the addition of light-weight and high-hardness B_4C (5 and 10 wt%) particles into Stellite 6 and Colmonoy 88 coating powders prior to the laser treatment. The aim was to disperse the B_4C particles without melting on the surfaces of both the coatings. Effects of B_4C reinforcement on microstructures of Stellite 6 and Colmonoy 88 coatings are presented and discussed in this section.

The effects of B_4C addition on microstructures of Stellite 6 and Colmonoy 88 coatings are given in Figs 7.12 and 7.13, respectively. As very high temperature is involved during LSA process, the complete dissolution of B_4C in Stellite 6 and Colmonoy 88 coatings was observed in all the microstructures due to its low melting point (2440 °C). The influence of B_4C reinforcement on microstructures of Stellite 6 and Colmonoy 88 coatings is two-fold. First,

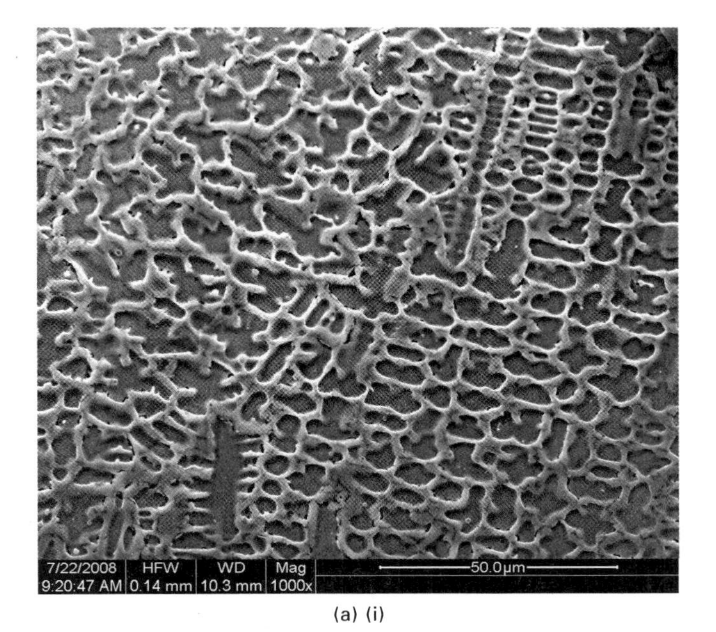

(a) (i)

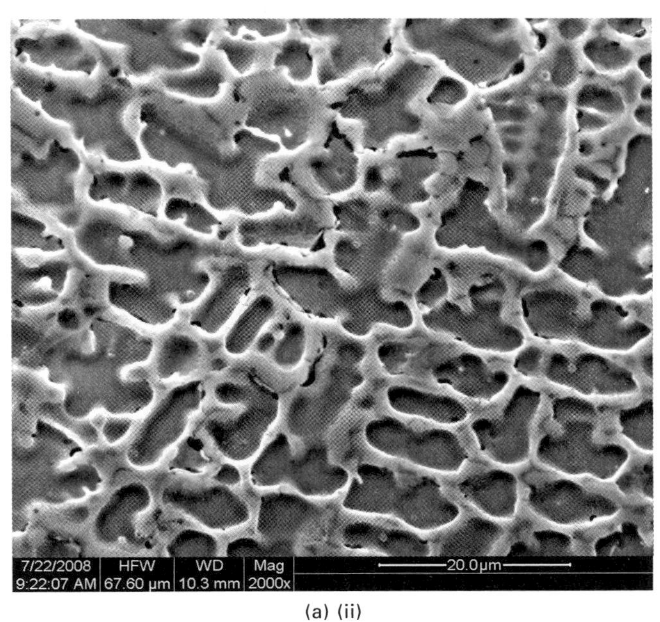

(a) (ii)

7.12 SEM microstructural features of various Stellite 6 coatings: (a) (i) at low magnification and (ii) at high magnification for Stellite 6 (without B_4C); (b) (i) at low magnification and (ii) at high magnification for Stellite 6 + 5 wt% B_4C; (c) (i) at low magnification and (ii) at high magnification for Stellite 6 + 10 wt% B_4C.

(b) (i)

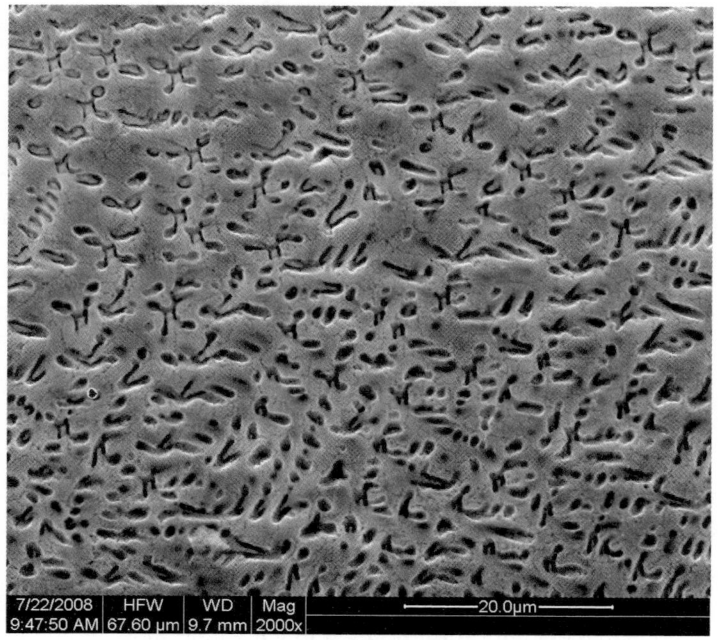

(b) (ii)

7.12 Continued

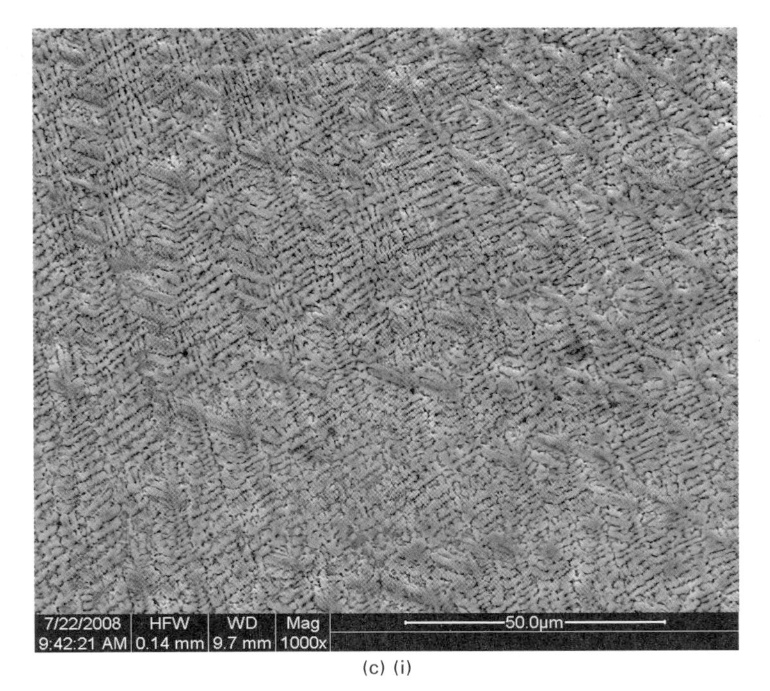

(c) (i)

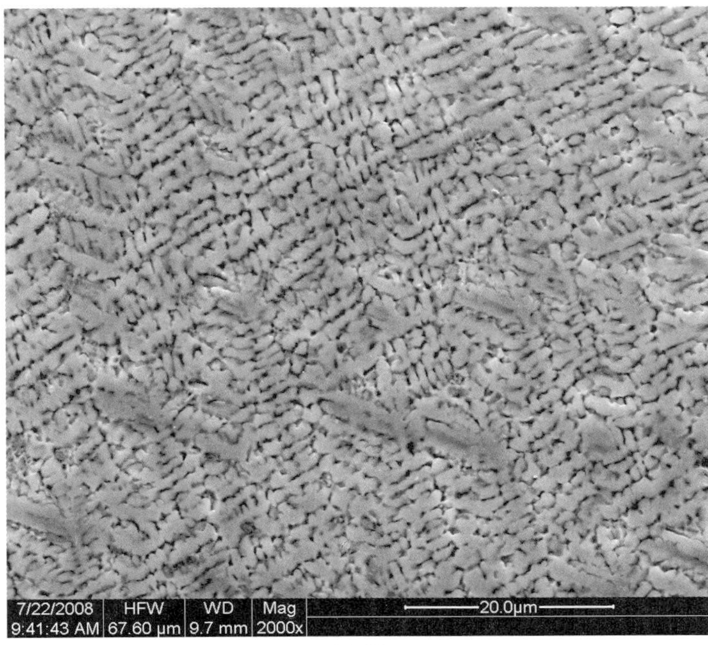

(c) (ii)

7.12 Continued

(a) (i)

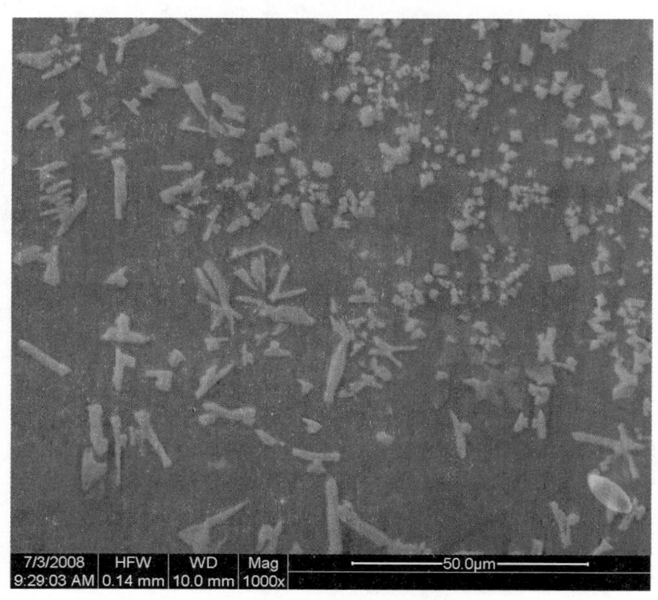

(a) (ii)

7.13 SEM microstructural features of various Colmonoy 88 coatings: (a) (i) at low magnification and (ii) at high magnification for Colmonoy 88 (without B_4C); (b) (i) at low magnification and (ii) at high magnification for Colmonoy 88 + 5 wt% B_4C; (c) (i) at low magnification and (ii) at high magnification for Colmonoy 88 + 10 wt% B_4C.

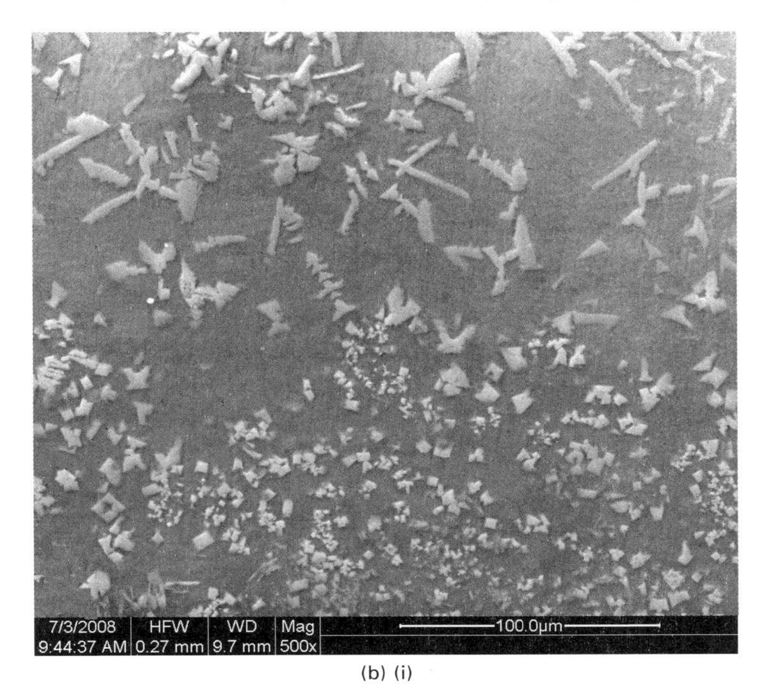

(b) (i)

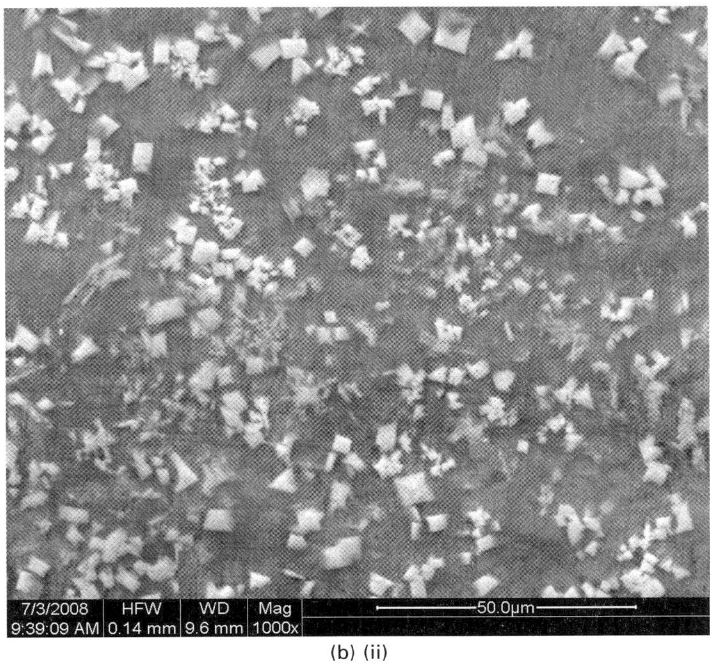

(b) (ii)

7.13 Continued

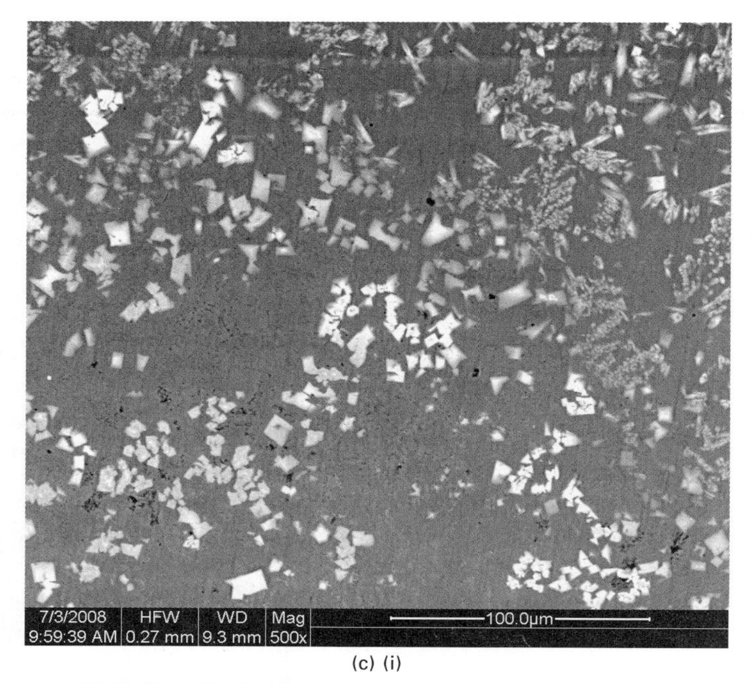

(c) (i)

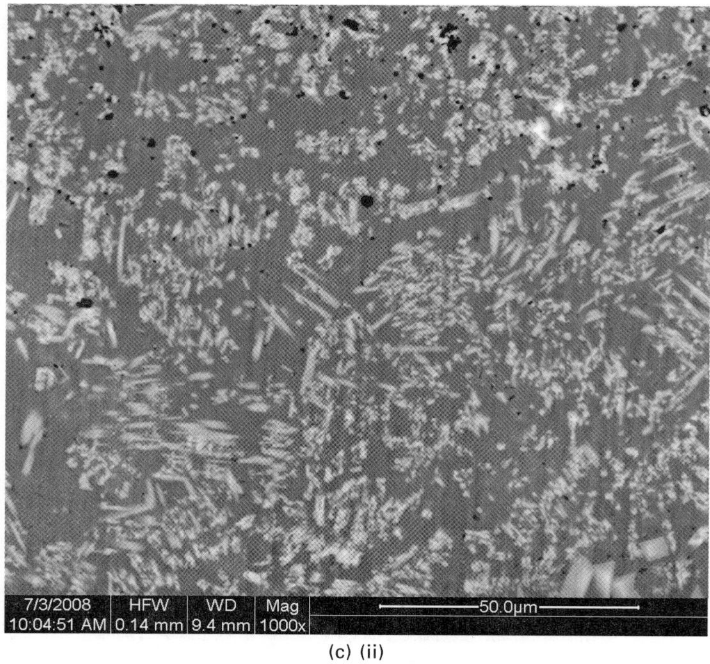

(c) (ii)

7.13 Continued

a significant effect is the refinement of Co-rich phases, and second is gradual increment of carbide-rich finer interdendritic regions observed for Stellite 6 coatings. Also, the morphology of eutectic was changed (Fig. 7.12). It is observed from the quantitative analysis that the fractions of Co-rich regions decreased from 65% to 45% and then to 35%, and also that the volume fractions increased from 35% to 55% and then to 65%, when added with 5 and 10 wt% of B_4C, respectively. For comparison, the analysis of Stellite 6 + 10 wt% B_4C coatings is given; these coatings do not bond well with substrate 13Cr–4Ni steel, and there are a few cracks as observed in Fig. 7.5(c)–(d). In the case of Colmonoy 88 coatings, changes in morphologies of secondary phases and decrease in volume fraction of Ni-rich matrix regions (in other words, increase in volume fractions of secondary phases) are observed (Fig. 7.13). The quantitative analysis confirmed the increase in volume fractions of secondary phases from 45% (only Colmonoy 88 coatings) to 65% (5 wt% B_4C reinforced Colmonoy 88 coatings) and then up to 75% when added with 10 wt% of B_4C. Colmonoy 88 + 10 wt% B_4C coatings showed few micro-porosities and were free from any types of cracks.

Melting and dissolution of B_4C powder particles during LSA process helped in nucleation of more and more secondary phases by restricting the growth of secondary phases, which are formed without B_4C addition. This led to enhancement of volume fractions of secondary phases or decreases in volume fractions of Co-rich and Ni-rich regions in both Stellite 6 and Colmonoy 88 coatings, respectively.

7.4.2 X-ray diffraction (XRD) analysis

To find out the various phases in the coatings, which are formed during LSA, X-ray diffraction (XRD) was performed using a Bruker Discover D8 diffractometer. This was done by using Cu-K_α radiation. Diffraction patterns were recorded and indexed using Joint Committee on Powder Diffraction Standards (JCPDS) software.

Stellite 6 coatings

The various phases of 13Cr–4Ni steel, Stellite 6 (with or without addition of B_4C) are identified from XRD and given in Fig. 7.14. XRD results confirmed the presence of solid solutions of Co, $Cr_{23}C_{26}$, Ni_3Si, Fe_3C and W_2C phases. The results are slightly different from earlier reports (Xu and Kutsuna, 2006), since they have laser clad with Stellite 6 coating powder on mild carbon steels. With the addition of 5 wt% B_4C, extra peaks corresponding to carbides are seen in Fig. 7.14. This also confirms the melting and dissolution of B_4C particles, which aids the enhancement of the volume fractions of chromium/tungsten carbides as explained on pages 200–7.

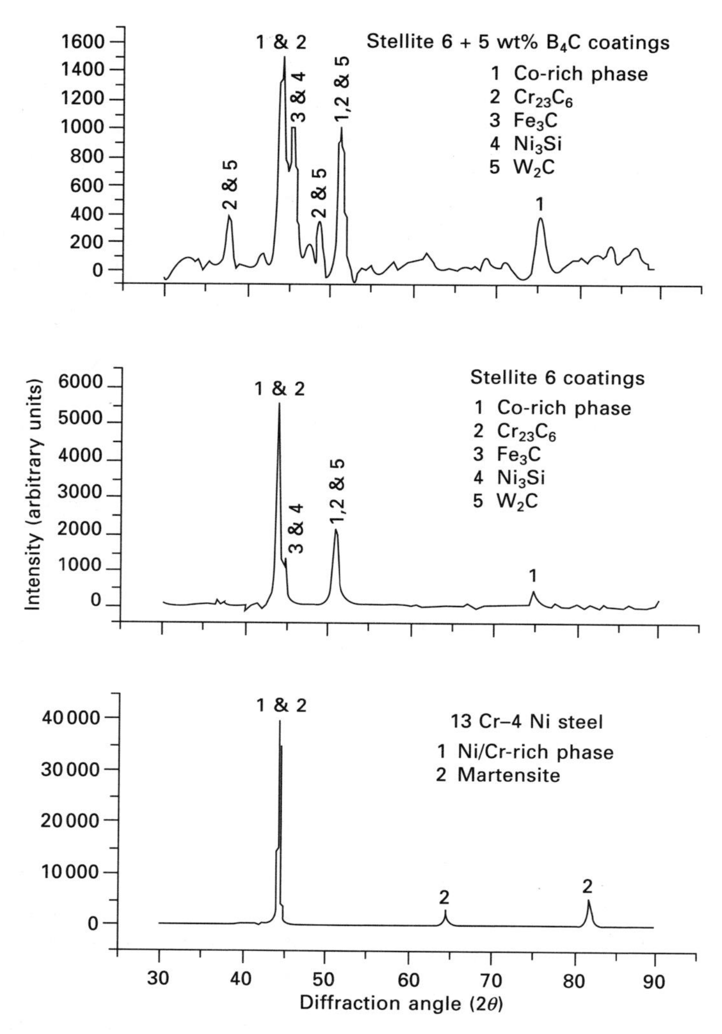

7.14 XRD patterns of substrate, Stellite 6 (with or without B_4C) coatings.

Colmonoy 88 coatings

Figure 7.15 depicts the XRD patterns of Colmonoy 88 and its 5 and 10 wt% B_4C reinforced coatings. The fcc γ-nickel peaks are most prominent along with carbide peaks of $Cr_{23}C_6$, WC, W_2C and boride types of (Ni, Fe)$_3$B, W_2B and Cr_5B_3 along with Ni_3Si phases. The work of Lim *et al.* (1998) and Conde *et al.* (2002) indicated the formation of complex carbides (W, Cr, Fe, Ni)C in addition to normal carbides ($Cr_{23}C_6$) and borides after laser cladding with Colmonoy 88 coating powder. Such types of carbide are not observed when

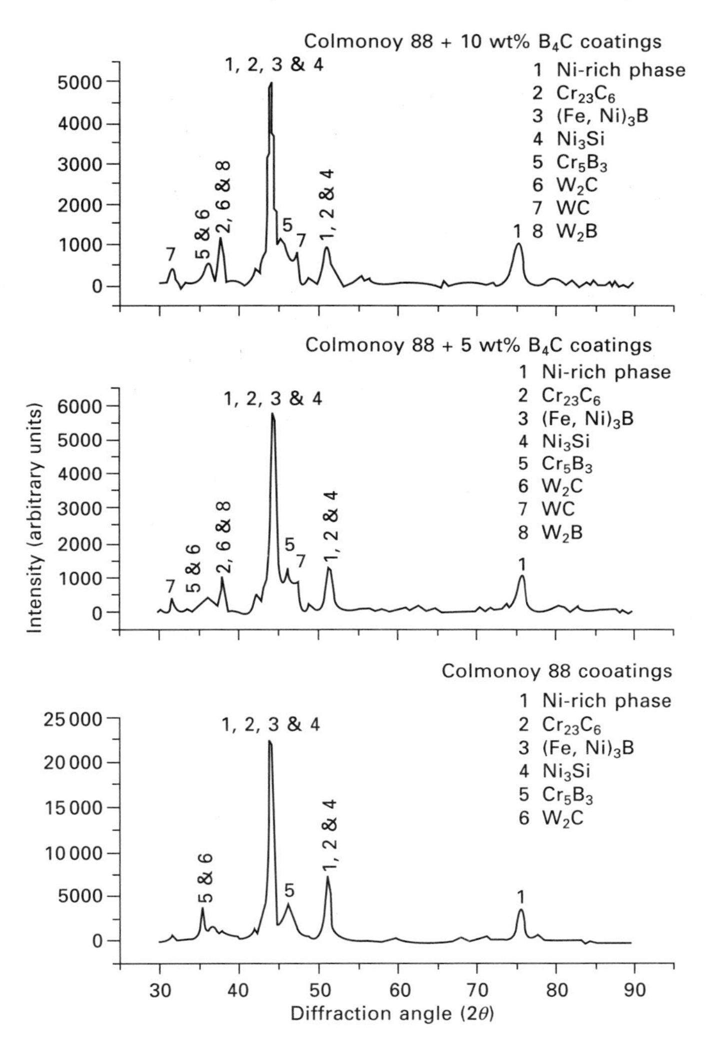

7.15 XRD patterns of Colmonoy 88 (0, 5 and 10 wt% B_4C) coatings.

the Colmonoy 88 powder particles are processed with LSA; instead, it was observed that W_2C was present along with other normal carbide and boride phases. The XRD analysis of Colmonoy 88 + 5 and 10 wt% B_4C coatings resulted in additional peaks (corresponding to respective carbides – WC, W_2C and $Cr_{23}C_6$ and borides – W_2B) compared with Colmonoy 88 (0 wt% B_4C) coatings (Fig. 7.15). These results agree well with the explanation given on pages 200–7.

7.4.3 Microhardness measurement

In order to obtain the microhardness of LSA coatings, the hardness survey across the interface and at the top of the coating surfaces was carried out using an MMT-7 MATSUZAWA microhardness tester at a load of 100 grams. A schematic illustrating the path of microhardness measurements across the interfaces is shown in Fig. 7.16.

The microhardness measurements have been carried out at three regions across the interfaces of LSA coatings. This study also aims to understand the effect of reinforcement of B_4C on hardness of Stellite 6 and Colmonoy 88 coatings. The results are plotted as 'average microhardness' against distance from top surface of the respective coatings. In both coatings, microhardness gradually decreased with distance measured from the top of the coatings to substrate (Figs 7.17 and 7.18). Substrate 13Cr–4Ni steel is used for this study in tempered conditions with average microhardness of 240-250 HV_{100}.

Stellite 6 coatings

Figure 7.17 gives the average microhardness profile across the interfaces of Stellite 6 (0 and 5 wt% B_4C) coatings. For the given LSA parameters, Stellite 6 (without B_4C) coatings exhibited maximum microhardness of 410 HV_{100} up to 0.3 mm, its 5 wt% B_4C reinforced coatings showed enhancement in microhardness up to 650–660 HV_{100}, up to 0.25 mm from the top surface of the coatings. It was observed from Fig. 7.17 that a total of 1.7-fold increase in hardness of Stellite 6 coatings was attained compared with 13Cr–4Ni steel. Further enhancement (about 1.6-fold) in hardness of Stellite 6 coatings was observed when 5 wt% B_4C is added to Stellite 6 coating powder. As explained on page 197, substrate 13Cr–4Ni steel has got only tempered needle shape martensite, whereas Stellite 6 and 5 wt% B_4C added coatings have fine carbides of Cr and W along with Co-rich phases. Also, volume fractions of fine carbides have been increased with the addition of 5 wt% B_4C. All these factors led to total of 2.6-fold enhancement in microhardness of Stellite 6 + 5 wt% B_4C coatings compared with 13Cr–4Ni steel.

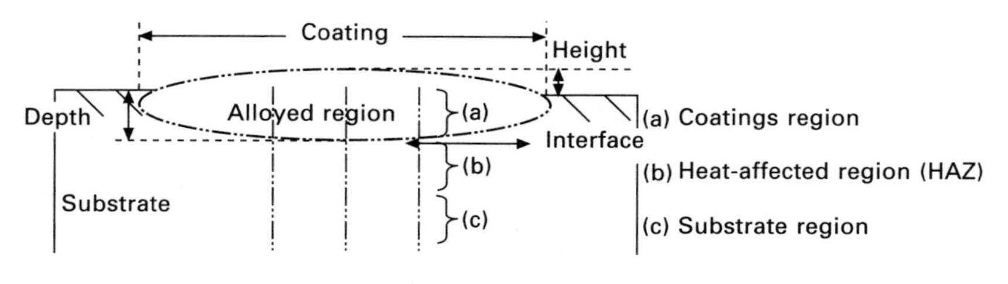

7.16 Schematic representing the paths of microhardness measurement.

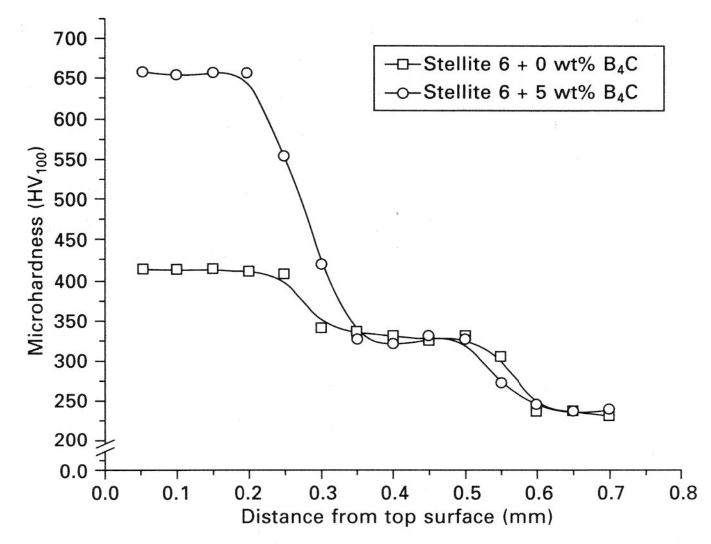

7.17 Hardness profiles of Stellite 6 (with and without B_4C) coatings.

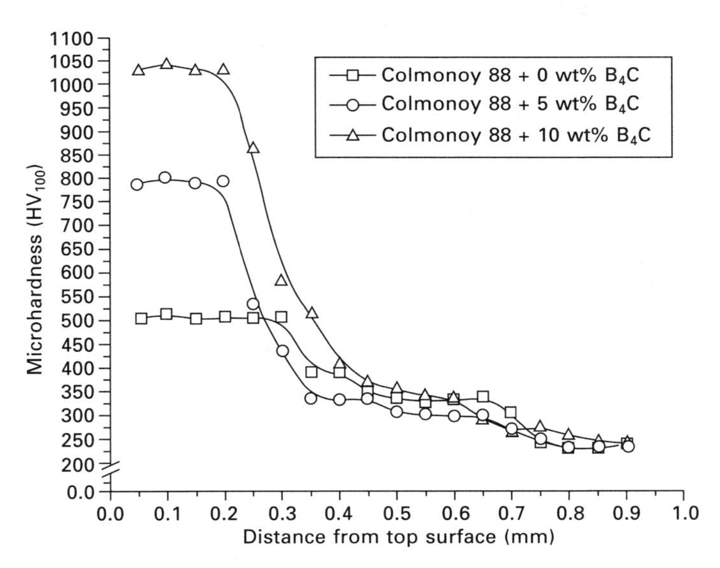

7.18 Hardness profiles of Colmonoy 88 (with and without B_4C) coatings.

Earlier, Tiziani *et al.* (1987), Oliveira *et al.* (2002), Ocelik *et al.* (2007) and Xu *et al.* (2006) reported the microhardness of Stellite 6 laser claddings at 600 HV_{300}, 525 HV_{500}, 500–600 HV_{200} and 580 HV_{50}, respectively, using CO_2 lasers. Higher hardness is attributed to either single layer or multilayer

Stellite 6 coatings obtained on a variety of steel substrates surfaces. However, Yellup (1995) showed a decrease in microhardness with percentage dilution, and, also reported that the hardness of Stellite 6 coatings is about 350 HV_{10} at 60% of dilution for a single-step process using CO_2 lasers. Xu *et al.* (2006) have shown marginal enhancement (from 580 HV_{50} to 650 HV_{50}) in microhardness when 7 wt% WC particles is added to Stellite 6. Here, WC particles are retained and distributed uniformly at the surface of Stellite 6 coatings, and not melted as described on pages 200–7.

Colmonoy 88 coatings

The microhardness profile taken across the interfaces of Colmonoy 88 and its reinforcement of 5 and 10 wt% B_4C coatings is shown in Fig. 7.18. From Fig. 7.18, maximum microhardness of 510 HV_{100} is observed for Colmonoy 88 (0 wt% B_4C) coatings, and this hardness is observed up to a length of 0.30 mm from the top surface of the coatings. When added with 5 and 10 wt% B_4C into Colmonoy 88, enhancement in microhardness of Colmonoy 88 coatings was observed about 1.58 and 1.85-fold, respectively. Colmonoy 88 + 5 wt% B_4C coatings showed maximum microhardness of about 790–800 HV_{100}, whereas Colmonoy 88 + 10 wt% B_4C coatings have exhibited higher hardness of 1020–1030 HV_{100}; in both cases, hardness has been attained up to 0.25 mm from top surface of the coatings towards substrate. A two-stage process of LSA with Colmonoy 88 and various percentages of B_4C coating powder on substrate 13Cr–4Ni steel yielded 2–4-fold enhancements in hardness. The steep increase in hardness is due to the formation of complex carbides/borides, and to the increased volume fractions of these phases in Colmonoy 88 coatings with the addition of B_4C particles (pages 200–7).

Ming *et al.* (1998) and Lim *et al.* (1998) reported that hardness of multipass clad layers of Colmonoy 88 coatings on AISI 1020 steel were about 850 to 920 HV_{200} and 890 to 930 HV_{200}, respectively. The Colmonoy 88 coatings have been developed using single step process of laser claddings. There were no reports found on the enhancement of hardness with addition of B_4C or WC particles which are processed via laser cladding. However, Przybylowicz and Kusinski (2000) and Wu *et al.* (2003) have reported an increase in hardness of Ni-based alloy powder with the addition of 50 wt% WC particles processed via two-stage process of laser cladding using a CO_2 laser.

The microhardness of LSA coatings is summarized and shown in Fig. 7.19. For given two-stage LSA process parameters, the hardness of Stellite 6 (0 and 5 wt% B_4C) coatings is inferior at least by 1.25-fold compared with Colmonoy 88 (0 and 5 wt% B_4C) coatings. The reason behind this difference is attributed to differences in microstructure, morphology and volume fraction of secondary phases of the coatings, as given on pages

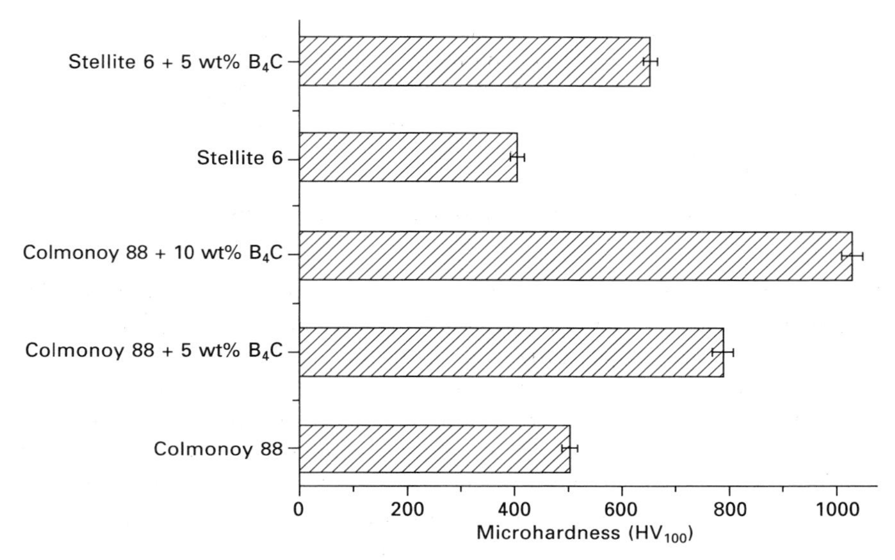

7.19 Comparison between microhardness of various LSA coatings (load: 100 g).

200–7. Higher hardness of Colmonoy 88 coatings has come from complex carbides as well as borides, whereas only carbide phases are contributing to the hardness of Stellite 6 coatings.

7.5 Slurry erosion performance of coatings: an overview

7.5.1 Details of slurry jet erosion tests and testing methodology

The slurry erosion tests were conducted as per ASTM G73 (2004). The LSA samples of size $20 \times 20 \times 7$–$8\ mm^3$ were machined and polished with alumina to a surface finish of 0.5 to 1 μm. After polishing, the samples were cleaned with acetone and water thoroughly to remove the surface contaminations and then dried. A schematic of erosion test samples is given in Fig. 7.20.

The slurry jet erosive wear test rig was used here to study the slurry erosion behaviour of substrate and of laser surface modified coating. The schematic of the slurry jet erosion test apparatus and slurry chamber is shown in Fig. 7.21. More details of the schematics can be found at Mann *et al.* (2006) and the laboratory slurry jet erosion test rig is as shown in Fig. 7.22. The erosion test rig consisted of a nozzle of diameter 6 mm, through which the slurry (tap water and erodents) in required concentration was directed

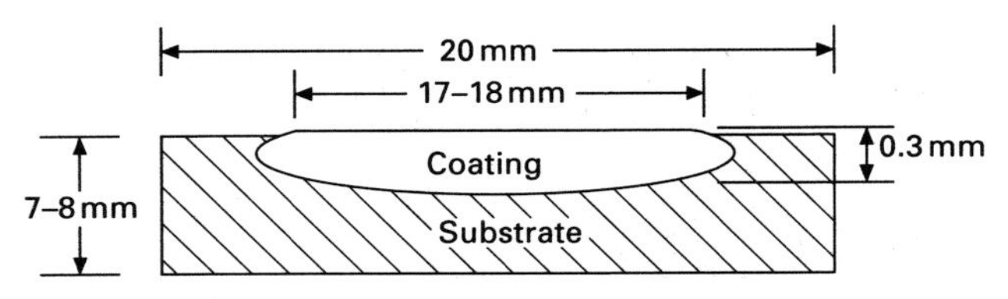

7.20 Schematic of erosion test specimen dimensions (not to scale).

towards the sample. The sample to be tested was kept in an adjustable holder by which the angle of the sample with respect to the jet can be varied. The required concentration of slurry was mixed in a conical vessel of 25 litres capacity and pumped through the impeller and directed through the nozzle. The flow of the slurry was controlled by varying the speed of the motor. The required velocity (Equation 7.1) of impingement was maintained through the flow of the slurry through the nozzle. The angle of impingement and the distance between the nozzle and sample were maintained as per the requirement. After testing for specified duration, the samples were removed and cleaned thoroughly with water and acetone and then dried. The weight measurements were carried out before and after the erosion test and the weight loss was determined. The erosion of coatings was determined through the weight loss. During each test, the distance between the nozzle tip and the sample was kept constant and the particle size, as well as the concentration of the slurry, was maintained to the most accurate level in order to ensure that the obtained results were reproducible and comparable. Also, the slurry was changed every 30 minutes during each test in order to overcome the problems of erodent particle blunting due to continuous impacts (Gandhi *et al.*, 1999). The cumulative weight loss was plotted versus the cumulative exposure time. Slurry erosion results of LSA modified and 13Cr–4Ni steel are reported as specific erosion rate (mass loss of the material eroded per unit mass of silica particles in the total amount of slurry used during a test) (Equation 7.2) for all test conditions.

The slurry velocity of erodents is assumed to be that of jet velocity and is calculated using the following equation:

$$V = \frac{Q}{A} \tag{7.1}$$

where V is slurry velocity (m/s), Q is the flow rate (L/s) of slurry and A is the area of the nozzle, πr^2, where r is the radius of nozzle (mm).

Specific slurry erosion rate (SSER) has been calculated using the following equation:

$$\text{SSER} = \frac{\text{Mass loss}}{\text{Amount of erodent} \times \text{time}} \tag{7.2}$$

Average erodent particle size and size distribution of pure silica and river

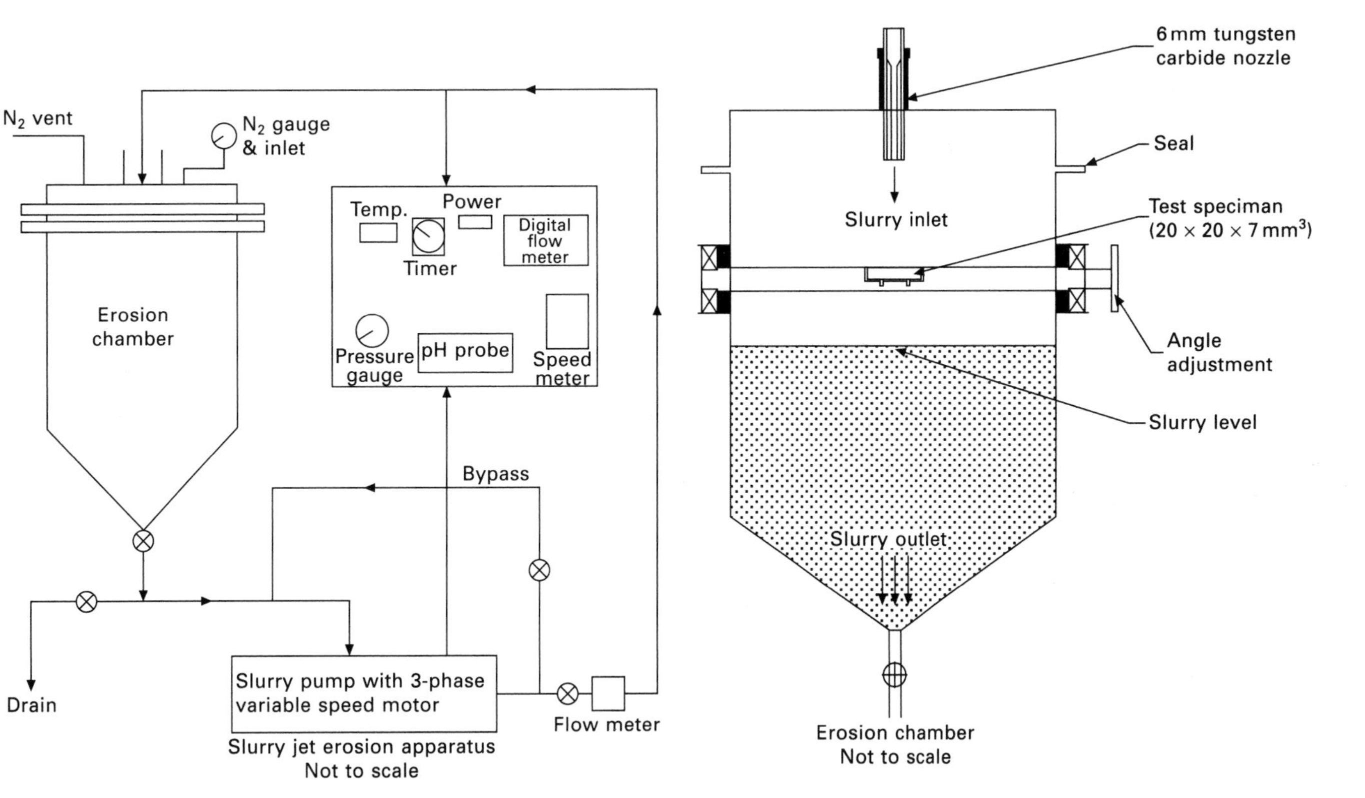

7.21 Schematic of slurry erosive wear test apparatus and erosion chamber.

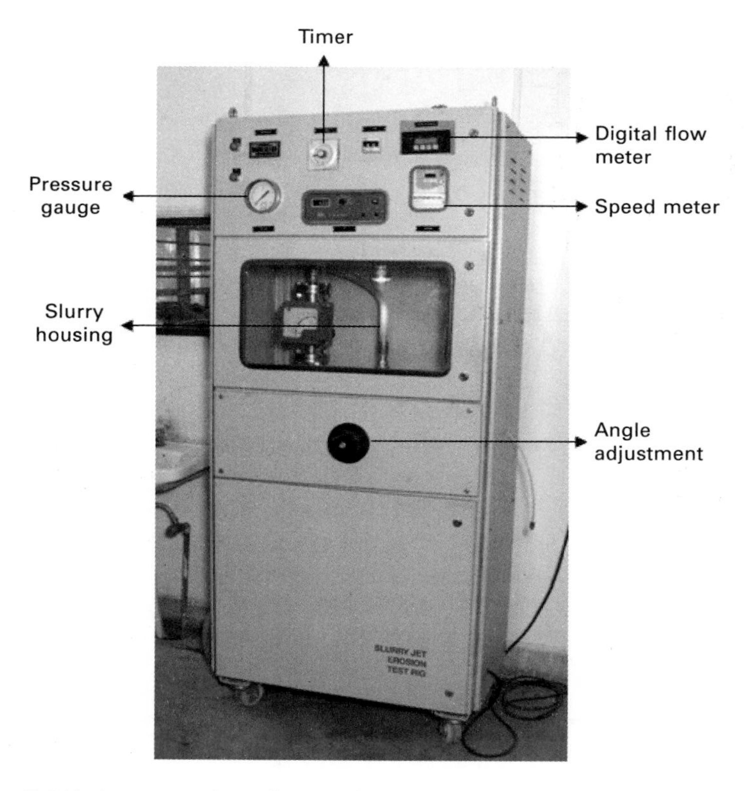

7.22 Laboratory slurry jet erosion test rig.

sand were measured using laser-assisted particle size analyser (Microtrac S-3500 model: Tri-laser technology). Table 7.4 lists the various erodent/particle sizes, which are used to prepare various slurries for erosion tests. The SEM features along with chemical analysis using EDS of different commercial SiO_2 sand and Manali river sand are given in Figs 7.23 and 7.24, respectively.

7.5.2 Factors influencing erosion/erosion rate

Erosive wear in hydroturbines, commonly known as sand/silt erosion, is caused by impacts of solid or liquid particles against the solid surface. These particles are contained in the flow medium and possess very high kinetic energy that will be sufficient to damage even metallic target surfaces. Therefore, studies on silt properties, operating conditions and substrate properties are essential for understanding the silt erosion mechanisms of hydroturbine components. In the literature, based on experimental test conditions (Clark, 1992, 2002;

Table 7.4 Different sizes of commercial SiO_2 and Manali river sand

Sieved erodent sizes (μm)	Measured average particle sizes (μm)	
	Commercial SiO_2 sand	Manali river sand
< 105	100	88
105–250	215	–
250–355	375	342
355–450	490	–

7.23 SEM morphology of SiO_2 of (a) <105 μm; (b) 105–250 μm; (c) 250–355 μm and (d) 355–450 μm (EDS analysis of various morphologies showed SiO_2 only).

and Finnie, 1995) and target materials (especially ductile materials; Finnie, 1972) one can find several factors which are influencing directly or indirectly the erosion rate and its mechanisms on target materials (Truscott, 1972; Stachowiak and Batchelor, 2001; Thapa, 2004). Hence, general factors/

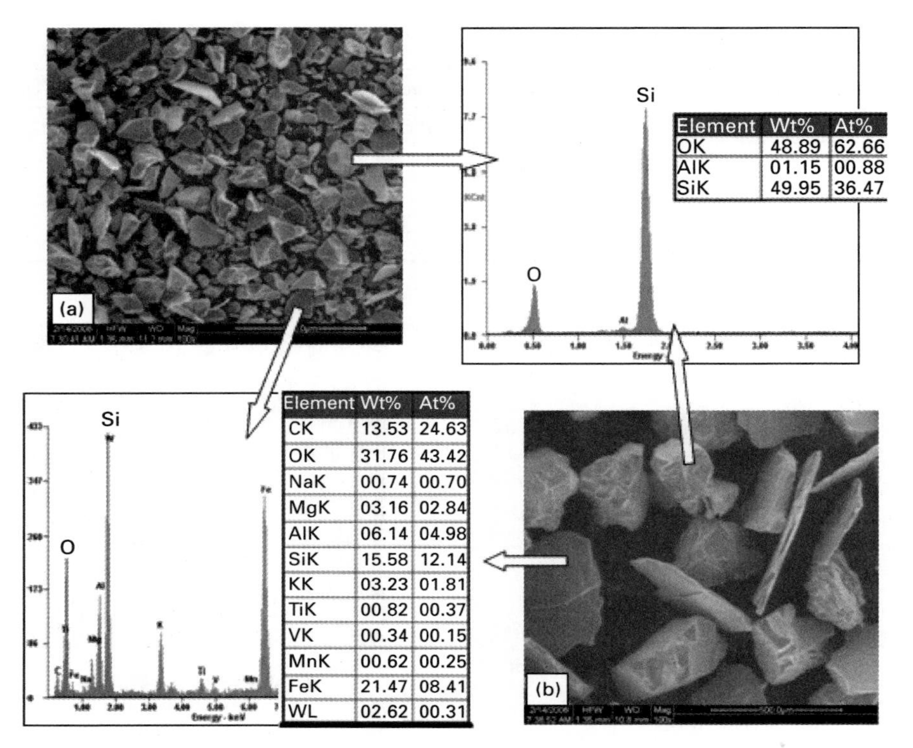

Element	Wt%	At%
OK	48.89	62.66
AlK	01.15	00.88
SiK	49.95	36.47

Element	Wt%	At%
CK	13.53	24.63
OK	31.76	43.42
NaK	00.74	00.70
MgK	03.16	02.84
AlK	06.14	04.98
SiK	15.58	12.14
KK	03.23	01.81
TiK	00.82	00.37
VK	00.34	00.15
MnK	00.62	00.25
FeK	21.47	08.41
WL	02.62	00.31

7.24 SEM morphologies of Manali river sand: (a) <105 μm and (b) 250–355 μm (EDS analysis confirmed the composition change for different morphologies).

parameters affecting erosion can be grouped in the following three distinct groups:

1. Operating conditions: impingement/impact angle (direction of impingement), particle impact speed, acceleration, flux rate or concentration, erosive medium nature and flow (density, viscosity and surface activity-lubricity), temperature, particle rotation at impingement, particle–particle interaction.
2. Eroding particles (sand or liquid droplets): size, density, shape/sharpness (form), hardness, strength (resistance to fragmentation), type (different erodent materials), friability.
3. Substrates (target materials): type (ductile/brittle), properties (melting point, work hardening, toughness, hardness, fatigue and structure), chemistry, elastic property (residual stress levels), surface morphology (size and shape of the surface, metallographic structure).

7.6 Impingement angle

According to Stachowiak and Batchelor (2001), the impingement angle is the angle between the target material's surface and the trajectory of the particle immediately before impact as shown in Fig. 7.25. Generally, the impingement angle can range between 0° and 90°. It is 0° when particles are striking parallel to the target material surface. When particles strike normal to the material, it is called '90° impact'. Hydroturbine components will usually show negligible or no erosion at '0° impact'. Based on historical research data, the effect of impingement angle on erosion of materials (irrespective of type of material) can be classified as brittle (wherein erosion will increases gradually and reaches maximum at normal impacts, 90°) and ductile (material shows maximum erosion at 10–30°). Depending on operating test conditions, particle/erodent size and shape, even ductile materials can show the brittle mode of erosion behaviour. A plot of erosion rate of target materials as a function of impingement angle is shown in Fig. 7.26; details of it can be found elsewhere (Stachowiak and Batchelor, 2001; Thapa, 2004).

The impingement angle is very important because, in a hydroturbine atmosphere, the water entrained with silt/sand strikes at high velocity at different angles over the turbine components.

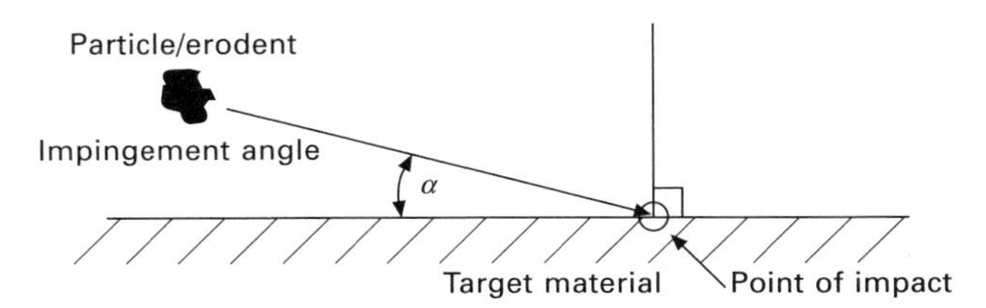

7.25 Schematic of impingement angle definition.

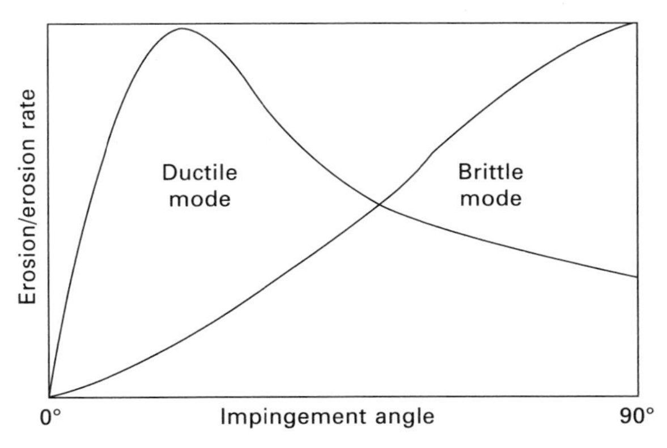

7.26 Illustration of ductile and brittle mode of erosion behaviour with respect to angle of impingement.

The erosive wear rate varies for different impingement angles for different materials, so the angle of impingement is one important parameter that decides the performance of materials in erosion. For instance, a material which is generally considered to be ductile undergoes severe erosion at low impingement angles, for example at 30°, whereas it experiences minimum erosion at high impingement angles. Hence, it becomes important to study the modified steels' performance at different angles of impingement.

In the present study, erosion tests were carried out at 30°, 45°, 60° and 90° impingement angles for substrate 13Cr–4Ni steel and for different laser surface modified coatings (Table 7.5). All tests were carried out for 1 h at room temperature and the erosion performance compared. The experimental test conditions are described in Table 7.6. Schematics illustrating four different impingement angles are shown in Fig. 7.27, which also illustrates the slurry jet distance between nozzle and target material (which was kept constant during all the experiments).

7.6.1 Stellite 6 and Colmonoy 88 coatings (without B_4C)

Macrographs of erosion-tested samples are shown in Fig. 7.28. The tests were done at room temperature with a slurry velocity of 12 m/s, impacted with

Table 7.5 Test parameters (variables) to study the effect of impingement angles on slurry erosion performances of various target materials

Type of slurry	Material	Angle of impingement (°)	Average erodent size (μm)
Commercial silica sand	1. 13Cr–4Ni steels 2. Stellite 6 coatings 3. Colmonoy 88 coatings	30, 45, 60 and 90	100 and 375
Commercial silica sand	1. Colmonoy 88+5 wt% B_4C 2. Colmonoy 88+10 wt% B_4C 3. Stellite 6 + 5 wt% B_4C	30, 45, 60 and 90	100
Manali river sand	1. 13Cr–4Ni steels 2. Stellite 6 coatings 2. Colmonoy 88 coatings	30, 45, 60 and 90	1. 88 and 342 2. 88 3. 88 and 342

Table 7.6 Experimental test conditions for substrate and LSA modified coatings

Slurry type	Erodent mixed in tap water
Concentration of the slurry	10 kg/m^3 *
Nozzle to sample distance	50 mm
Nozzle diameter	6 mm
Velocity of the slurry flow	12 m/s **
Test duration	60 min
Test temperature	Room temperature

* Concentration was varied to study the effect of slurry concentration.
** Velocity was varied to study the effect of such variations.

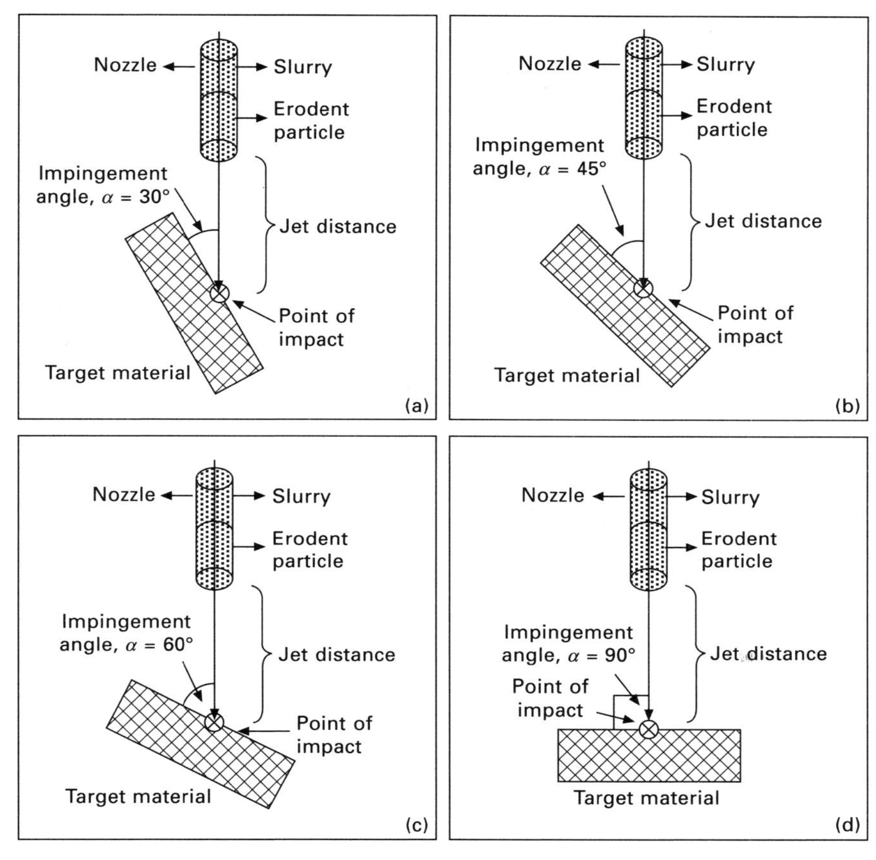

7.27 Schematics of various impingement angles: (a) 30°; (b) 45°; (c) 60°; and (d) 90°.

commercial silica sand of average size 100 μm at 30, 60 and 90° impingement angles. The qualitative surface morphology obtained is deep in 13Cr–4Ni steel (Fig. 7.28(a)) compared with Stellite 6 (Fig. 7.28(b)) and Colmonoy 88 coatings (Fig. 7.28(c)) in the range between 30 and 90°. A deeper impression implies more erosion and vice versa. Visual inspection of Colmonoy 88 coatings showed lower erosion. However the exact (quantitative) erosion (mass loss) was calculated by the weight loss method.

Cumulative mass loss plots

Characteristic plots of erosion test results (ASTM G73, 2004) are plotted as cumulative mass loss against time of exposure for substrate 13Cr–4Ni, Stellite 6 and Colmonoy 88 coatings, as shown in Fig. 7.29. All target materials

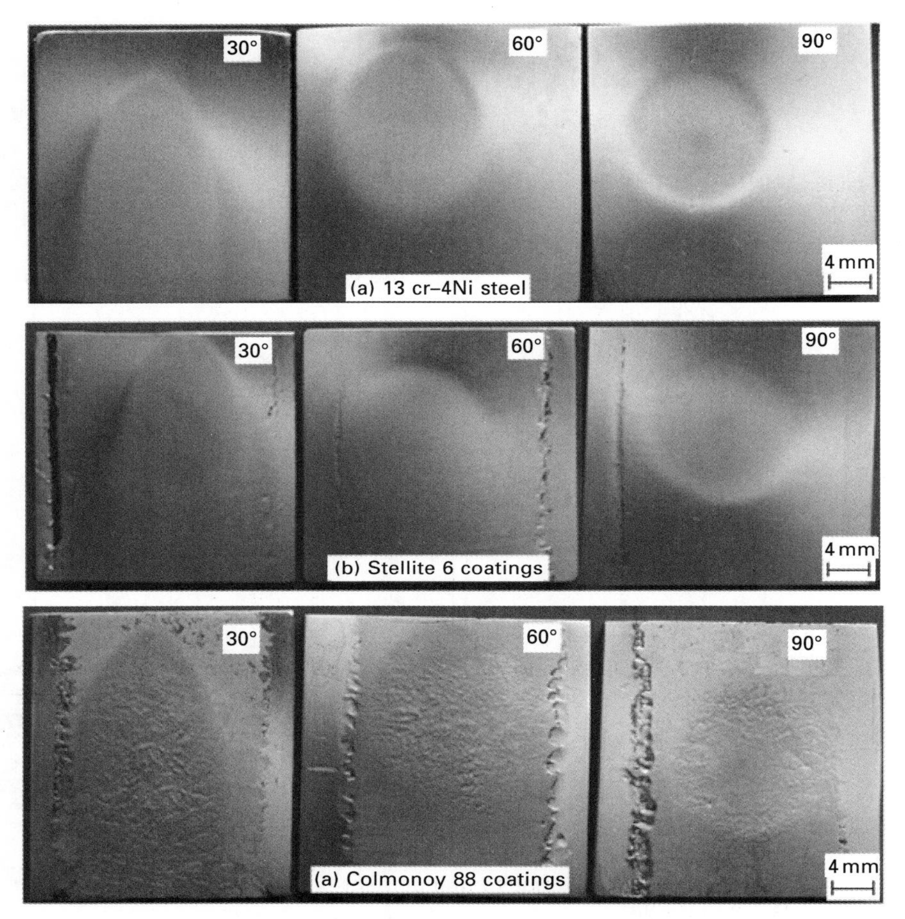

7.28 Representative slurry impressions of (a) 13Cr–4Ni steel; (b) Stellite 6 coatings; and (c) Colmonoy 88 coatings at three different impingement angles after 1 h.

showed very little or negligible mass loss at initial stages ('incubation period'). Because of the mild test conditions, LSA coatings exhibited very small mass losses at initial test time (1 or 2 min) of tests. Hence, for simplification and comparison purposes, the starting time of erosion has been measured after 5 min for all target materials and at all test conditions. Consequently, the absence of a run-in period (until which mass loss will be constant, with a subsequent sudden increase in mass loss) was observed.

The slopes of erosion test curves given in Fig. 7.29 are almost constant for individual target materials. Variations in the slope of erosion curves (depending on target materials and also on impingement angle) may indicate variations in the mechanism of erosion damage during the erosion test.

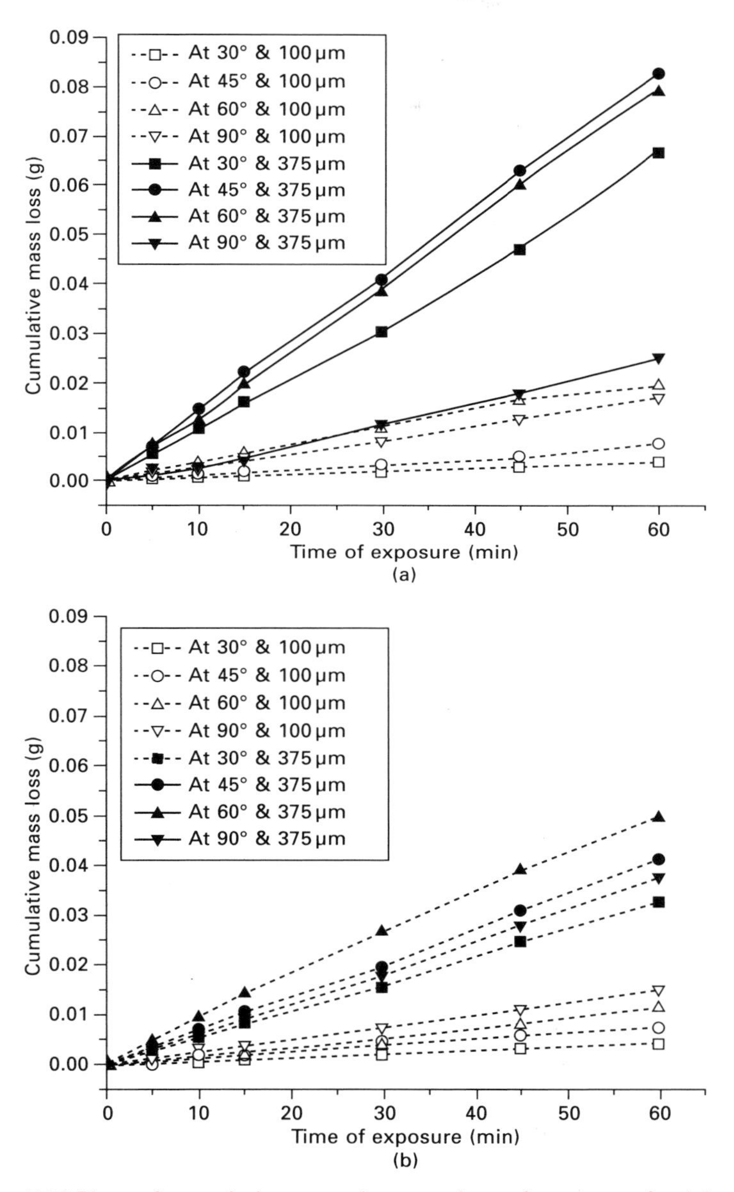

7.29 Plots of cumulative mass loss vs. time of exposure for (a) 13Cr–4Ni steel; (b) Stellite 6; and (c) Colmonoy 88 coatings.

Irrespective of the target material and test conditions, cumulative mass loss increased linearly with increase in exposure time. This indicates that the erosion mechanism for particular target materials does not change noticeably, implying a steady erosion damage during the impingement process, regardless

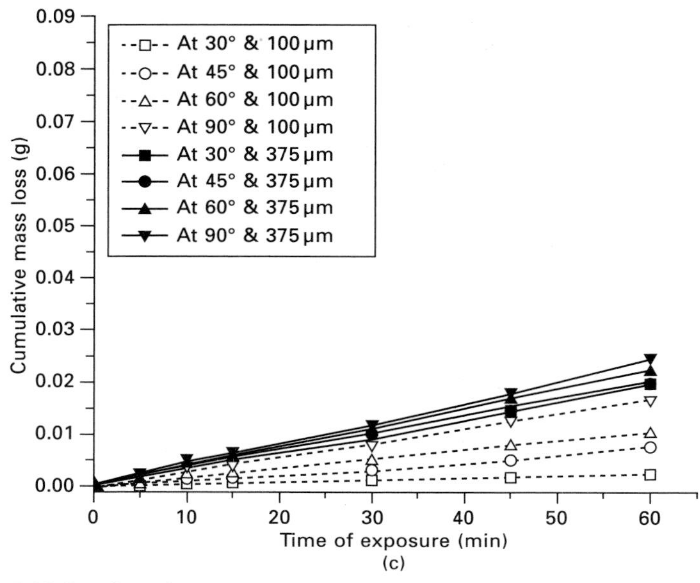

7.29 Continued

of the variation in impinged angle and erodent size. For a particular time of exposure and irrespective of erodent size, LSA coatings showed minimal mass loss compared with that of substrate 13Cr–4Ni steel. Among LSA coatings, significant erosion resistance was observed for the cases of Colmonoy 88 coatings (Fig. 7.29(c)). Similar findings have been reported by Santa *et al.* (2009), when studying the erosion performances of 13Cr–4Ni steel, thermal sprayed Ni, WC/Co–Ni, and CrO_3 coatings.

Erosion and SSER plots

Before reviewing SSER plots, the processes of erosion (or mass loss) occurring with time at different impingement angles for individual target materials are discussed here. The plots of erosion curves when impacted with 100 and 375 µm erodents are given in Fig. 7.30 and Fig. 7.31, respectively. This investigation is essential to identify impingement angles wherein maximum erosion occurs, and thereby to identify the mode of erosion behaviour for specific test conditions. Individual target materials showed unique erosion behaviour with respect to impingement angles at different time intervals. There was no qualitative shift in maximum erosion point from 5 to 60 min of erosion tests, which indicates that only an increase in erosion magnitude occurs with test time.

Figures 7.30(a) and 7.31(a) reveal the case of substrate 13Cr–4Ni steel impacted with two different erodent sizes. When impacted with erodents of

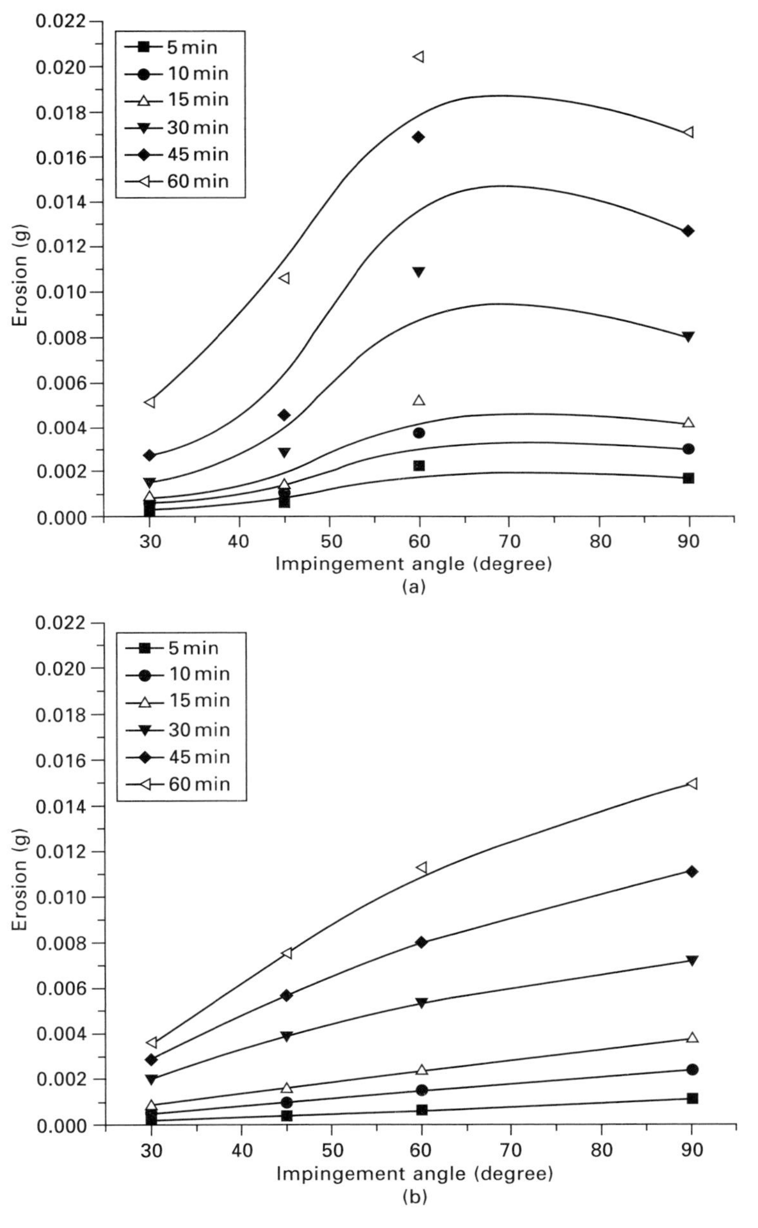

7.30 Erosion curves at different time intervals when impacted with erodents 100 μm: (a) 13Cr–4Ni steel; (b) Stellite 6; and (c) Colmonoy 88 coatings.

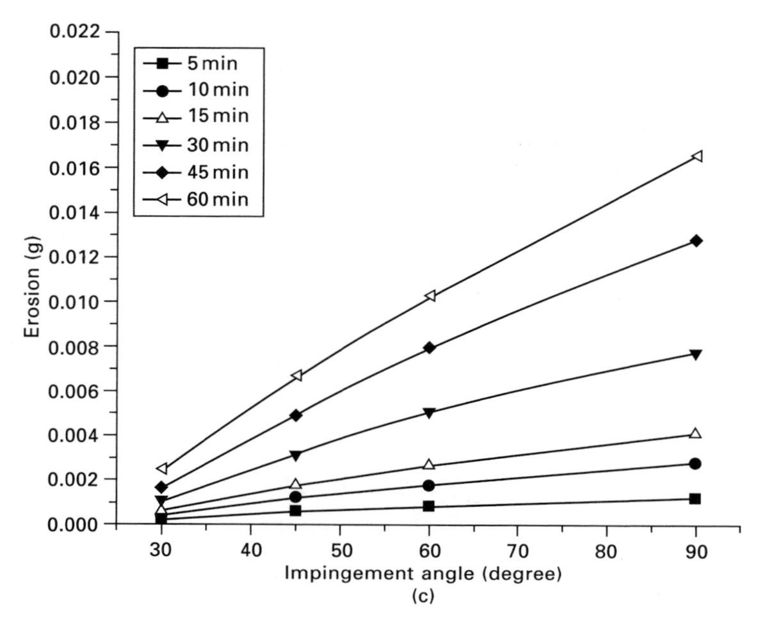

7.30 Continued

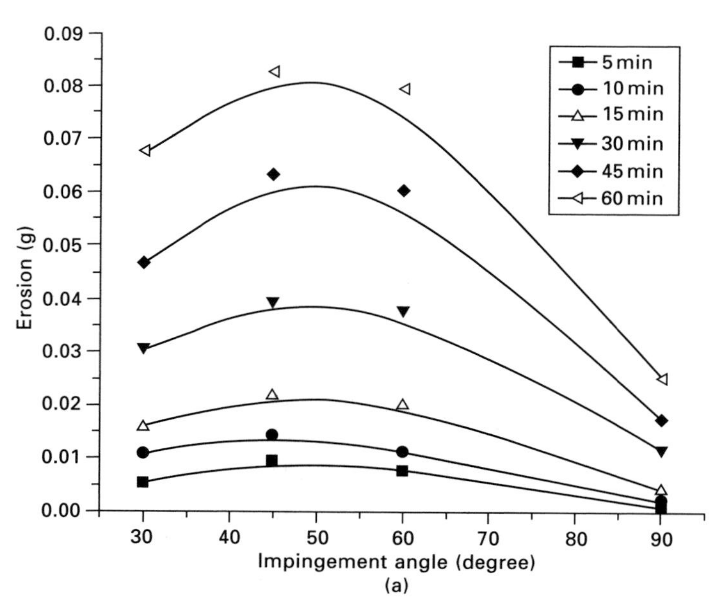

7.31 Erosion curves at different time intervals when impacted with erodents 375 μm; (a) 13Cr–4Ni steel; (b) Stellite 6; and (c) Colmonoy 88 coatings.

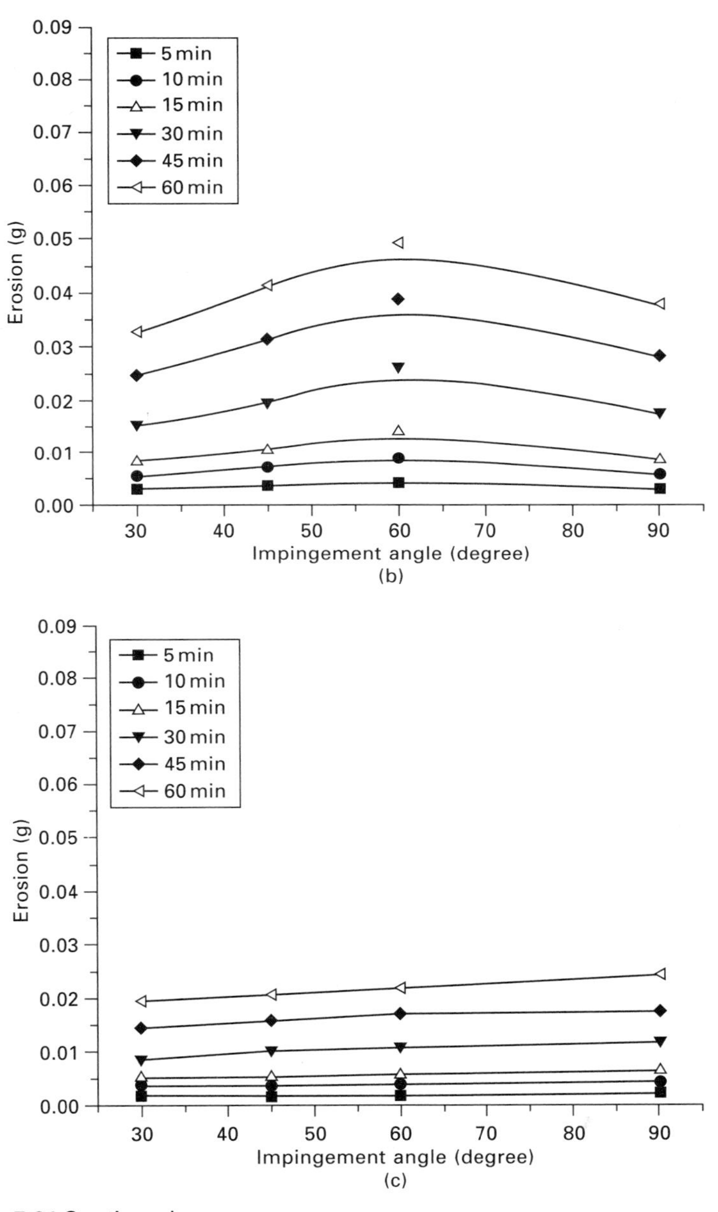

7.31 Continued

100 μm, erosion continuously increases with increasing impingement angle and reaches a maximum at 60°. Afterwards, erosion decreases with angle of impingement, which is not in line with theory of ductile (maximum erosion

rate at 20–45°) or brittle (maximum erosion rate at 90°) mode of erosion behaviour (Stachowiak and Batchelor, 2001). This behaviour is called 'mixed mode' (neither ductile nor brittle) erosion behaviour. Erosion approaches a maximum at 45° and then starts to decrease up to 90° for impacts of 375 μm sized erodents. At these test conditions, 13Cr–4Ni steel showed the theoretical ductile mode of erosion behaviour. 13Cr–4Ni steel showed a change in mode of erosion behaviour, i.e. mixed to ductile, when impacted erodent size changed from 100 to 375 μm. Similar findings have been reported by Thapa (2004) when martensitic stainless steels such as 13Cr–1Ni (peak erosion rate at 45°) and 13Cr–4Ni (peak erosion rate at 60°) steels are impacted with Baskarp-15 sand particles of average size 150 μm at relatively high slurry velocity (53 m/s). Manisekaran (2005) reported that maximum rate for the case of as-cast 13Cr–4Ni steel is at 30°, when impacted with commercial white silica sand of less than 150 μm a 1 h test duration. Hawthorne *et al.* (1999) reported the change in erosion mechanisms from brittle to ductile of 316L steel when impacted with alumina erodents of 35 μm and 200 μm for 1 h of exposure time.

Stellite 6 coatings (Figs 7.30(b) and 7.31(b)) showed brittle (peak erosion rate at 90°) and mixed (peak erosion rate at 60°) mode of erosion behaviour when impacted with erodents of 100 and 375 μm, respectively. Similar to 13Cr–4Ni steel, a change in erosion behaviour (brittle to mixed) was observed when erodent size changed from 100 to 375 μm, however, the results are exactly opposite to that of substrate 13Cr–4Ni steel erosion behaviour. The change in erosion behaviour may be attributed to differences in microstructure, and in turn, surface hardness of 13Cr–4Ni steel and Stellite 6 coatings. Although there was no literature about the use of Stellite 6 LSA coatings for erosion applications, some related references are reported here. Hawthorne *et al.* (1999) have reported the brittle nature of Co–Cr–Ni–Mo–W–Fe and Ni–W–S–Cr–Si–Fe–B–C coatings obtained on 316L steels using HVOF spraying. They have also reported that the latter coatings showed less erosion than the former due to greater volume fractions of carbide and boride. Target materials are impacted with alumina of average size 200 μm for 1 h test duration at a slurry velocity 15 m/s. Haugen *et al.* (1995) also showed the brittle mode of erosion behaviour of Stellite 6 (having hardness 400 HV) when impacted with river sand of 200–250 μm at 45 m/s slurry velocity.

Erosion of Colmonoy 88 coatings continuously increases with impingement angle (Figs 7.30(c) and 7.31(c)), reaching a maximum at 90°. Hence, according to the theory of brittle erosion behaviour, these coatings exhibit only the brittle mode of erosion behaviour for both impacts (erodents of size 100 and 375 μm). Also, impingement angle dependence of erosion behaviour is not significant for Colmonoy 88 coatings when impacted with 375 μm erodents (Fig. 7.31(c)), as only a slight change in erosion was observed with impingement angles, i.e. the differences in erosion at high and low impinging angles are quite small. Similar kinds of results have been published by Lin *et al.* (2006). The main difference between their work and the present work is with chemical composition (only NiCrBSi) of coating powder and the coating process (HVOF technique) (viz. NiCrWBSi (Colmonoy 88) and LSA process). According to Zu *et al.* (1991) and Hawthorne *et al.* (1999), the characteristic mode of erosion behaviour is well known in dry erosion, wherein ductile materials usually show greater erosion at low impingement angles compared with the brittle materials, and vice versa at higher impingement angles. However, they have reported that such trends are less clear in slurry erosion.

To compare slurry erosion performances among target materials, SSER (Equation 7.2) is plotted against impingement angles at two different erodent sizes as shown in Fig. 7.32. First, individual target materials have been compared with respect to SSER and then relative to LSA coatings. SSER values of 13Cr–4Ni steel impinged with erodents of 100 μm are at least 1.5–13 times lower than those when impinged with erodents of 375 μm at impingement angles 30–90°. For the cases of Stellite 6 coatings, differences in SSER values of 2.5–10 times are observed for impacts by erodents of size 100 and 375 μm. SSER values of Colmonoy 88 coatings impinged with 375 μm are at least 1.5–8.5 fold higher when compared with impingement with 100 μm sized erodents. The SSER values of Colmonoy 88 coatings are much lower than that of 13Cr–4Ni steel and Stellite 6 coatings. Higher values of SSER with erodent size are attributed to an increase in erodent particle

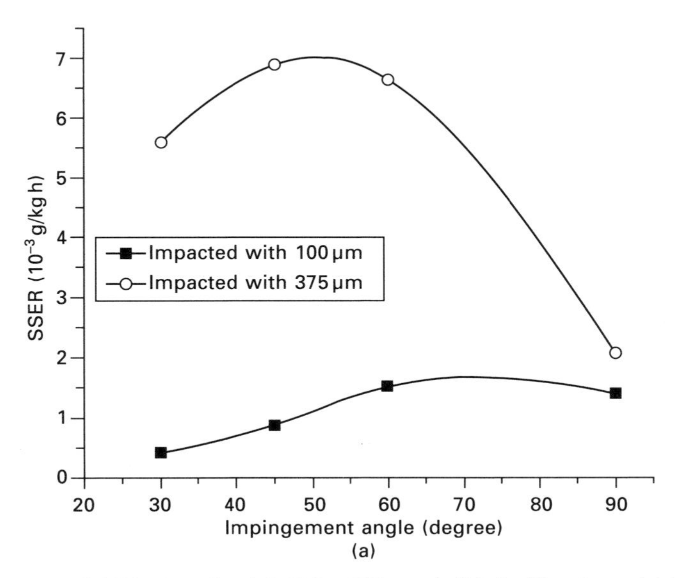

7.32 SSER plots for (a) 13Cr–4Ni steel; (b) Stellite 6; and (c) Colmonoy 88 coatings.

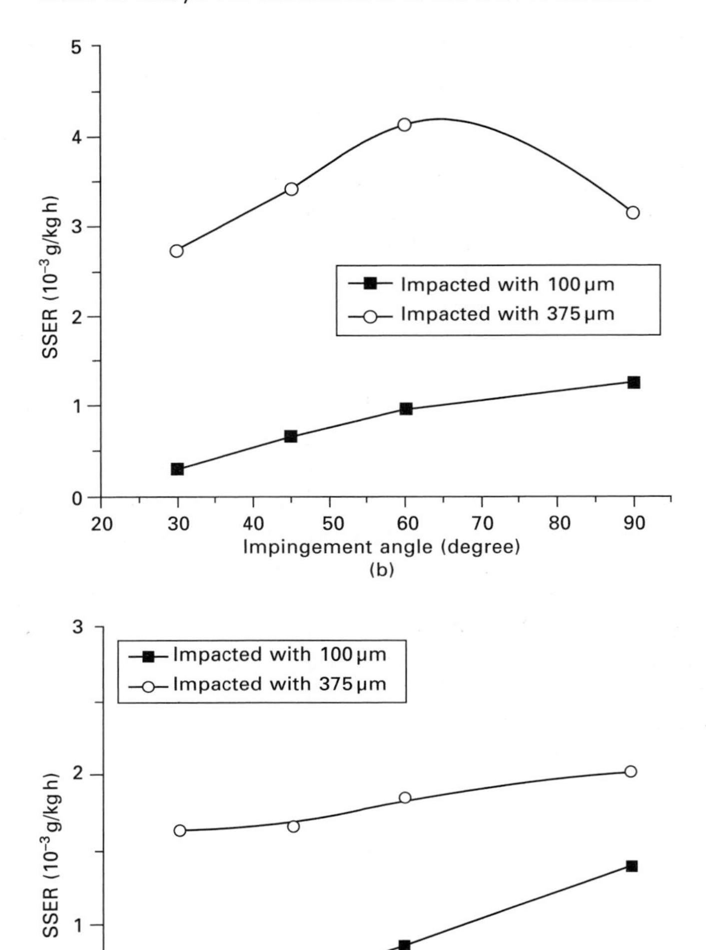

7.32 Continued

impact energy (as erodent size is increased from 100 to 375 μm). Also the finer erodents have fewer multiple cutting edges than larger particles; therefore, material removal by cutting is difficult (Manisekaran *et al.*, 2007). Moreover, Clark and Hartwich (2001) have reported that less energy is sufficient to remove unit volume of material for larger erodent particles than for smaller erodent particles. With an increase in erodent impact energy, target materials tend to lose resistance to erodent penetration, thereby leading to more mass losses.

When impacted with erodents of 100 μm size, relative SSER (SSER of 13Cr–4Ni steel/SSER of LSA coatings) of Stellite 6 and Colmonoy 88 coatings are 1.4–1.6 and 1.6–2.2-fold, respectively, at impingement angles of 30° to 60°. However, at 90° impacts, they were in the range 1.1 to 1 for Stellite and Colmonoy 88 coatings, respectively. For impacts of erodents of size 375 μm, relative SSER values of Stellite 6 and Colmonoy 88 coatings are 2.1 and 3.4-fold, respectively, at an impingement angle of 30°. At impingement angles from 45° to 60°, erosion rates of Stellite 6 and Colmonoy 88 coatings are 2–1.6 and 4.2–3.6-fold, respectively. But at normal impacts (90°), both substrate 13Cr–4Ni steel and Colmonoy 88 coatings showed almost same erosion rates, whereas Stellite 6 coatings showed a relatively higher erosion rates compared with other two target materials. Depending on the test conditions and target material properties, Bhushan (2002) has reported higher erosion rates of brittle materials compared with ductile materials, and Matsumura and Chen (2002) have shown higher erosion rates of ductile materials compared with brittle materials. The slurry erosion test results showed the following sequence of ranking of laser surface modified and substrate 13Cr–4Ni steel based on increasing erosion rate:

Colmonoy 88 coatings < Stellite 6 coatings < substrate 13Cr–4Ni steel

The above sequence holds good for all impingement angles except normal impact (90°), where the sequence is as follows:

Colmonoy 88 coatings < substrate 13Cr–4Ni steel < Stellite 6 coatings (for 375 μm sized erodents)

Stellite 6 coatings < Colmonoy 88 coatings < substrate 13Cr–4Ni steel (for 100 μm sized erodents)

Substrate and Stellite 6 coatings exhibited a change in erosion behaviour with change in erodent size. The shift in erosion behaviour was reported in the literature. Stachowiak and Batchelor (2001) reported that with the increase in erodent size, the mode of erosion changed from ductile to brittle even for hardened steel. They have also explained the alteration of ranking of target materials based on wear resistance. According to them, erosive wear rate depends on target material hardness at small erodent size impacts, whereas toughness of the target materials plays an important role during impacts with larger erodents. Materials, which are neither tough nor hard, e.g. 13Cr–4Ni steel, show inferior erosion resistance.

The impingement angle is a very important parameter, which alters the erosion performance of materials during jet impingent erosion tests. For instance, at low impingement angles, material removal takes place by micro-cutting as a result of oblique shear force, thus increasing the erosion rate. This is due to the fact that more plastic deformation results when the stresses act at low angles on certain planes in the material. At high impingement angles, the normal compressive force will mainly produce accumulated damage from fatigue, shear localization, microforging and extrusion processes (Lin *et al.*, 2001). In such cases, the ductile material undergoes plastic deformation to accommodate the repeated impact stresses, resulting in lesser removal of material. These processes will produce less erosion damage compared to that of cutting removal at low impingement angles. Hence, there appears to be a maximum erosion rate at about 20–45° impinged angle for ductile materials, such as tempered 13Cr–4Ni, as-cast 13Cr–4Ni, 12Cr martensitic, SUS304 stainless steel, etc. The high erosion resistance (Fig. 7.13(b) and (c)) of the LSA coatings at low impingement angles is attributed to the high hardness (due to complex carbides and borides) as well as to high toughness (due to Co and Ni matrices), wherein material removal by micro-cutting at low impinged angles is difficult. These LSA coatings can also have a less plastic deformation at high impingement angles and exhibit a lower erosion rate depending on the test conditions. All these features make the LSA coatings exhibit a different erosion behaviour from the substrate 13Cr–4Ni martensitic stainless steel. However, these coatings showed mostly brittle nature at high impingement angles due to the fact that erodents impinge with impact stresses where the extremely brittle precipitates, such as complex carbides and borides, cannot withstand the stresses; hence, these secondary phases will be removed by fracturing with successive impacts and matrix is removed by micro-cutting.

The secondary phases, such as complex carbides and borides of laser modified coatings are tough enough to survive the particle impact energies; hence, they show lower slurry erosion than substrates. According to Turenne *et al.* (1989), the protection effect at small impingement angles is because reinforcing particles give rise to protruding particles, thereby reducing the shearing effect on matrices of impinging particles. High erosion at large impingement angles is due to fracturing of these reinforcing particles resulting from extreme brittle nature of reinforcements. In LSA coatings, complex hard carbides/borides, which are distributed uniformly over the modified surfaces, protect the matrix by acting as reinforcements. The differences in erosion results among LSA modified coatings are due to the change in morphology and chemistry/types of phases. Colmonoy 88 coatings contained both carbides and borides with average sizes of 2 to 10 μm, whereas Stellite 6 coatings contained only carbide phases with average sizes of less than 1 μm (Figs 7.10 and 7.12). Jiang and Kovacevic (2004) have demonstrated the improved erosion behaviour of laser-clad chromium carbides when compared with tungsten carbide coatings and substrate AISI 4140 steel when impacted with silica sand in water. Although first indications are that the morphology and size of carbides and borides play a crucial role in erosion resistance of LSA coatings, the volume fraction and type of these phases are very vital in determining erosion resistance. Similar findings have also been reported by Tucker *et al.* (1984). Hence, for the cases of Stellite 6 coatings, chromium and tungsten carbides enhance erosion resistance, whereas additional phases of chromium boride along with chromium/tungsten carbides lead to enhanced erosion resistance of Colmonoy 88 coatings compared with Stellite 6 coatings.

7.6.2 Stellite 6 and Colmonoy 88 coatings (with B_4C)

To study the erosion performances of B_4C added LSA coatings, silica sand of average size 100 μm was impacted at room temperature and at 12 m/s slurry velocity on Stellite 6 + 5 wt% B_4C and Colmonoy 88 + 5 and 10 wt% B_4C coatings for about 1 h and erosion results obtained. Typical slurry eroded impressions are shown in Fig. 7.33. Interestingly, for the given slurry erosion test conditions, Colmonoy 88 + 10 wt% B_4C coatings did not show noticeable eroded impressions similar to other target material cases. Hence, these results have been omitted here.

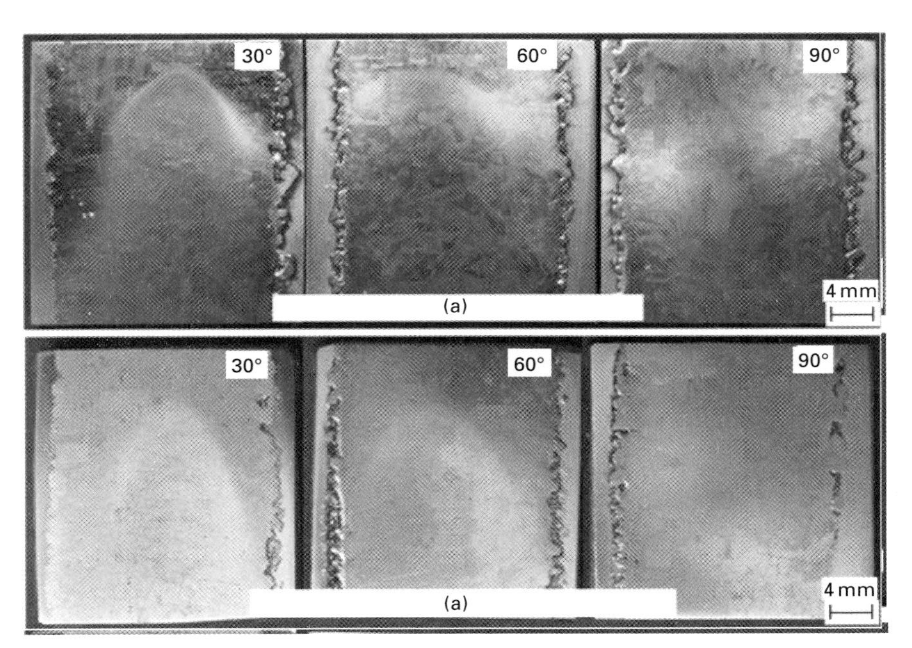

7.33 Typical slurry impressions after 1 h test for (a) Stellite 6 + 5 wt% B_4C and (b) Colmonoy 88 + 5 wt% B_4C coatings.

Cumulative mass loss plots

Typical cumulative mass loss plots for Stellite 6 + 5 wt% B_4C and Colmonoy 88 + 5 and 10 wt% B_4C coatings are given in Fig. 7.34. Similar to substrate 13Cr–4Ni steel and basic LSA (0 wt% B_4C) coatings, here also an increasing

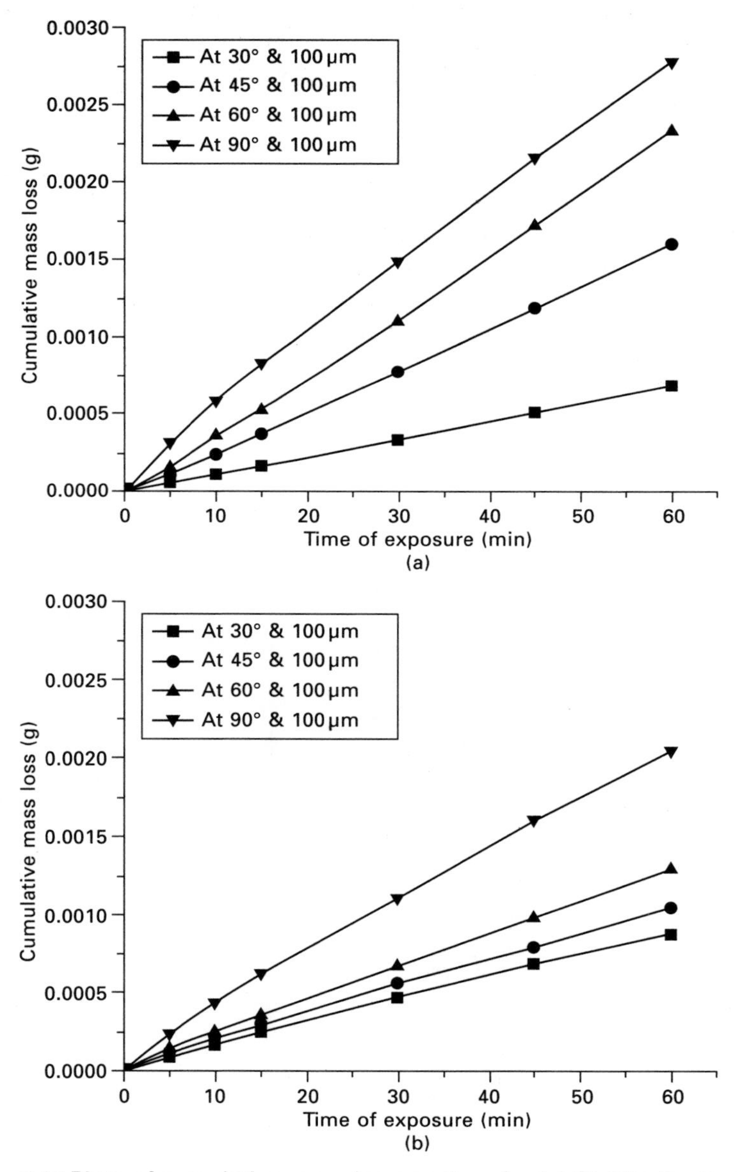

7.34 Plots of cumulative mass loss vs. time for (a) Stellite 6 + 5 wt% B_4C; (b) Colmonoy 88 + 5 wt% B_4C: and (c) Colmonoy 88 + 10 wt% B_4C coatings.

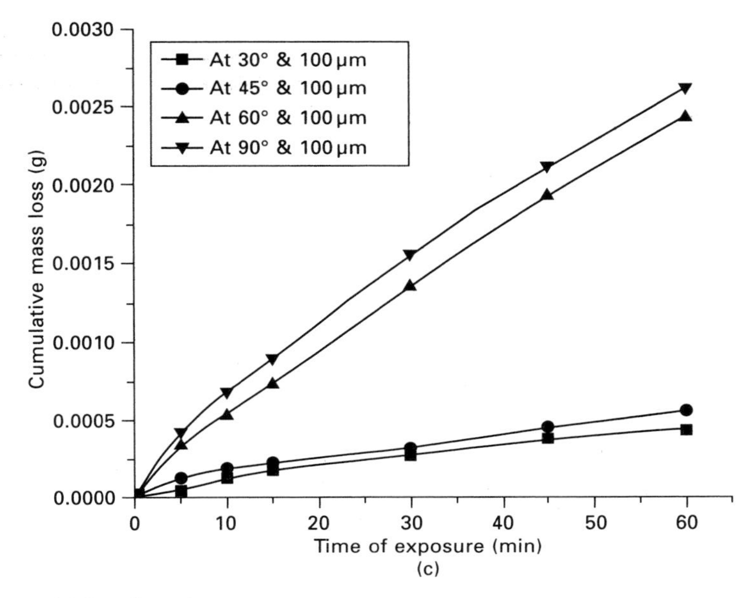

7.34 Continued

trend of cumulative mass losses was observed with increasing time of exposure at all impingement angles. At given test conditions, an order of magnitude lower mass loss (10 times) was observed for B_4C added coatings. Although 5 wt% B_4C coatings do not show any significant change in slopes, Colmonoy 88 + 10 wt% B_4C showed change in slopes when tested at higher impingement angles, implying a change in erosion mechanisms.

Erosion and SSER plots

Erosion curves plotted with erosion (mass loss) against angle of impingement angles for B_4C coatings are presented in Fig. 7.35. Here also, erosion curves indicated that the erosion behaviour is independent of time of exposure, i.e. only the actual quality of erosion or mass loss of target materials increased with time, but time of exposure does not alter mode of erosion behaviour.

The effect of impingements angles on SSER values of B_4C added LSA coatings, in comparison with substrate 13Cr–4Ni steel and basic LSA coatings, is shown in Fig. 7.36. SSER values of B_4C added coatings showed increasing trend with impingement angles and resulted in typical brittle mode of erosion behaviour (wherein maximum erosion is observed at 90°). Also, SSER values of B_4C added coatings do not increase significantly with increasing impingement angles as compared with 13Cr–4Ni steel and basic LSA coatings, which implies that impingement angles do not significantly affect erosion performances of B_4C added coatings. Stellite 6 + 5 wt%

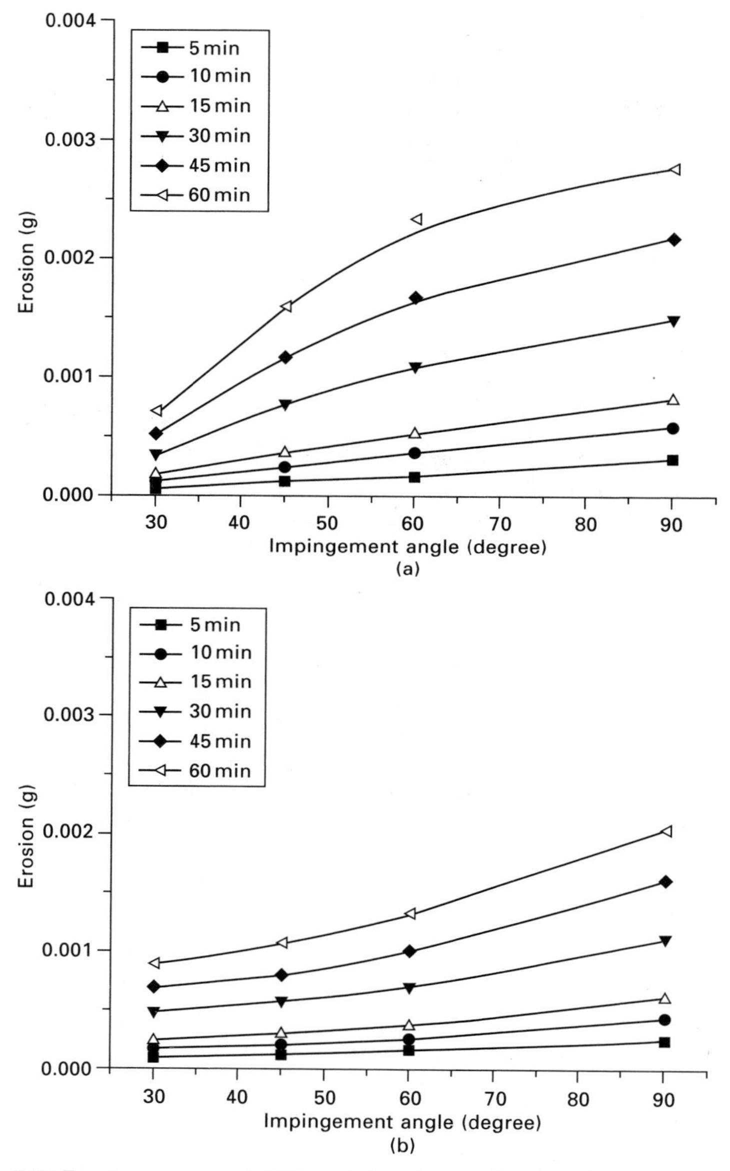

7.35 Erosion curves at different time intervals when impacted with erodents 100 μm: (a) Stellite 6 + 5 wt% B_4C; (b) Colmonoy 88 + 5 wt% B_4C; and (c) Colmonoy 88 + 10 wt% B_4C coatings.

B_4C coatings (Fig. 7.36(a)) showed at least 8-fold lower SSER values than Stellite 6 coatings with respect to angle of impingement. So, for the given test conditions, the ranking of target materials based on erosion resistance is as follows:

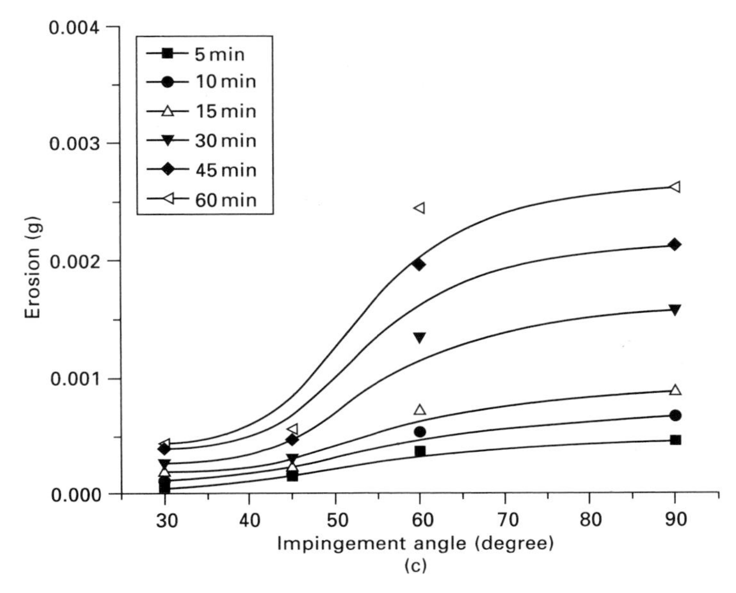

7.35 Continued

13Cr–4Ni Steel < Stellite 6 < Stellite 6 + 5 wt% B_4C coatings

At 30° impacts, 10 wt% B_4C reinforced Colmonoy 88 coatings (Fig. 7.36(b)) had the lowest SSER compared with the rest, whereas 5 wt% B_4C reinforced Colmonoy 88 coatings showed the lowest SSER compared with all target materials. For the case of 10 wt% B_4C coatings, crossover of SSER values is observed after 45° impacts. This is attributed to increased hardness due to increase in carbide and boride phases, or decrease in matrix area, compared with other target materials. At high impingement angles, erodent particles can erode the matrix faster than secondary phases. At least 6–8-fold differences in SSER values were observed between Colmonoy 88 and 5 and 10 wt% B_4C reinforced Colmonoy 88 coatings. For specific test conditions, slurry erosion tests resulted in the following sequence of ranking of the target materials based on increasing erosion rates:

10 wt% < 5 wt% < 0 wt% B_4C added Colmonoy 88 coatings

< 13Cr–4Ni steel (at 30°)

5 wt% < 10 wt% < 0 wt% B_4C added Colmonoy 88 coatings

< 13Cr–4Ni steel (at 90°)

For comparison purposes, SSER values of B_4C added coatings are plotted and shown in Fig. 7.37. 10 wt% B_4C added Colmonoy 88 coatings exhibited at least 2-fold lower erosion rates than 5 wt% B_4C reinforced Colmonoy 88 coatings. This trend was continued up to 45°, and then onwards, 5 wt% B_4C

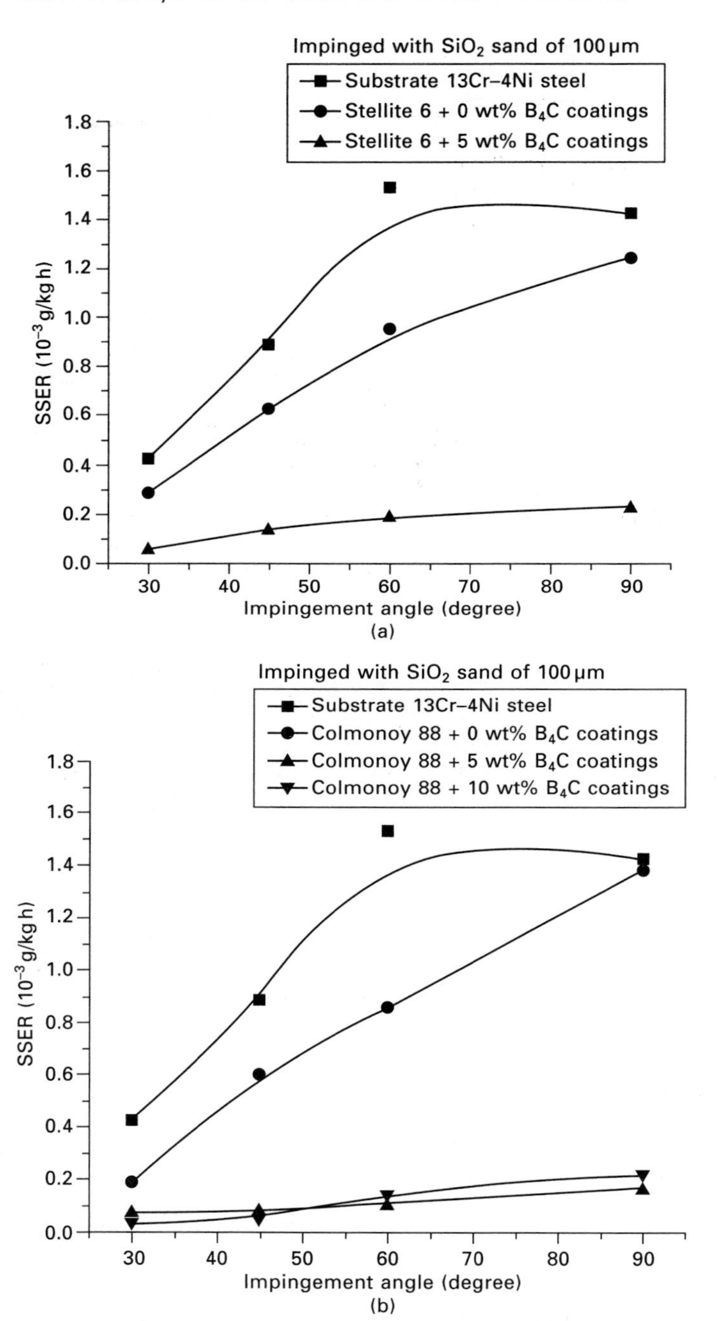

7.36 SSER comparison between substrate 13Cr–4Ni steel and LSA (with and without B_4C) coatings; (a) stellite 6 coatings and (b) Colmonoy 88 coatings.

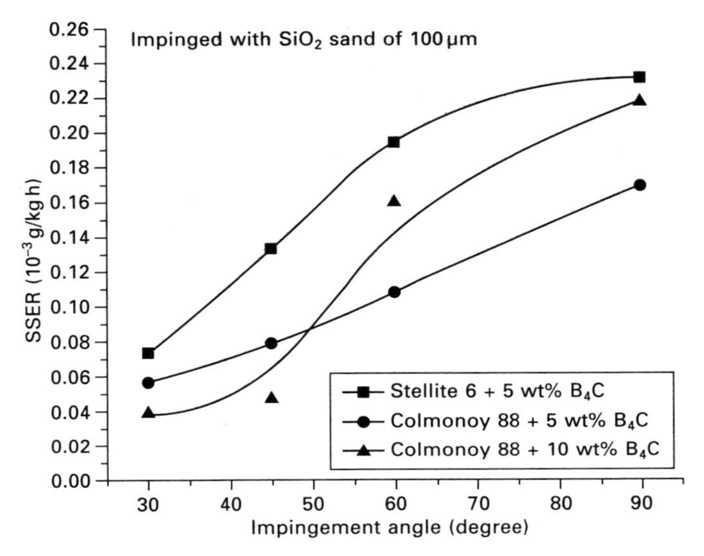

7.37 SSER comparison between B_4C added coatings.

reinforced Colmonoy 88 coatings showed minimum erosion rates up to 90° impacts. Under identical test conditions and based on erosion resistance, the following sequence of ranking of target materials has been obtained:

13Cr–4Ni steel < Stellite 6 < Colmonoy 88
< Stellite 6 + 5 wt% B_4C < Colmonoy 88 + 5 wt% B_4C
< Colmonoy 88 + 10 wt% B_4C (at 30°)
13Cr–4Ni steel < Colmonoy 88 < Stellite 6
< Stellite 6 + 5 wt% B_4C < Colmonoy 88 + 10 wt% B_4C
< Colmonoy 88 + 5 wt% B_4C (at 90°)

The higher erosion resistance of B_4C added coatings is due to their uniform distribution and high hardness compared to other target materials. The difference in SSER among Stellite 6 and Colmonoy 88 coatings is attributed to increased volume fractions of secondary phases.

7.6.3 Erosion mechanisms

The slurry-eroded samples were analysed under SEM in order to study the characteristic features of erosion surfaces, and to understand the operative mechanisms of slurry erosion. Several researchers (Bitter, 1963; Hutchings, 1993; Magnee, 1995; Mann and Arya, 2002; Singal and Singh, 2006) have described two important erosion mechanisms (different at low- and high-impingement angles) by which the erodent particle can impinge the target. When particles impact at the surface at low angles of incidence, chip formation or cutting plays a role in the material removal. Erosion at, or close to, normal impact involves the formation of crater 'lips' of highly deformed material due to the fatigue process; these are subsequently removed as platelets after a number of successive particle impacts (Manisekaran *et al.*, 2007). Many researchers (Wood, 1999; Stachowiak and Batchelor, 2001) have shown that a ductile material generally undergoes mass loss by a process of direct microcutting or plastic deformation, followed by cutting action. In brittle materials, repeated particle impacts result in a fatigue process.

For simplification purposes, only maximum and minimum erosion rates for different materials at a particular angle of impingement and for a fixed erodent size are presented in Figs 7.38 to 7.43. In order to understand the erosion mechanisms of various target materials, an attempt has been made to analyse the SEM micrographs taken at impact areas, and at leading edges (or away from the impact area).

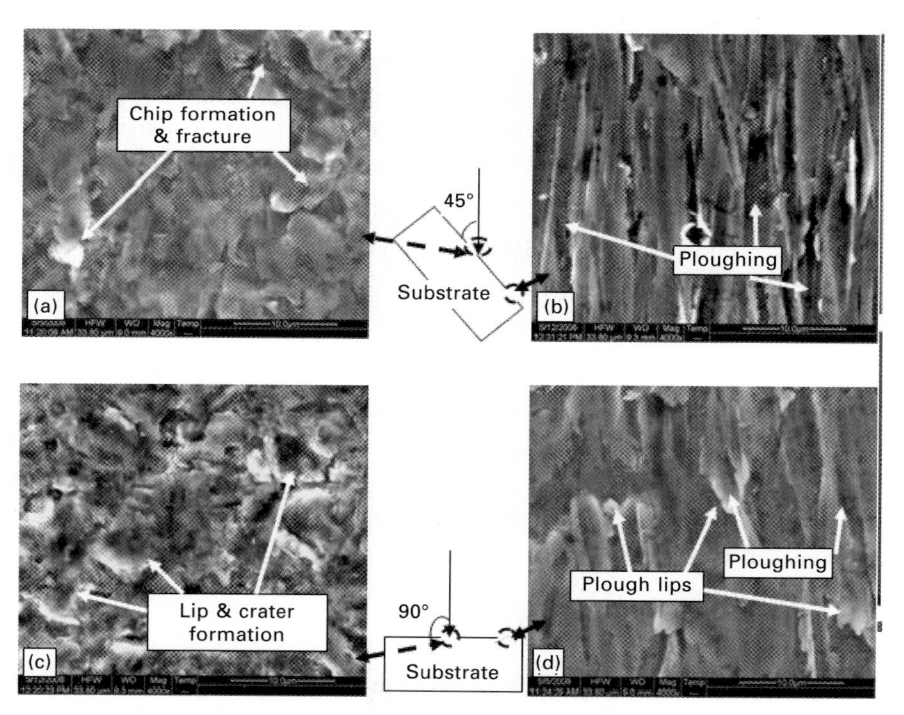

7.38 SEM features of eroded surfaces of substrate 13Cr–4Ni steel (with 375 μm): (a) and (b) impacted at 45° (maximum erosion rate); (c) and (d) impacted at 90° (minimum erosion rate); and (c) at the point of impact; (b) and (d) at the leading edge/away from the impact region.

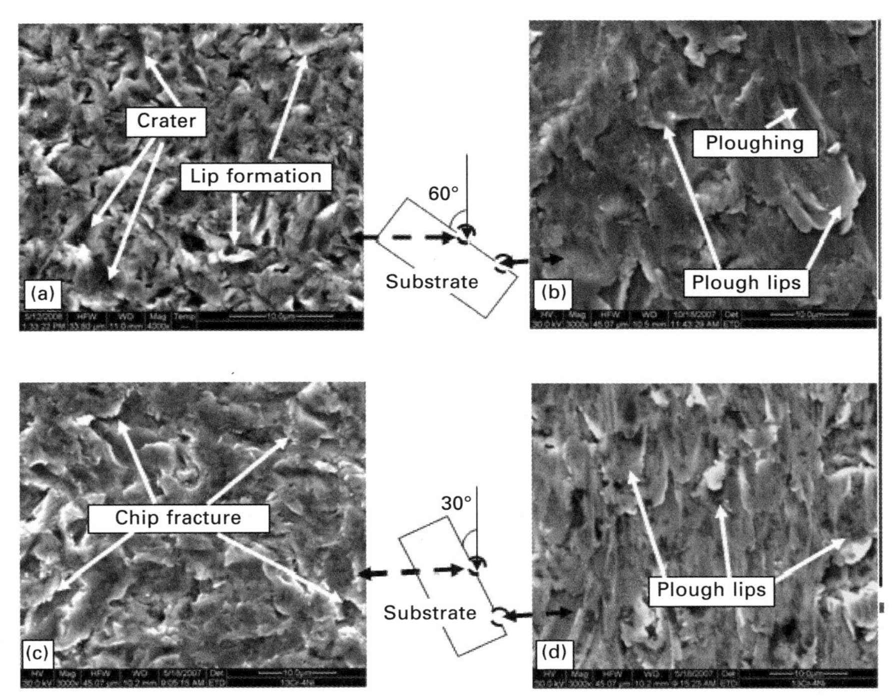

7.39 SEM features of eroded surfaces of substrate 13Cr–4Ni steel (with 100 μm): (a) and (b) impacted at 60° (maximum erosion rate); (c) and (d) impacted at 30° (minimum erosion rate); and (c) at the point of impact; (b) and (d) at the leading edge/away from impact region.

The SEM features of the eroded surfaces of the 13Cr–4Ni steel eroded with silica erodents with an average erodent size of 375 μm are shown in Fig. 7.38. It reveals that chip formation, the subsequent fracturing of chips (Fig. 7.38(a)), and several deep ploughing marks at the leading edge (Fig. 4.19(b)) are responsible for the observed maximum erosion rate at 45°. The minimum erosion rate occurred at 90° and the corresponding eroded surface showed crater lip formation (Fig. 7.38(c)), and subsequent material removal as platelets, as well as shallow discontinuous ploughing marks along with plough lips similar to wedge formation (Hutchings, 1993) of abrasive wear (as shown in Fig. 7.38(d)).

Figure 7.39 shows the SEM micrographs of eroded surfaces of the substrate 13Cr–4Ni steel eroded with particles with an average size of 100 μm. Figures 7.39(a) and (b) correspond to features of the maximum erosion rates that occurred at 60°, in which lip formation (Fig. 7.39(a)), microcutting action, and very few discontinuous ploughing marks along with the plough lips (Fig. 7.39(b)) are the main operative mechanisms. The SEM features of the surfaces that showed minimum erosion rates at 30° are shown in Figs 7.39(c) and (d). Only chip formation, chip fracture and wedge formation (plough lips) are observed at the point of impact and at the leading edge of the specimen, respectively. Figures 7.38 and 7.39 also illustrate the significantly reduced damage in eroded surfaces, along with the changes in the mechanisms, when impacted with particles with an average size that changed from 375 to 100 μm.

The SEM micrographs of the eroded surfaces of the Stellite 6 coatings exhibited erosion mechanisms that are more or less similar to those exhibited by the substrate 13Cr–4Ni steel when impacted with particles of 375 μm. However, the maximum erosion rate occurred at 60° (Figs 7.40(a) and (b)), at which both chip formation and subsequent chip fracture, crater lips formation, lip fracture and shallow ploughing marks were predominant. The minimum erosion observed at the 30° eroded surfaces showed mainly cutting action and deep micro-tunnels at the point of impact (Fig. 7.40(c)) and discontinuous ploughing marks away from the point of impact region (Fig. 7.40(d)).

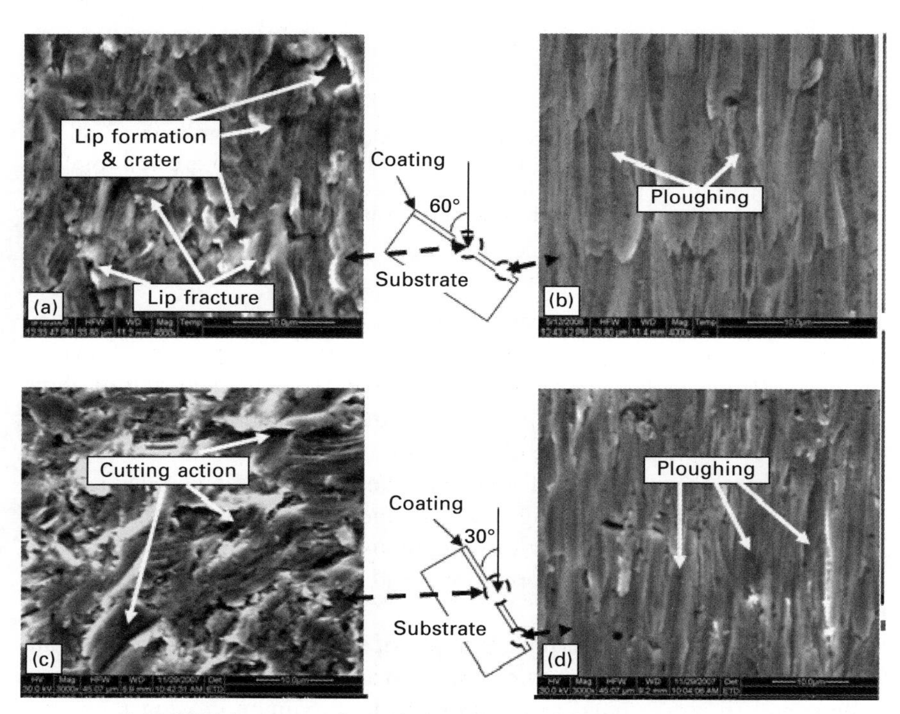

7.40 SEM features of eroded surfaces of Stellite 6 coatings (eroded with 375 μm): (a) and (b) impacted at 60° (maximum erosion rate); (c) and (d) impacted at 30° (minimum erosion rate); and (c) at the point of impact; (b) and (d) at the leading edge/away from impact region.

The effect of the impingement of particles with a size of 100 μm on the eroded surface features of Stellite 6 coatings is presented in Figs 7.41(a) to (d). Eroded surfaces, which correspond to a maximum erosion rate at 90°, showed a higher number of craters, evidence of lip deformation (Fig. 7.41(a)), and a lower number of shallow ploughing marks along with plough lips (Fig. 7.41(b)). However, only a higher number of microcutting regions (Fig. 7.41(c)) and a lower number of deep and discontinuous ploughing marks (Fig. 7.41(d)) were observed at the eroded surfaces which exhibited a minimum erosion rate at 30°.

Figures 7.42 and 7.43 show the SEM features of Colmonoy 88 eroded surfaces, along with the EDS analysis of the fractured and intact secondary phases, when impacted with erodent particles of size 375 and 100 μm, respectively. It is evident from Figs 7.42 and 7.43 that the material removal mechanisms were entirely different from those of the substrate and the Stellite 6 coatings. Here, the progressive fracture of the secondary hard phases, such as carbides/borides pullout (i.e selective removal of fractured phases from the impacted regions of coatings), and intact carbide/boride phases along

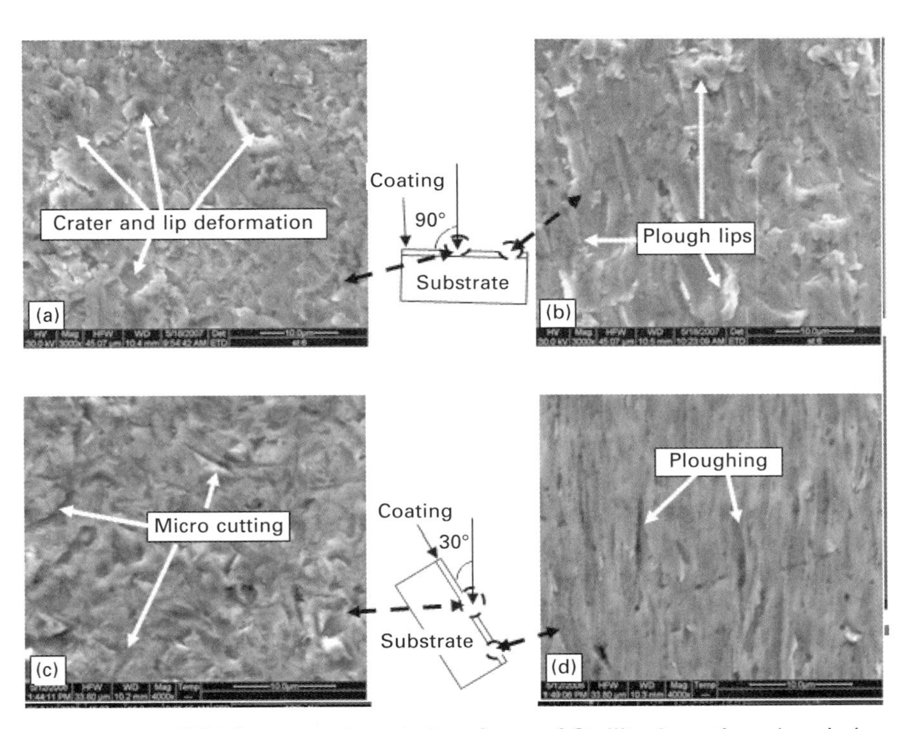

7.41 SEM features of eroded surfaces of Stellite 6 coatings (eroded with 100 μm): (a) and (b) impacted at 90° (maximum erosion rate); (c) and (d) impacted at 30° (minimum erosion rate); and (c) at the point of impact; (b) and (d) – at the leading edge/away from impact region.

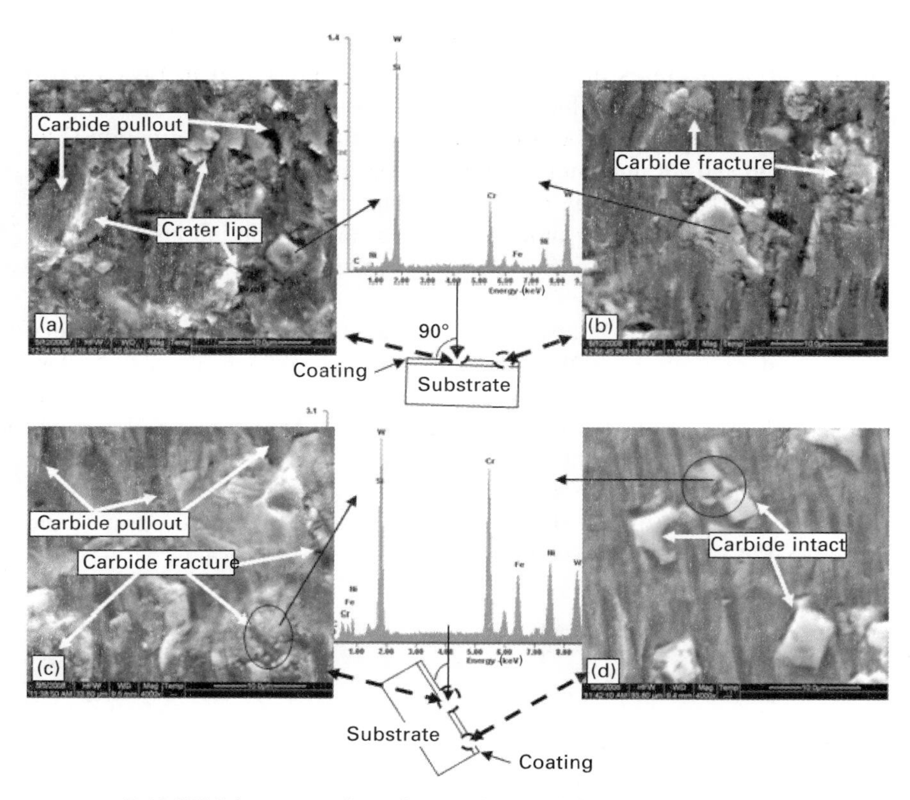

7.42 SEM features of eroded surfaces of Colmonoy 88 coatings (with 375 μm): (a) and (b) impacted at 90° (maximum erosion rate); (c) and (d) impacted at 30° (minimum erosion rate); and (c) at the point of impact; (b) and (d) at the leading edge/away from impact region.

with crater lips were taken into consideration when describing the erosion mechanisms of Colmonoy 88 coatings.

When the Colmonoy 88 coatings are impacted with 375 μm particles at 90° (corresponds to the maximum erosion rate), more carbide pullout and crater formations were observed at the point of impact (Fig. 7.42(a)), and carbide/boride fracture and intact carbides/borides were found away from the impact region (Fig. 7.42(b)). The SEM features of the eroded surfaces (minimum erosion rate observed at 30°) showed fracture of hard phases and less carbide/boride pullout from the impact region (Fig. 7.42(c)). There was no effect on carbides/borides at regions away from the impact; only deep ploughing marks were observed (Fig. 7.42(d)).

When Colmonoy 88 is impacted with particles of size 100 μm, more carbide/boride fracture, less carbide/boride pullout and more intact carbides/borides (Fig. 7.43(a) and (b)) were observed at eroded surfaces that correspond to

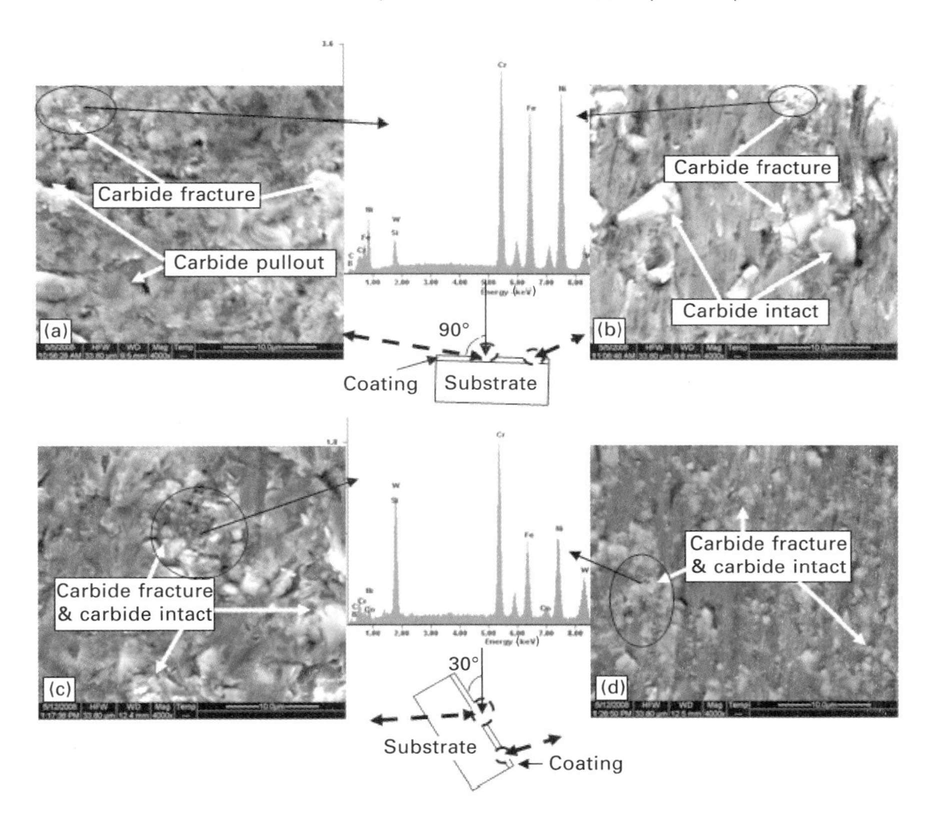

7.43 SEM features of eroded surfaces of Colmonoy 88 coatings (with 100 μm): (a) and (b) impacted at 90° (maximum erosion rate); (c) and (d) impacted at 30° (minimum erosion rate); and (c) at the point of impact; (b) and (d) at the leading edge/away from impact region.

a maximum erosion rate at 90°. Minimum eroded surfaces showed only carbide fracture and intact carbide at both the regions of impact and away from impact regions (Figs 7.43(c) and (d)).

More or less similar erosion mechanisms were observed for the cases of B_4C added coatings. From the quantitative and qualitative analysis results, an attempt has been made to explain all the observed mechanisms in schematic representation (Figs 7.44 and 7.45). Figure 7.44 illustrates the material removal mechanisms for the substrate 13Cr–4Ni steel and Stellite 6 coatings at low- and high-impingement angles.

In Figs. 7.44(a) and (b), it is observed that a fractured chip is formed by repeated impingement of erodents at an angle less than 45° from the target materials. The erodent particles first generate a plastically deformed built-up edge for the subsequent erodent particles to fracture the chip. Figures 7.44(c) and (d) show failure mechanisms at high angles of impingement. Here, the

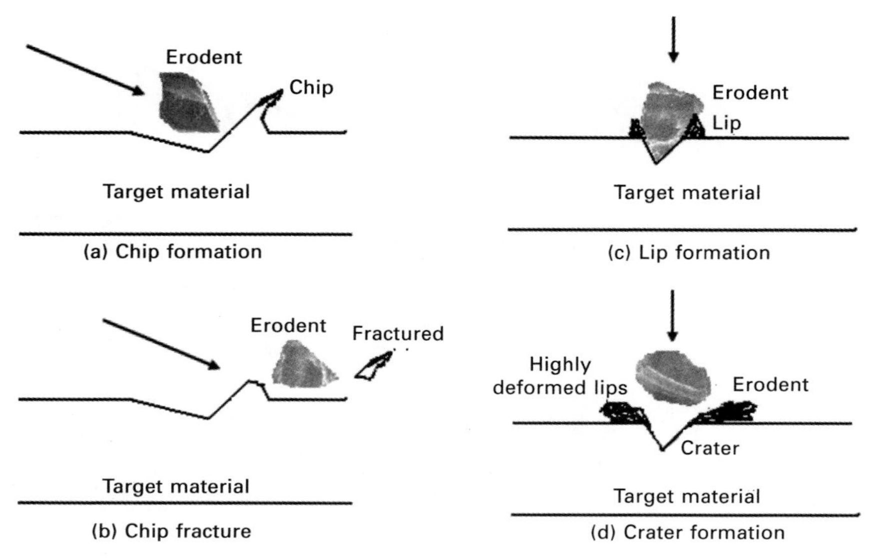

7.44 Material removal mechanisms for 13Cr–4Ni steel and Stellite 6 coatings at low impingement angles: (a) chip formation and (b) chip fracture at high impingement angles; (c) lip and (d) crater formation.

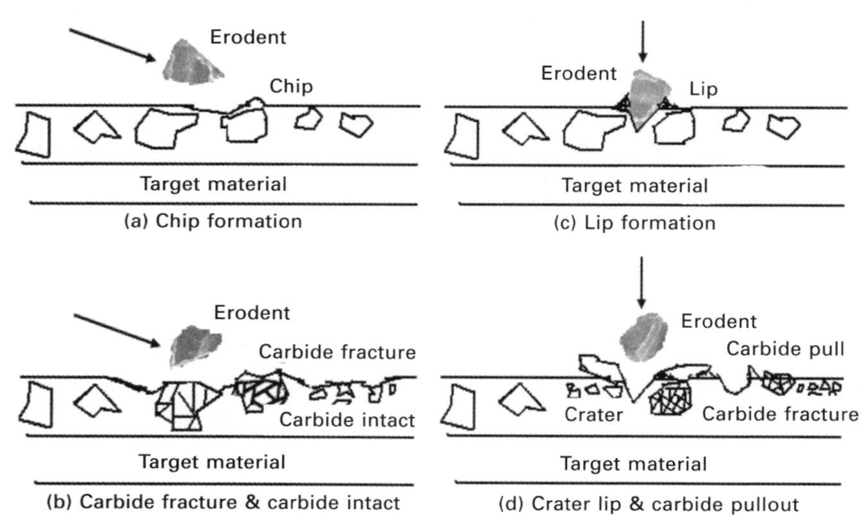

7.45 Mechanisms of progressive fracturing of secondary phases for Colmonoy 88 at low impingement angles: (a) chip formation and (b) fracture and intact at high impingement angles; (c) lip and (d) crater and secondary phase pull out.

crater formation is followed by a lip from the impingement of the first few erodent particles. Then, built-up lips further deform or elongate along the surface by the subsequent impingement of erodents, leading to a further widening of the crater and removal of the elongated lips as platelets. These types of failure mechanisms are observed in the substrate 13Cr–4Ni steel and Stellite 6 coating (Figs 7.38 to 7.41).

Figure 7.45 explains the mode of the material removal mechanisms for the Colmonoy 88 coatings at low- and high impingement angles. As observed in Figs 7.45(a) and (b), the erodent particles impinging at low angles initially strike the surface, creating shallow craters. The successive impact of erodent particles would tend to fracture the secondary hard phases within the coating, but without removing it. Figures 7.45(c) and (d) show the mode of material removal at high-angle impingements. As can be seen, the erodent particles tend to form craters with built-up lips, while subsequent impingements result in a small fractured carbide pullout from the coating. On first impact, the secondary phases can withstand the impact of erodent particle, as the hardness of secondary phase is greater than erodent particles. However, with subsequent impact of erodent particles, secondary phases can break or fracture, as these are brittle in nature. Pullout of these fractured phases will be much easier than for un-fractured phases. These types of failure mechanisms are displayed in Figs 7.42 and 7.43.

Table 7.7 gives the summary of the experimental results obtained from the slurry erosion test results and SEM studies of the eroded surfaces of the substrate and the laser surface alloyed Stellite 6 and Colmonoy 88 coatings. The maximum and minimum erosion rates are related to the average particle size and to the impact angle, with specific erosion mechanisms. The erosion modes, as described in Table 7.7 for the substrate and LSA coatings, are derived from Figs 7.32(a) and (b).

7.7 Effect of erodent size

The particles causing erosion of turbine components in hydropower plants are river sediments, which are in the form of clay, silt, sand and gravel. Sediments are made of fragments of rock formed from chemical and mechanical weathering. Basically, it is the sand fraction of the sediment, which causes turbine erosion. The sand fraction can be further classified into fine (0.06–0.2 mm), medium (0.2–0.6 mm) and coarse (0.6–2 mm). Together with particle properties, particle transport mechanisms also play a role in erosion models. The movement of such particles basically depends on particle characteristics (density, shape, size) and fluid characteristics (velocity, turbulence, viscosity). Many investigators (Lynn *et al.*, 1991; Gandhi and Borse, 2004; Thapa, 2004) have reported that the material loss due to erosive wear is directly related to the kinetic or impact energy of

Table 7.7 Effect of test parameters on slurry erosion rates and erosion mechanisms of three different target materials

	Substrate 13Cr–4Ni steel				Stellite 6 coatings				Colmonoy 88 coatings			
Erosion rate	Maximum		Minimum		Maximum		Minimum		Maximum		Minimum	
Ave. particle size (μm)	375	100	375	100	375	100	375	100	375	100	375	100
Impact angle	45°	60°	90°	30°	60°	90°	30°	30°	90°	90°	30°	30°
Mechanisms	Chip fracture and ploughing	Micro cutting, lips formation and ploughing	Lips formation	Cutting action	Chips fracture, lips and crater lips	Crater and Carbide pull out	Cutting and ploughing	Cutting and ploughing	Crater lips and carbide fracture	Carbide pull out and fracture	Ploughing and carbide intact	Ploughing, carbide fracture and intact
Mode of erosion	• Ductile when impacted with 375 μm erodents • Mixed mode when impacted with 100 μm erodents				• Ductile when impacted with 375 μm erodents • Brittle when impacted with 100 μm erodents				• Brittle when impacted with 375 and 100 μm erodents			

solid particles. The kinetic energy of a particle is directly proportional to its mass. The relationship between the particle size and the erosion rate can be defined as erosion rate ∝ (particle size)^n. The variation in the exponent value is normally attributed to different experimental conditions and fluid dynamic effects.

The de-silting arrangement (settling basin) is designed to trap sediment particles of up to 200 µm in size during hydropower plant operation. But during a monsoon, it is impossible to control the passing of higher erodent sizes. Hence, erodent size is one of the important parameters to characterize erosion behaviour of LSA coatings with 13Cr–4Ni steel substrate. In practice, erodents of size up to 500 µm can impact on turbine components. So, four different sizes of silica sand are impacted on five different target materials under identical test conditions and results are discussed in the following sections. Only low impingement angle (30°) was considered here due to the fact that turbine components are designed to get maximum efficiency (by keeping in view that at low angles, water can have greater duration of sliding contact while impacting the turbine components). The particle size of the erodent is another important factor which alters the performance of hydroturbine components, especially during monsoon. Since the properties of solid particles are of great importance in erosion studies, a single source of solid particles (erodent) was used throughout the experiments. Fresh silica sand was used for every test to avoid any erosion effect due to the degradation of the impacting erodents during erosion tests. The tests were conducted (Table 7.8) for four different sizes of erodent for 1 h duration at constant slurry velocity of 12 m/s. Constant parameters, as described in Section 7.6, were also utilized to study the effect of erodent size on erosion performance of laser surface modified coatings and 13Cr–4Ni steel.

Although some test conditions have been studied for the same target material to assess effect of impingement angle (Section 7.6), due to uncertainties and problems with reproducibility, fresh target materials are utilized at all test conditions. Hence, erosion/SSER values may or may not be different from earlier sections. To minimize the number of experiments, tests have not been conducted on Colmonoy 88 + 10 wt% B_4C coatings.

Table 7.8 Test parameters to study effect of erodent size

Type of slurry	Material	Angle of impingement (°)	Average erodent size (µm)
Commercial silica sand	1. 13Cr–4Ni steels 2. Stellite 6 coatings 3. Colmonoy 88 coatings 4. Stellite 6 + 5 wt% B_4C 5. Colmonoy 88 + 5 wt% B_4C	30	100, 215, 375 and 490

7.7.1 Cumulative mass loss plots

To simplify and for comparison purposes, cumulative mass loss plots for surfaces impacted by erodents of sizes 100 and 490 μm are given in Fig. 7.46. The plots clearly indicate an increasing trend of cumulative mass

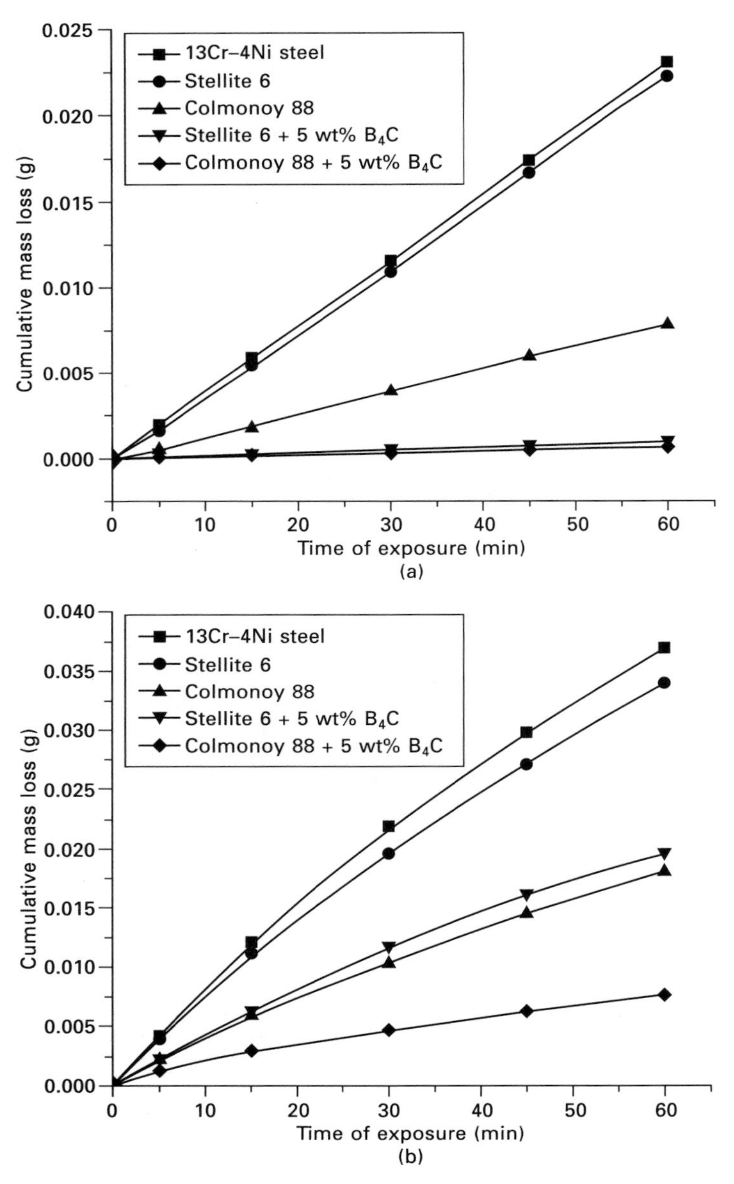

7.46 Cumulative mass loss vs. time of exposure plots at (a) 100 μm and (b) 490 μm.

loss with time of exposure. No changes in slopes were observed for all the target materials, indicating an absence of change in respective erosion mechanisms with time of exposure. However, magnitudes of mass losses increased with increasing erodent particle sizes due to increase in kinetic or impact energy.

7.7.2 Erosion and SSER plots

Standard erosion curves plotted as erosion (mass loss) against various erodent sizes at a given impingement angle at different time intervals for different target materials (obtained under identical erosion test conditions) are discussed in this section. For the purposes of simplification and depending on the nature of erosion curves, only two different plots have been shown in Fig. 7.47.

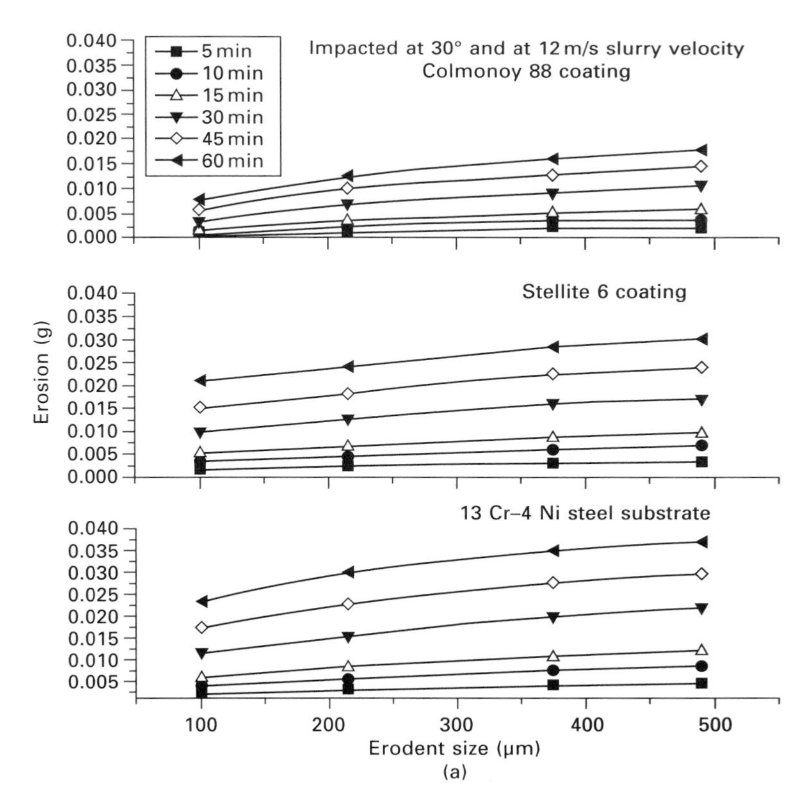

7.47 Standard erosion curves for (a) 13Cr–4Ni steel substrate, Stellite 6 and Colmonoy 88 coatings and (b) 5 wt% B_4C added Stellite 6 and Colmonoy 88 coatings.

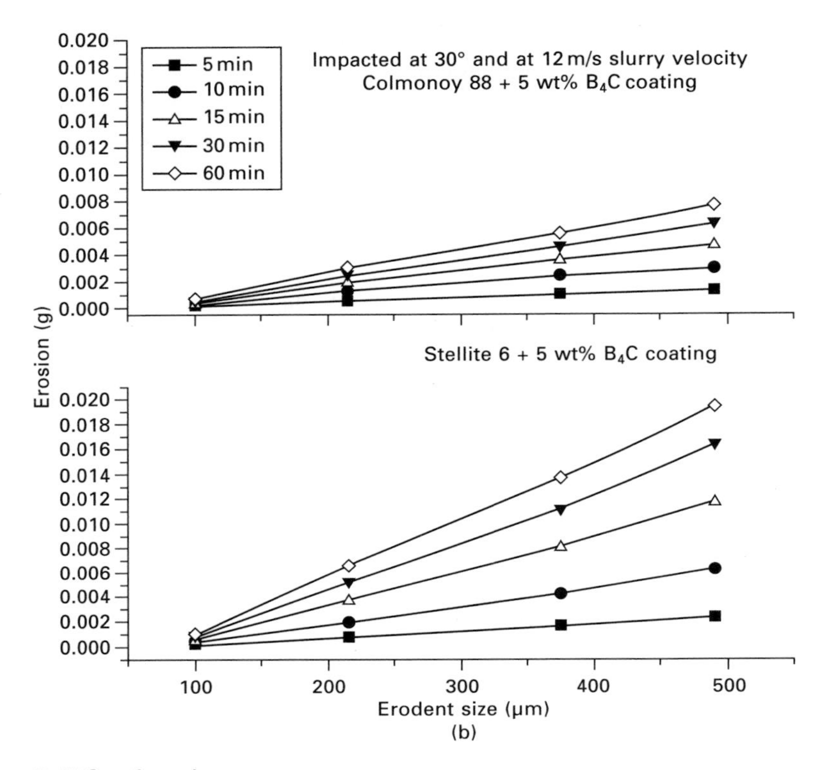

7.47 Continued

The erosion curves (irrespective of target materials) show that erosion behaviour is independent of time of exposure, i.e. only erosion or mass loss of target materials increased with time, but time of exposure does not alter the qualitative erosion characteristics of individual materials. Two distinct erosion profiles have been observed for different target materials under identical test conditions. Target materials such as 13Cr–4Ni steel, Stellite 6 and Colmonoy 88 coatings showed an increasing trend in respective erosion values up to a certain erodent size (say, 375 μm), and afterwards a decrease in rate of erosion when impacted with erodent of size 490 μm (Fig. 7.47(a)). Although this observation is not very obvious at initial time interval (up to 15 min) for 13Cr–4Ni steel and Stellite 6 coatings, measurements after test duration of 1 h showed a clear trend of decreasing rates of erosion with the impacts of erodents of size 490 μm. B_4C added LSA coatings, however, showed a monotonically increasing trend with erodent size at all time intervals under identical test conditions (Fig. 7.47(b)).

Turenne *et al.* (1990) explained the phenomenon of increase in slurry erosion with increasing erodent size, citing two important effects. Firstly, the normal component of velocity for coarse abrasive erodents is higher, and secondly, their masses are also higher than those of fine erodents, resulting in higher impact energy for the material removal process. Erodent particle size can be characterized mainly in two basic dimensions, such as mass and length. For a fixed velocity, kinetic energy of erodent is directly proportional to mass. But in the present study, mass of erodent is constant. However, mass of spherical particle is proportional to (diameter)^3. Wood (1999) and Thapa (2004) have utilized the following general equation to calculate impact energy or kinetic energy of erodents, E_k:

$$E_k = \frac{1}{2} M_P V_P^2 = \frac{2}{3} \rho \pi R^3 V_P^2 \quad [7.3]$$

where M_P is mass of erodent particles, V_P is erodent particle velocity, ρ is erodent particle density and R is radius of erodent particle. From the above equation, the calculated kinetic energy of finer erodents (average erodent size 100 μm) and coarser erodents (average erodent size 490 μm) are 0.0829 and 9.76 μJ, respectively. The steep increase in kinetic energy of the erodent particle is responsible for the increase in erosion of individual target materials. So, the type of damage resulting from an impact is related to the energy of a single erodent particle. For an erodent particle having a low energy (finer erodent), the micro-cutting of the matrix is the operating mechanism accompanied by the protection effect (these erodents will take more time to fracture the secondary phases, compared with coarser erodents). For high-energy values (coarser erodent), the fracture of secondary phases, such as carbides and borides, is also activated along with the micro-cutting of the matrix. Hence, finer erodents have more cutting effect, while coarser erodents will deform material by elastic deformation and fatigue.

Clark (1992) also explained that impact and flow conditions changed with change in erodent size from coarser to finer. If the erodents are finer, they have less irregular multiple cutting edges than coarser erodents and, hence, erosive wear is also reduced significantly. The less sharp edges in the finer erodents may not remove the material by the so-called micro-cutting mechanisms. Besides, the impact energy imparted to the surface on each impact is reduced with finer erodents. In addition, the finer erodent can only coincide with a smaller defective area in the surface, thereby reducing the amount of erosive wear.

The low rate of erosion for coarser erodents (say, 490 μm) compared with finer erodents (100 μm) is described by the roundness factor, which was calculated by Bukhaiti *et al.* (2007) using quantitative analysis of silica sand. According to them, coarser erodents showed a roundness factor very close to 1 (1.06), which means regular and round, whereas finer erodents had a roundness factor above 1.5. As described by Stachowiak (2000), the erodent particle angularity plays a crucial rule in erosion of the target materials. Stachowiak has explained that the more angular erodent results in more material removal due to its sharp nature. Since B_4C added LSA coatings have lower fractions of matrix regions than other normal LSA coatings and 13Cr–4Ni steel substrate, all the kinetic energy of erodents is spent on fracturing carbide and borides. Hence, linear increases in erosion curves were observed for these coatings (Fig. 7.47(b)).

Relative erosion performances of LSA coatings and 13Cr–4Ni steel substrate have been compared using SSER bar charts, which are plotted against erodent size. To simplify, first the same family of LSA coatings was compared with the substrate (Fig. 7.48(a) and (b)), and then of erosion performances were compared among LSA coatings as shown in Fig. 7.48(c). Typical SSER plots for Stellite 6 (with and without addition of B_4C) coatings compared with that of substrate 13Cr–4Ni steel are shown in Fig. 7.48(a).

Erosion resistance of Stellite 6 coatings (without B_4C) was between 1 and 1.5, whereas the differences between with and without B_4C added Stellite 6 coatings were in the range of 1.75–18 times. Overall, 5 wt% B_4C added Stellite 6 coatings showed erosion resistance in the range of 1.9 to 20 times that of substrate. The lower erosion resistance of Stellite 6 coatings is due to low enhancement in hardness (in turn due to change in microstructure) compared to substrate. Moreover, the finer carbides were low in volume fraction with Stellite 6 coatings. These carbides were not able to withstand

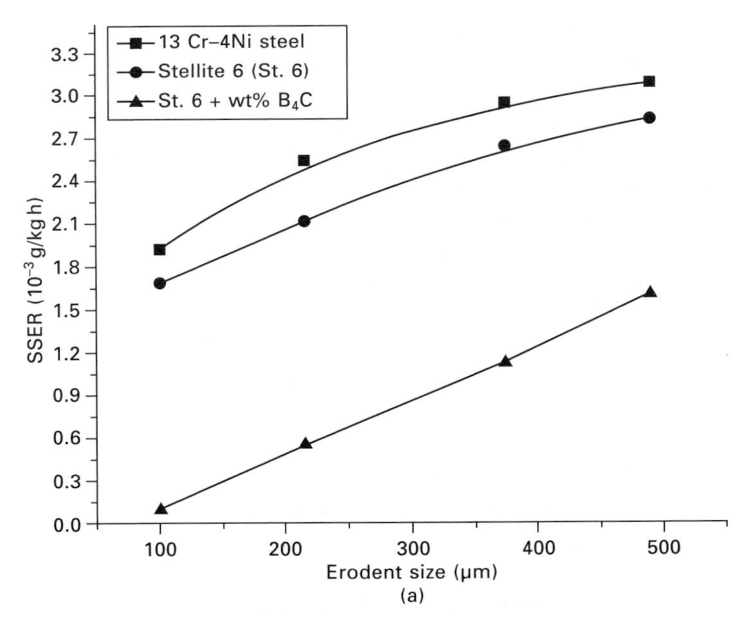

7.48 SSER comparison between substrate and LSA coatings: (a) Stellite 6 coatings; (b) Colmonoy 88 coatings; and (c) only LSA coatings.

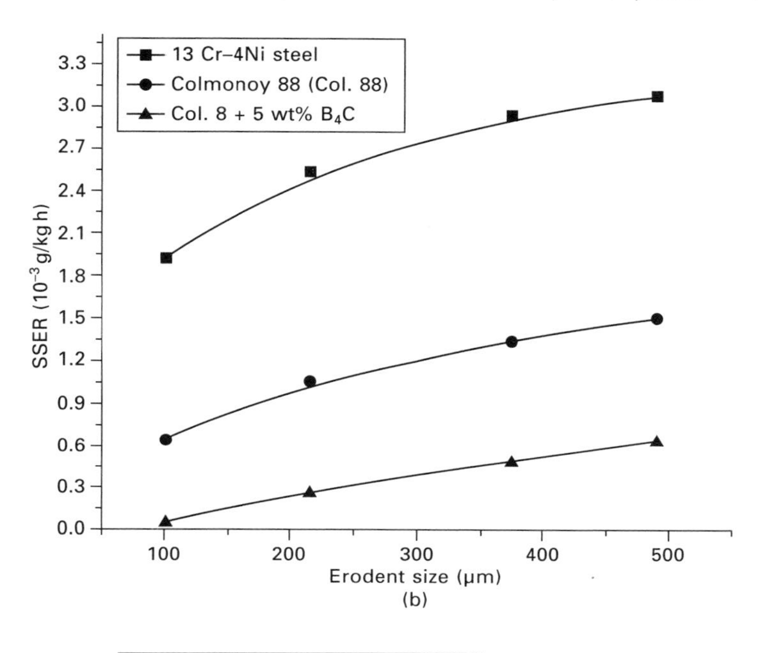

(b)

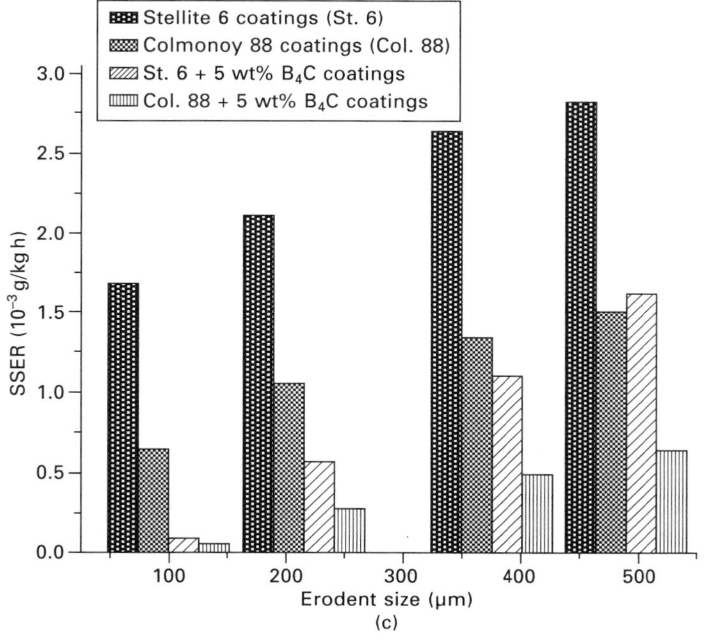

(c)

7.48 Continued

and protect the matrix from high impact energy of erodents. With the addition of B_4C, the increase in volume fraction of carbides leads to an increase in matrix toughness and, thereby, an increase in erosion resistance.

Figure 7.48(b) represents the comparison between substrate and Colmonoy 88 (with and without addition of B_4C) coatings. Colmonoy 88 coatings (without B_4C) showed at least 2 to 3-fold increase in erosion resistance. About 3 to 11 times differences in SSER values between with and without B_4C added Colmonoy 88 coatings were seen. With the addition of 5 wt% B_4C, Colmonoy 88 coatings showed very high erosion resistance (5 to 33-fold) compared with substrate. Change in microstructure and increase in volume fractions of carbides and borides are responsible for higher erosion resistances of Colmonoy 88 (with and without B_4C) coatings.

Among LSA coatings, Colmonoy 88 (with or without addition of B_4C) coatings showed high erosion resistance (in the order of 1.6 to 2.6-fold) compared with Stellite 6 (with and without addition of B_4C) coatings (Fig. 7.48(c)). This may be attributed to the presence of different types of secondary phases. As described (on pages 200–7), Stellite 6 coatings contained only carbides, whereas borides along with carbides are present in Colmonoy 88 coatings. This study also helped in understanding the effect of volume fractions (rather than size and shape) of secondary phases on erosion performances. Even though Stellite 6 coatings showed finer secondary phases than Colmonoy 88 coatings, they have shown almost equivalent erosion resistance when 5 wt% B_4C is added. The erosion test results indicate the following sequence of ranking of the target materials with respect to erosion resistance:

- 13Cr–4Ni steel substrate < Stellite 6 < Colmonoy 88 < Stellite 6 + 5 wt% B_4C < Colmonoy 88 + 5 wt% B_4C (impacted with average size of 100 μm)
- 13Cr–4Ni steel substrate < Stellite 6 < Stellite 6 + 5 wt% B_4C < Colmonoy 88 < Colmonoy 88 + 5 wt% B_4C (impacted with average size of 490 μm).

7.7.3 Erosion mechanisms

SEM analysis is very important for observing the erosion pattern and determining which mechanism plays a role in the removal of material. For comparison, only test samples which are impacted with two different erodent sizes (average sizes of 100 and 490 μm) are discussed. Figure 7.49 shows features of slurry erosion mechanisms, which are observed by SEM of the eroded surfaces of substrate and basic LSA (without addition of B_4C) coatings. It is evident from Figs 7.49(a) and (b) that cutting action, chip formation and fracture are operative mechanisms for substrate. Stellite 6 coatings

(a)

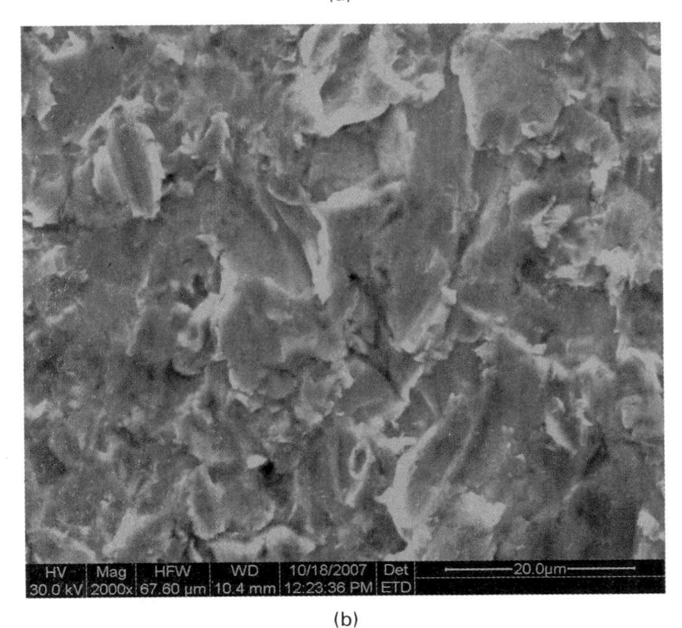

(b)

7.49 SEM features of different eroded samples (at 30° impacts at 12 m/s): (a) with 100 μm and (b) with 490 μm cutting and chip fracture, substrate; (c) with 100 μm and (d) with 490 μm, micro-cutting action, Stellite 6 coatings; (e) with 100 μm and (f) with 490 μm, fracture and intact of secondary phases, Colmonoy 88 coatings.

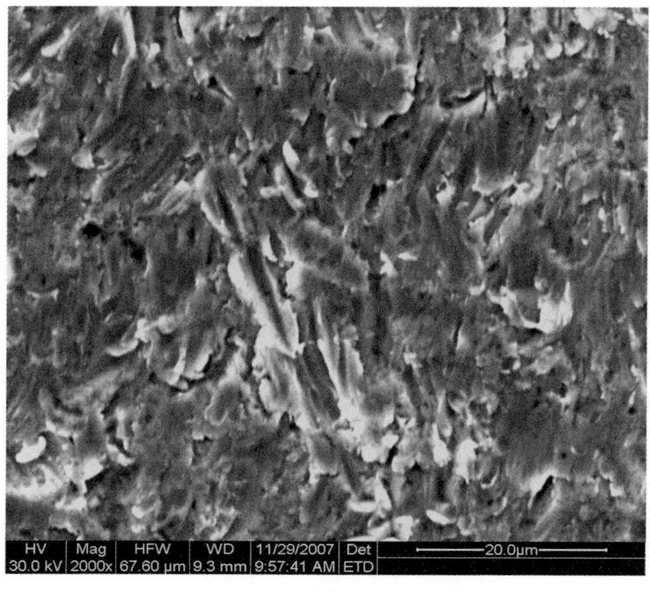

(c)

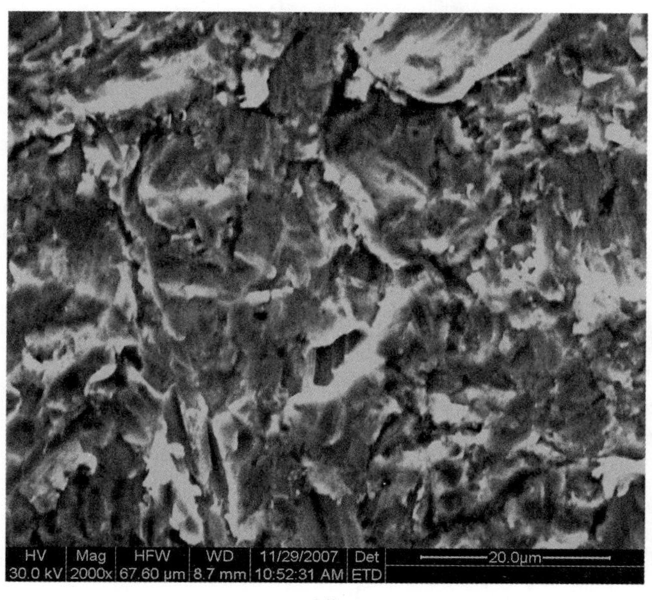

(d)

7.49 Continued

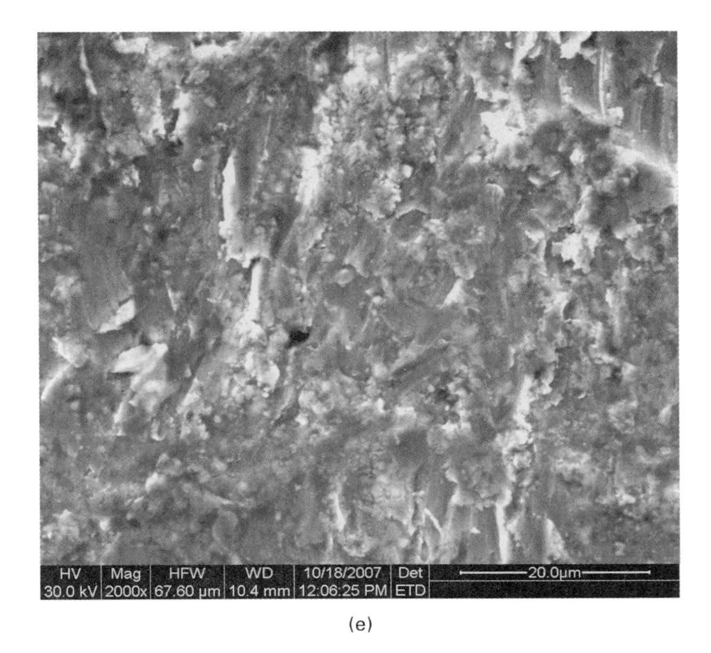

(e)

(f)

7.49 Continued

showed that micro-cutting and cutting actions were main material removal mechanisms as shown in Figs 7.49(c) and (d). Secondary phase fracture, intact nature of those fractured phases and pullout from the surfaces are main operative erosion mechanisms for Colmonoy 88 coatings (Figs 7.49(c) and (d)). Figures 7.49(b), (d) to (f) also revealed the increase in large material removal at the point of impact with increase in erodent size. Similar erosion mechanisms of basic Stellite 6 (without addition of B_4C) coatings prevailed in B_4C added coatings too, but with lower distortion (indicative of lower mass losses) (Figs 7.50(a) and (b)). The irregular sharp erodents of average size 100 μm were not able to fracture the secondary phases at a few places (Fig. 7.50(c)), whereas these phases were fractured (but intact with matrix) when impacted with erodents of average size 490 μm.

7.8 Effect of slurry velocity

In the literature, slurry velocity is referred to as the speed of erosive particles or particle velocity, etc. The relation between erosion and velocity of particle is $E = kV^n$, where k and n are constant and velocity exponent, respectively. The n value varies depending on properties of target materials and test conditions.

To develop a relation between erosion rate and velocity, slurry erosion tests were carried out at four different slurry velocities with all other parameters kept constant. The relations obtained are useful in predicting the erosion of target material operating at similar conditions of actual practice. To simplify, test samples have been removed to weigh after 30 mins (not at regular intervals) for the entire test duration of 1 h. Generally after 30 min, slurry has been changed to maintain the sharpness of commercial silica erodent. Tests have been conducted at low impingement angle (30°) in order to simulate the actual applications in hydroturbine atmosphere.

Velocity of the particle flow is also one of the important parameters in assessing turbine erosion behaviour. Applications of components in hydroturbines decide the velocity of particle impacts. So, effects of slurry velocity (Table 7.9) on 13Cr–4Ni steel and laser surface modified coatings are studied at 30° impingement angle by using commercial SiO_2 sand of average size 100 μm (slurry concentration of 10 kg/m^3).

7.8.1 SSER plots

Figure 7.51 shows the SSER plots as a function of slurry velocity for Stellite 6 (with and without added B_4C) coatings and Colmonoy 88 (with and without addition of B_4C) coatings. In the same figures, erosion performances of LSA coatings are compared with that of 13Cr–4Ni steel. The erosion rates of all target materials are found to increase substantially with slurry velocity as

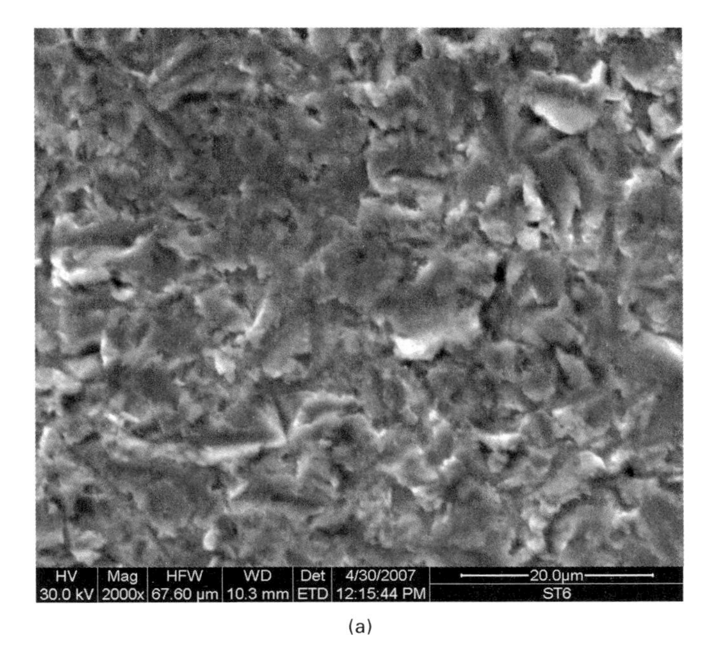

(a)

(b)

7.50 SEM features of B_4C added coatings (at 30° impacts at 12 m/s): with 100 μm and (b) with 490 μm, plastic deformation and micro-cutting, Stellite 6 + 5 wt% B_4C coatings; (c) with 100 μm and (d) with 490 μm matrix damage, fracture and intact of secondary phases, Colmonoy 88 + 5 wt% B_4C coatings.

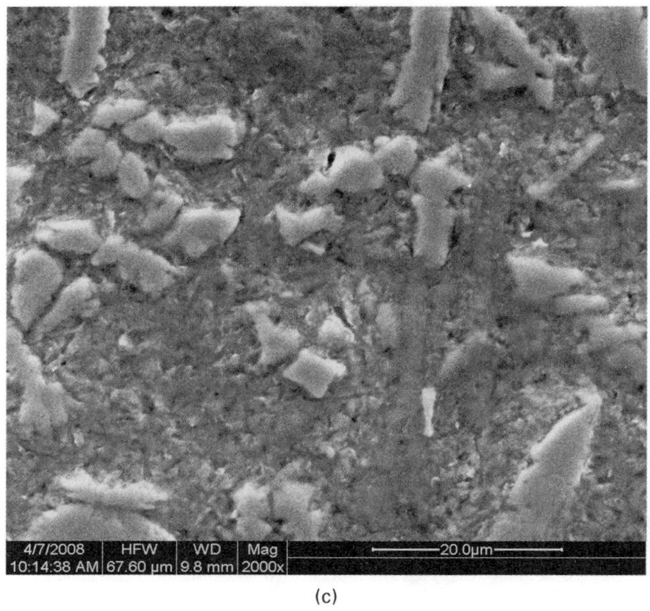

(c)

(d)

7.50 Continued

Table 7.9 Test parameters to study the effect of slurry velocity

Type of slurry	Material	Angle of impingement (°)	Slurry velocity (m/s)
Commercial silica sand	1. 13Cr–4Ni steels 2. Stellite 6 coatings 3. Colmonoy 88 coatings 4. Stellite 6 + 5 wt% B_4C 5. Colmonoy 88 + 5 wt% B_4C	30	8, 12, 16 and 20

shown in Figs 7.51(a) and (b). As is the case with erodent size effects on SSER, here too SSER is dependent on the erodent particle kinetic energy or impact energy (E), i.e. $E = ½mV^2$, where m is mass of erodent and V is slurry or erodent velocity. The equation describes how impact energy is directly related to square of velocity. Using the equation, the calculated kinetic energies for tested slurry velocities (8 to 12 m/s) ranged between 0.04 and 0.23 μJ.

Higher impact energy leads to more severe damage during the process of erosion. Hence, the erosion rate should be higher at higher slurry velocity. These results confirmed the dependence of SSER on the square of the slurry velocity. For given test conditions, the velocity exponent (n) for each target material is given in Table 7.10. It is observed that the velocity exponent for a given test condition is in the range of 1.04 to 2.25. Thapa (2004) reported that n values are between 1.5 and 5 for various target materials with or without surface modifications. Large differences between literature and observed values of n are due to differences in the range of slurry velocities (generally, 40–160 m/s slurry velocity used in literature, whereas present investigation was confined to 8–20 m/s) and different coating methods (thermal spray technique, e.g. HOVF compared to LSA method used in this work). Clark (1992) described how the large differences in observed n are due to various reasons. According to him, velocities referred to are the nominal test speeds, which are almost certainly not representative of actual particle impact velocities. Also, in some test set-ups, an increase in velocity results in particles impacting more frequently, and values of velocity exponent must be adjusted accordingly. Data obtained from high slurry concentration and experiments for brittle materials may also yield high velocity exponents. However, for low slurry velocities (less than 5 wt% erodents), the exponent was about 2.5 for ductile target materials.

The SSER differences between substrate and Stellite 6 coatings are in the ranges of 1.3 to 1.5 and 1.9 to 3.1 for Colmonoy 88 coatings. Though B_4C added LSA coatings show better erosion resistance at low slurry velocities from 8 to 12 m/s, cross-over of SSER values was observed after certain velocities (i.e. after 16 and 12 m/s for B_4C added Stellite 6 and Colmonoy 88 coatings, respectively). B_4C added Stellite 6 coatings showed enhanced

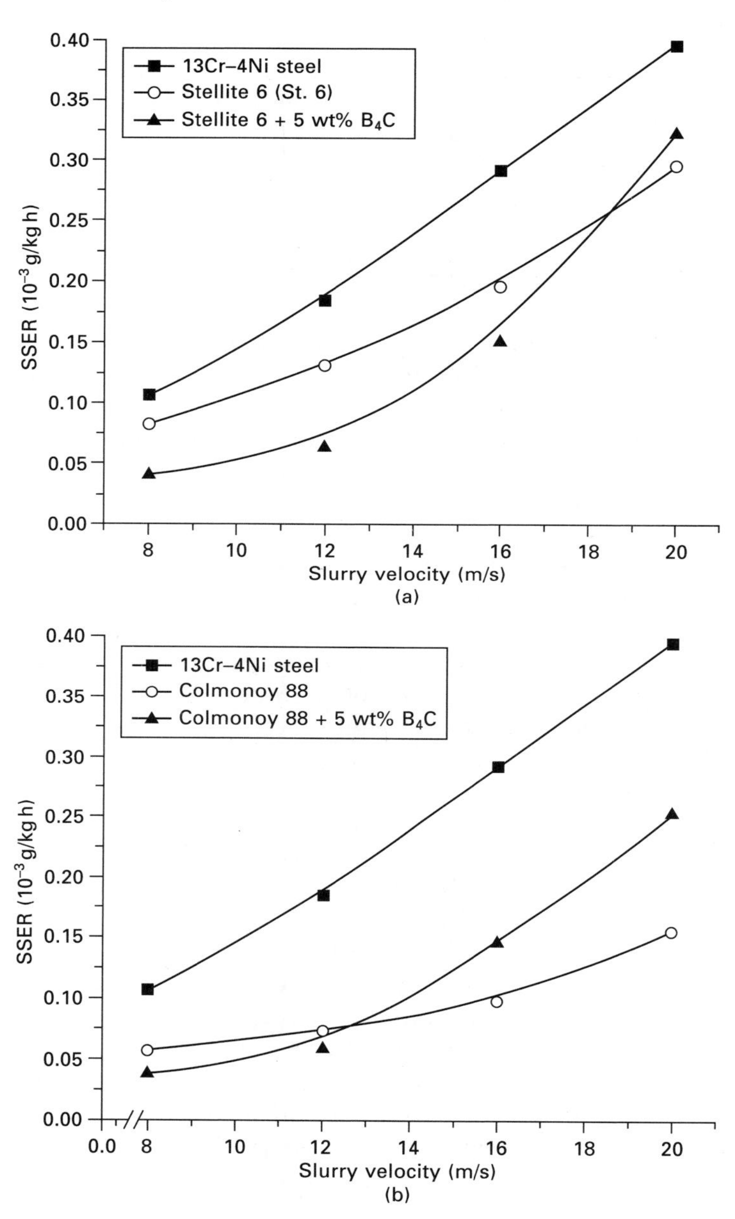

7.51 Slurry velocity dependence of SSER plots at 30° impacts for (a) Stellite 6 and (b) Colmonoy 88 (with and without addition of B_4C) coatings.

erosion resistance in the range of 1.2 to 2.9 compared with that of the substrate, and 1.3 to 2.1 compared to that of Stellite 6 coatings (up to 16 m/s velocity). Similarly, Colmonoy 88 + 5 wt% B_4C coatings showed erosion resistance up

Table 7.10 Velocity exponent *n* for different target materials

Target material	13Cr–4Ni steel	Stellite 6 (St. 6)	Colmonoy 88 (Col. 88)	St. 6 + 5 wt% B_4C	Col. 88 + 5 wt% B_4C
Velocity exponent, *n*	1.44	1.37	1.04	2.25	2.11

from 1.6 to 3.2 compared with that of the substrate, and 1.3 to 1.5 compared with that of Colmonoy 88 coatings (up to 12 m/s velocity). After 12 m/s, Colmonoy 88 coating showed better erosion resistance (at least by 1.5 times) than B_4C added Colmonoy 88 coating. Similar findings have been reported by Tu *et al.* (1999), that the slurry erosion resistance of aluminium alloys AC4C and $Al_{18}B_4O_{33}$ whisker-reinforced AC4C Al alloy matrix composites. The high erosion resistances of B_4C added LSA coatings at lower slurry velocity are due to higher volume fractions of secondary phases, which in turn act like reinforcements (taking responsibility for protecting the matrix from damage by deflecting the erodents). At lower velocities, the carbides are tough enough to survive direct particle impacts in the environment of the test rig. Hence, the mode of material loss in B_4C added coatings would be related strongly to the behaviour of the matrix. The binding matrix of phases wears out preferentially, leading to subsequent loss of phases. In contrast, at higher velocities, the secondary phases are also scratched to an extent and are removed, together with their surrounding matrix phase.

The differences in cross-over of SSER values between Stellite 6 and Colmonoy 88 (with and without B_4C) coatings are attributed to the morphology, type and volume fractions of secondary phases. As observed in the microstructure of Stellite 6 (with and without addition of B_4C) coatings, only very fine secondary phases are formed with volume fractions of 35% for Stellite 6 coatings and 55% for B_4C added Stellite 6 coatings (Figs 7.10 and 7.12). It is assumed that fine carbides are physically too small to resist and deflect the erodents. Hence, their contribution would be limited to enhancing hardness of the matrix to reduce scratch depth and related erosion. On the other hand, Colmonoy 88 coatings showed coarse precipitates (carbides as well as borides) along with increasing volume fractions. Hence, erosion resistance of Colmonoy 88 coating is better than that of Stellite 6 coatings.

To compare the erosion performances all five target materials, SSER values have been plotted against slurry velocity and are shown in Fig. 7.52. which illustrates the better erosion resistance of LSA (with or without B_4C) coatings compared with the substrate at all slurry velocities for a given test condition. Except for the cross-over between Colmonoy 88 (without addition of B_4C) and Stellite 6 (with addition of B_4C) coatings, Colmonoy 88 coatings showed better erosion performance (lower by 1.1 to 2-fold) than to Stellite 6 (with and without B_4C) coatings. Otherwise, no cross-over of SSER values was observed between substrate and LSA coated target materials.

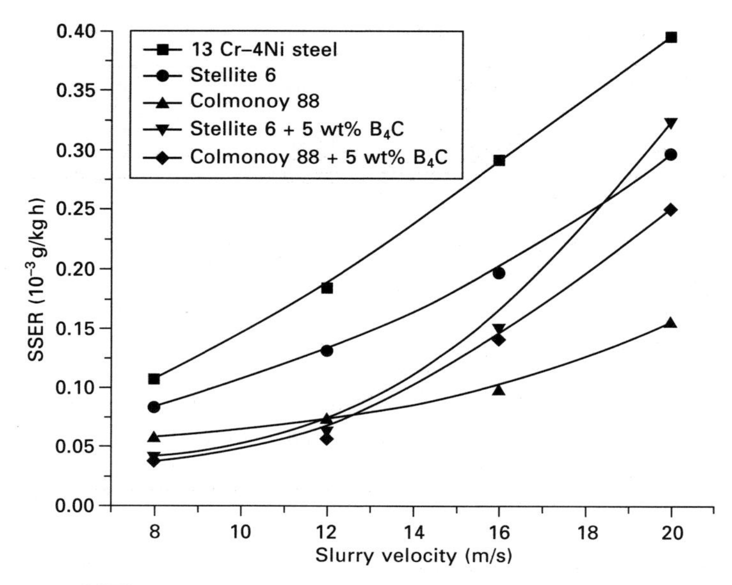

7.52 SSER comparison among all LSA coatings at 30° impacts after 1 h tests.

According to power law curves (Fig. 7.52), the beneficial effect of B_4C added LSA coatings on erosion resistance of turbine components appear to be more pronounced at lower slurry velocities. At a velocity of 8 m/s, B_4C added Stellite 6 and Colmonoy 88 coatings resist approximately 2.4 and 2.8 times better, respectively, than to 13Cr-4Ni steel. At higher velocities, Colmonoy 88 (without addition of B_4C) coatings showed better erosion resistance at least by 3 times compared with 13Cr–4Ni steel. These results confirmed the suitability of B_4C added LSA coatings for low-head turbines, wherein low velocity is involved. Otherwise, Colmonoy 88 coatings are useful for high-head turbine applications.

7.8.2 Erosion mechanisms

According to Thapa (2004), material damages due to plastic deformation and cutting occur simultaneously in actual practice. The ratio of these damage mechanisms depends on slurry (erodent) velocity and impingement angle together with other parameters. Up to a certain velocity (referred to as the 'critical velocity'), the erodents cannot skid in the surface due to friction and so cutting action does not take place easily. As the velocity increases higher beyond critical velocity, both cutting and plastic deformation components increase, which enhances the erosion rate drastically. The modes (ductile or brittle) of erosion behaviour also vary depending on erodent velocity. At low velocity, the erodents do not have enough energy to erode the material by cutting action, but elastic deformation or fatigue effect can be observed (Thapa, 2004).

The eroded surfaces of the substrate and LSA coatings were examined by SEM in order to understand the characteristic features of erosion mechanisms. At lower (8 m/s) slurry velocity, SEM features similar to those observed for impacted by 100 μm erodent size (Figs 7.49(a), (c) and (e) and 7.50(a) and (c)) were observed and, hence, results of lower velocity are not provided. Figure 7.53 shows typical SEM features of the substrate and different LSA coatings eroded at a slurry velocity of 20 m/s. As seen in Fig. 7.53, for the given test conditions, the eroded surfaces showed a higher degree of damage compared with lower velocity impacts.

The substrate steel is characterized by a high degree of chip formation and fracture (Fig. 7.53(a)). The extent of damage was less in Stellite 6 (with and without addition of B_4C) coatings (Fig. 7.53(b) and (c)). The extent of damage was uniform throughout the eroded surfaces of substrate and Stellite 6 (without addition of B_4C) coatings, whereas a scoop of material has been

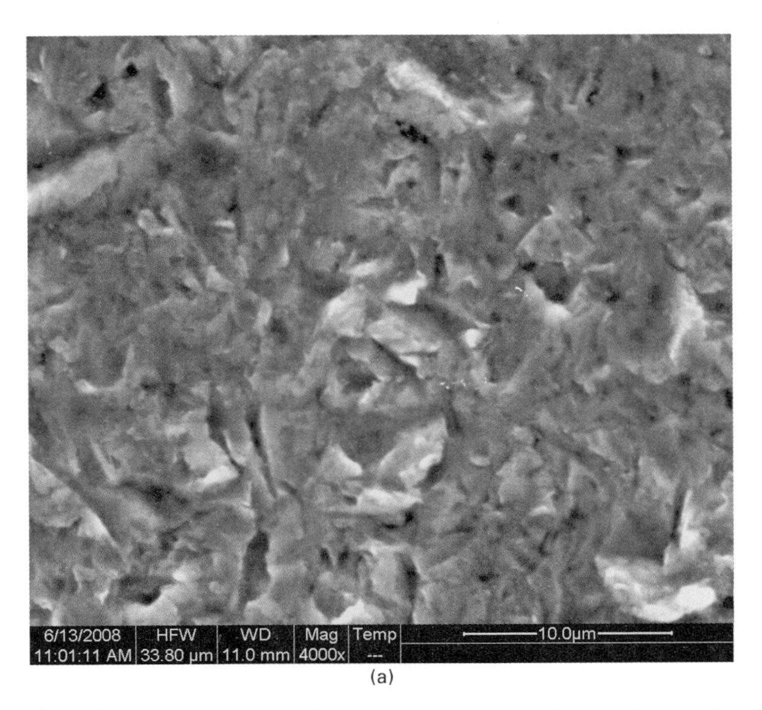

(a)

7.53 SEM morphological features of eroded surfaces of (a) 13Cr–4Ni steel; (b) Stellite 6 (without B_4C) coatings; (c) Stellite 6 + 5 wt% B_4C coatings; (d) Colmonoy 88 (without B_4C) coatings and (e) Colmonoy 88 + 5 wt% B_4C coatings. Impacted at a slurry velocity of 20 m/s.

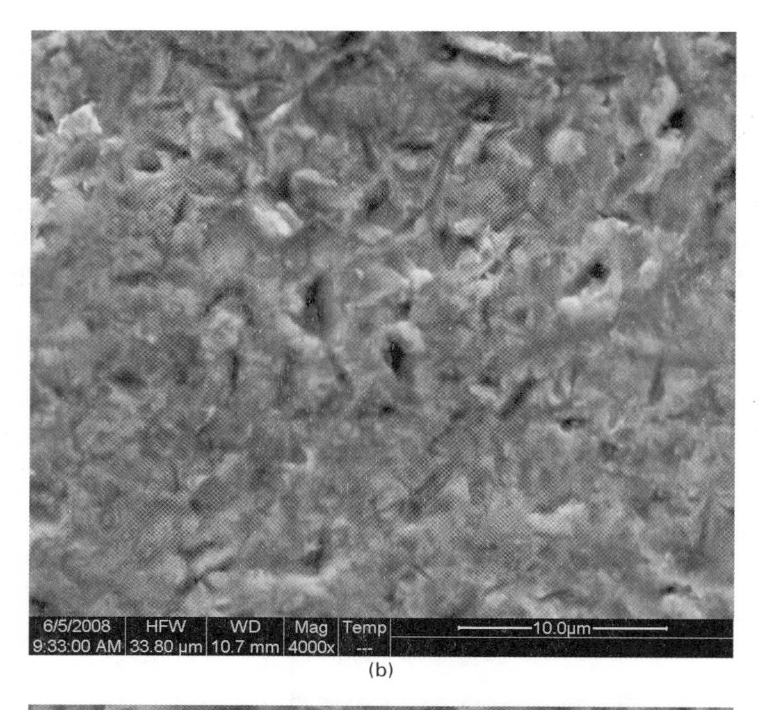

(b)

(c)

7.53 Continued

(d)

(e)

7.53 Continued

removed (indication of severe erosion) at few places for the cases of Stellite 6 + 5 wt% + B_4C coatings (Fig. 7.53(c)). This might be the reason for its low erosion resistance compared with Stellite 6 (without addition of B_4C) coatings. Similarly, Figs 7.53(d) and (e) show that the secondary phases (carbides and borides) are partially and completely dislodged along with severe damages at matrixes for Colmonoy 88 (with and without addition of B_4C). Less fracture of phases were observed and severe deformation was seen at matrix for Colmonoy 88 (without addition of B_4C) coatings (Fig. 7.53(d)). Colmonoy 88 + 5 wt% B_4C coatings, as shown in Fig. 7.53(e), exhibited more severe and uniform damages over the eroded surfaces.

7.9 Effect of slurry concentration

Slurry concentration, also referred to as particle flux rate, is defined as the mass of impacting particles per unit area and time. It is the mass (or volume) of particles present in unit mass (or volume) of fluid. It can also be represented in terms of percentage of particles in a given fluid mass (or volume). Especially for river sedimentation, concentration is presented in term of ppm (parts per million), which is equivalent to mg/litre or kilogram of particles in 1000 m^3 of water (1000 ppm is equivalent to 0.1%). Erosion rate is considered linearly proportional to concentration up to a certain limit, and thereafter reduces due to interference between re-bounding particles and arriving particles. The Himalayan rivers contain very high sediment concentrations, especially during the monsoon season. Although a settling arrangement has been designed according to requirements, a huge quantity of fine sand is impacting turbines each year. Thapa (2004) and Bajracharya *et al.* (2008) have reported that 35–40% of sediment would impact turbine components during flood conditions, and annual average sediment load passing through a turbine would be approximately 1.83 million tonnes, thus causing severe erosion of the turbine components. The maximum concentration recorded during the flood condition (or monsoon) is 2000–60 000 ppm, depending on the place of the hydropower plant. Hence, erosion tests have been conducted on 13Cr–4Ni steel and Colmonoy 88 (without B_4C addition) coatings at four different slurry concentrations (10 to 40 kg/m^3) at very high slurry velocity of 12 m/s using commercial silica erodent of average size 100 μm.

In order to evaluate the effect of slurry concentration on slurry erosion, Colmonoy 88 coatings and substrate 13Cr–4Ni steel were tested at various slurry concentrations at constant slurry velocity (20 m/s) and at 30° impingement angle by using SiO_2 particles of average size 100 μm (10 kg/m^3). Slurry concentration is a very important parameter to understand the erosion performance of components during monsoon times as it is expected that particle concentration will be very high. The test parameters are given in Table 7.11.

Table 7.11 Test parameters to study the effect of slurry concentration

Type of slurry	Material	Slurry concentration (kg/m^3)
Commercial silica sand	1. 13Cr–4Ni steels 2. Colmonoy 88 coatings	10, 20, 30 and 40

7.9.1 SSER plots

The quantitative results of erosion are plotted as SSER against slurry concentration and shown in Fig. 7.54. Higher slurry concentration implies a high possibility of large numbers of erodent particles coming in contact with the surface. The increasing trend of SSER values is due to an increase in kinetic or impact energy of erodent particles ($E = ½\ mV^2$). For a given slurry velocity and erodent size, impact energy is directly proportional to mass of the erodents, and the relation between erosion and concentration is expected to be a linear one. The erosion of actual hydropower turbine components increases linearly with increase in sediment/silt concentration. However, trends of power law or SSER remaining constant with respect to slurry concentration were observed for both the target materials during laboratory tests. This is because, above a certain concentration (critical concentration limit, 30 kg/m^3 for the present work), interference (inter-collision) between oncoming and rebound erodent particles takes place within the erosion chamber. According to Turenne *et al.* (1989), the laminar flow of the slurry jet becomes turbulent near the surface of the target material and the resulting flow conditions lead to an erodent particle cloud, which in turn leads to a reduction in the erosion efficiency corresponding to a limited protection of the surface. In such cases, they suggested that the erosion rate should be measured on the basis of weight of striking erodent particles rather than time duration.

Figure 7.54 also reveals lower SSER values (at least by 2.2 to 3.75) compared with substrate 13Cr–4Ni steel under identical test conditions. Once again, the higher hardness due to presence of various carbides and borides of LSA coatings is attributed to the enhanced erosion resistance observed. Exponent values of substrate and Colmonoy 88 coating are 0.94 and 0.59, respectively.

7.9.2 Erosion mechanisms

Erosion mechanisms of both target materials when tested at 10 kg/m^3 at slurry velocity 8 m/s, were similar to those depicted in Figs 7.49(a), (c) and (e). At 40 kg/m^3 slurry concentration, the extent of damage increased as shown in Fig. 7.55. Commonly observed erosion mechanisms were extensive plastic

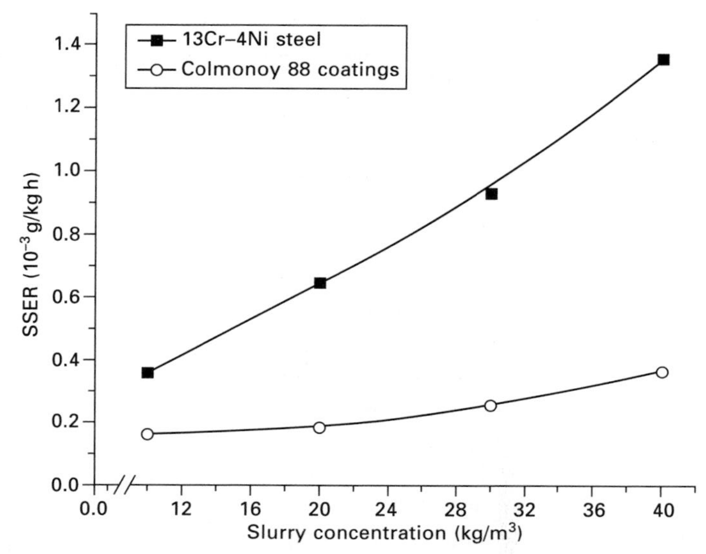

7.54 Slurry concentration dependence of SSER values of 13Cr–4Ni steel and Colmonoy 88 (without B_4C) coatings. Impinged at an angle of 30° for 1 h.

deformation (micro-cutting and cutting) with characteristic chip formation, chip fracture and scratches or plough marks for substrate 13Cr–4Ni steel (Fig. 7.55(a)); progressive fracturing of secondary hard phases/precipitates along with material removal as platelets, pull outs of fractured carbide/boride phases and intact nature of hard phases in a few places were observed for Colmonoy 88 coatings (Fig. 7.55(b)).

7.10 Erosion tests with river sand

In the laboratory, conduction of erosion tests by using sediment (sand collected from actual hydropower plant) slurry is not possible for various reasons. Hence, commercial silica sand, which is representative of river sand, has been utilized in slurry preparation for the laboratory tests, and it is assumed that the finer fraction of these sand particles represents the erodent size distribution of actual river sand particles. Thapa (2004) and Bajracharya *et al.* (2008) have analysed the sediment sand for its composition, and also to determine the percentage by volume of the minerals in the water being carried up to the turbine. They classified the minerals based on Mohs hardness scale. According to this, the minerals that are having 6 to 7 Mohs scale are quartz (average content is 50 to 76%), feldspar (2 to 8%) and other hard minerals such as sillimanite, garnet, ilminite, amphibole and hematite (7%).

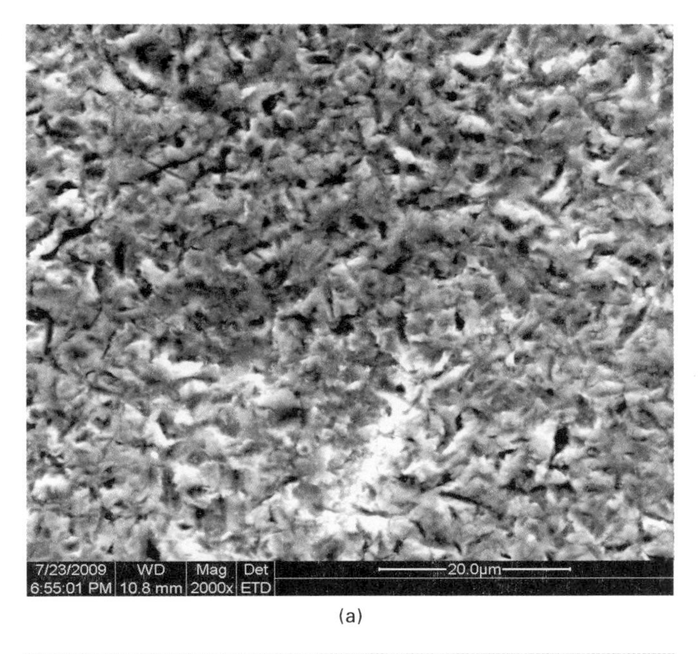

(a)

(b)

7.55 Typical SEM features of eroded surfaces of (a) 13Cr–4Ni steel and Colmonoy 88 (without B_4C) coatings. Slurry velocity of 20 m/s and slurry concentration of 40 kg/m^3.

The remaining constituents of the sediment samples are minerals softer than 5 Mohs scale, which are rock fragments, clay minerals and others. They also reported that minerals harder than 7.5 Mohs scale are not found in the collected river sand samples. The river sand particles also range in shape from the angular to the sub-rounded shape. Hence, impacts with commercial silica sand may or may not represent the actual erosion performances of turbine components.

7.10.1 SSER plots

To understand the effect of river sand on SSER values, erosion tests have been conducted on substrate 13Cr–4Ni steel and basic LSA coatings, i.e. Stellite 6 and Colmonoy 88 (without addition of B_4C) coatings using river sand particles (Table 7.5). The results of river sand impacts are compared with silica sand under identical test conditions and comparison plots are presented in Fig. 7.56.

Two different observations can be made from Fig. 7.56: first, on the mode of erosion behaviour (with respect to impingement angle) and, secondly, on the magnitude differences of SSER values. It is clearly evident that the shape of the erosion curve (mode of erosion behaviour) does not alter even with the impact of river sand, implying that mode of erosion behaviour of these target materials depends mainly on the erodent size and other test conditions.

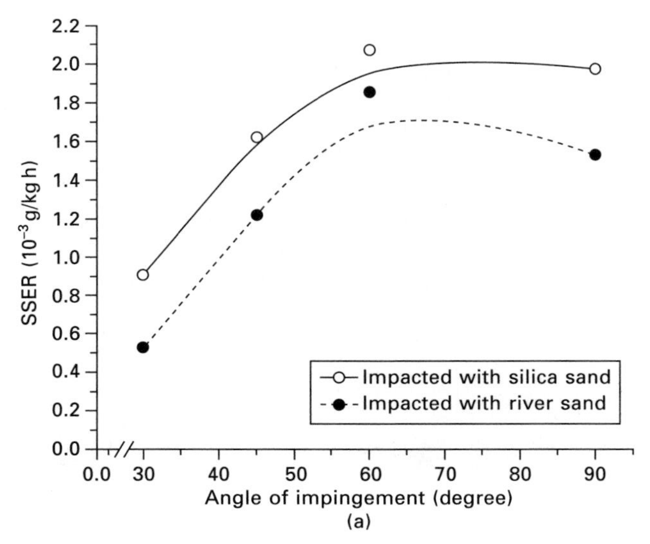

7.56 Effect of impingement angle on SSER values of (a) substrate 13Cr–4Ni steel; (b) Stellite 6; and (c) Colmonoy 88 (without addition of B_4C) coatings. Impacted with two different erodents of sizes < 105 μm for 1 h test duration.

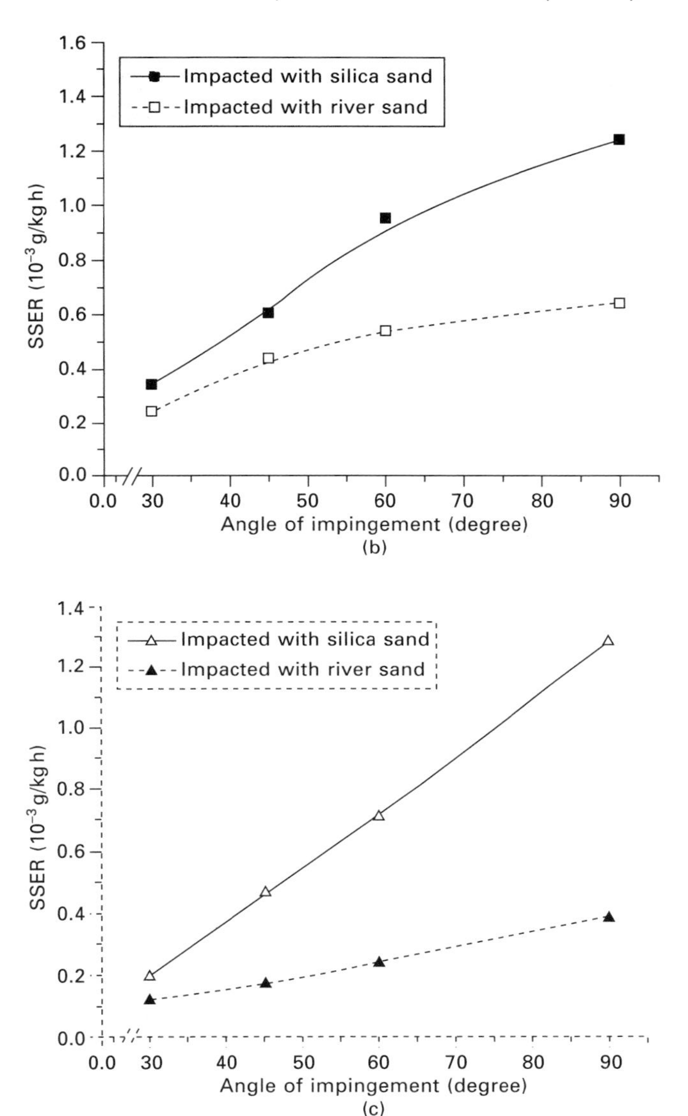

7.56 Continued

For example, the substrate showed the mixed (neither ductile nor brittle) mode of erosion behaviour, since maximum SSER value was observed at 60° for both types of erodent impacts (Fig. 7.56(a)). SSER of LSA coatings approached a maximum at 90°, implying brittle mode of erosion behaviour for the both types of erodent impacts (Figs 7.56(b) and (c)). It is noted from Fig. 7.56(a) that the difference in SSER between silica and river sand was about 1.11 to 1.75 times. Similarly, river sand impacts result in almost 1.37

to 1.92 and 1.64 to 3.35-fold lower SSER values for Stellite 6 and Colmonoy 88 coatings, respectively, compared with commercial silica sand (Figs 7.56(b) and (d)).

To cross-check erosion behaviour of substrate and Colmonoy 88 coatings (as observed in Figs 7.32(a) and (c)), tests have been conducted by impacting with river sand of average size 242 μm and results were compared with commercial silica of average size 375 μm. The effect of impingement angle on SSER is shown in Fig. 7.57.

There was a change in erosion mode, i.e. mixed (neither ductile nor brittle) mode of erosion behaviour for the river impacts compared with the ductile mode of erosion behaviour when impacted with silica sand, as observed for substrate 13Cr–4Ni steel, whereas the brittle mode of erosion behaviour was observed for Colmonoy 88 coatings. The differences in SSER values, measured between 30 to 90° for Colmonoy 88 coatings, were negligible and indicate that SSER is not affected significantly by impinging material for the given test conditions. Similar findings have already been obtained for Colmonoy 88 coatings, when impacted with silica sand of average size 375 (Fig. 7.32(c)).

The change in magnitude (lower values) of SSER for river sand impacts is attributed to two main factors, i.e. shape and size of erodents. Very hard silica sand has a sharp, irregular shape, unlike angular-shaped river sand (Figs 7.23 and 7.24). Relatively, the average size of silica sand was also

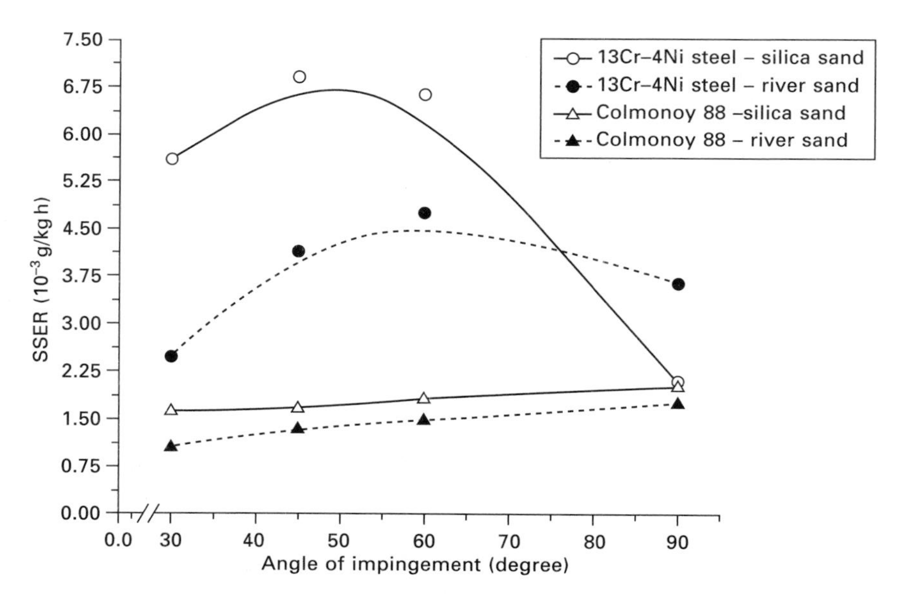

7.57 SSER plots comparison when impacted with two different erodent types (of size 250–355 μm) on substrate 13Cr–4Ni steel and Colmonoy 88 coatings.

greater (Table 7.4). As explained in Sections 7.5 to 7.11, the kinetic or impact energy of erodent increases with increase in erodent size, and hence leads to an increase in the rate of erosion.

From the results of erosion tests conducted with river sand, it is understood that the mode of erosion behaviour of target materials is independent of erodents. However, magnitudes of SSER may be lower with the impacts of river sand than with silica.

7.10.2 Erosion mechanisms

By and large, erosion mechanisms similar to those shown in Figs 7.38 to 7.40 are observed when impacted with river sand, but the degree of deformation or extent of damages is slightly reduced. For example, the deep ploughing marks (indicative of more material loss) were absent in substrate eroded surfaces, whereas presence of fractured phases of carbides and borides was observed for Colmonoy 88 coatings.

7.11 Development of correlation for erosion rate

From the literature, it is noted that power law relations between erosion rate and test parameters have been well accepted because they have been cited most frequently in published literature and, hence, indicate widespread acceptance in the technical community. Typical erosion equations emphasized that erosion rate is a function of various operating parameters, as well as erodent and target properties.

Thapa (2004) and Padhy and Saini (2008) have reported many generalized predictive equations based on experimental and field observations, one such example being $E = aV^b d^c C^d$, where E is erosion rate, V is velocity of particle, d is particle size, C is erodent particle concentration, a is proportionality constant, b, c and d are values of exponents for velocity, erodent particle size and erodent particle concentration, respectively.

It has been observed from erosion experiments that operating parameters such as slurry velocity (V), slurry concentration (C) and impingement angle (α), erodent properties including erodent size (d) and target material properties such as hardness (H), play critical roles in erosion performances of LSA coatings and substrate. The simplified correlation is of following type:

$$E = f\ (V, d, C, \alpha, H) \qquad [7.4]$$

Given the availability of large number of erosion test data, it is possible to produce an empirical relationship (based on above example) between erosion test parameters and erosion rate. The following generalized form of correlation is proposed for the present work:

$$E = a \times (V)^b \times (d)^c \times (C)^d \times (\sin \alpha)^e \times (H)^f \qquad [7.5]$$

where E is erosion rate expressed in g/h, a, b, c, d, e and f are model fitting parameters. H is expressed in terms of Vickers hardness number (HV). To normalize the erosion rate values, obtained erosion rate data have been divided by maximum erosion values. Similarly, erosion test parameters such as V, d and C have been divided by reference values. From the literature and their own measurement, Thapa (2004) and Bajracharya *et al.* (2008) have reported that, depending on the head and type of turbine in actual practice, velocity (V) can vary between 10 and 150 m/s. Also depending on de-silting arrangements and monsoon conditions, size of silt (d) impacting on various turbine components is in the range of 50–125 μm and annual sediment concentration (C) through turbine is almost about 5000–60 000 ppm, respectively. To simplify, the average values of V, d and C are considered while normalizing.

In the present investigation, data obtained from five different target materials (substrate 13Cr–4Ni steel, Stellite 6 (without and with 5 wt% B_4C) and Colmonoy 88 (without and with 5 wt% B_4C) coatings) have been considered while developing a correlation for normalized erosion rate. From the regression analysis of collected test data, it has been found that the relation is of first order between test parameter and erosion rate, with interaction effects being of secondary importance. The values of coefficient (a) and each exponent (b, c, d, e and f) of each erosion parameter in proposed correlation (Equation 7.5) have been evaluated from the regression data in log–log scale. The final form of the correlation for normalized erosion rate is as given in Equation 7.6.

$$E = \frac{2.66 \times 10^4 \times (V)^{1.89} \times (d)^{151} \times (C)^{0.21} \times (\sin \alpha)^{1.47}}{(H)^{1.49}} \qquad [7.6]$$

From Equation 7.6, it is observed that normalized erosion rate is inversely proportional to hardness and directly related to rest of the parameters. The obtained velocity exponent value matches the experimental observations. The normalized erosion rate data generated using Equation 7.6 and experimental normalized erosion rate data are compared and presented in Fig. 7.58.

A good agreement between experimental and theoretical normalized erosion rate was observed. For all target materials, the correlation coefficient (R^2) was about 91% or, in other words, theoretical normalized erosion rate values are lying within a band of ±9% from experimental erosion rate values.

7.12 Conclusions

7.12.1 Summary

13Cr–4Ni martensitic stainless steels were surface modified with two-stage processes of laser surface alloying (LSA) with commercial Co-based Stellite

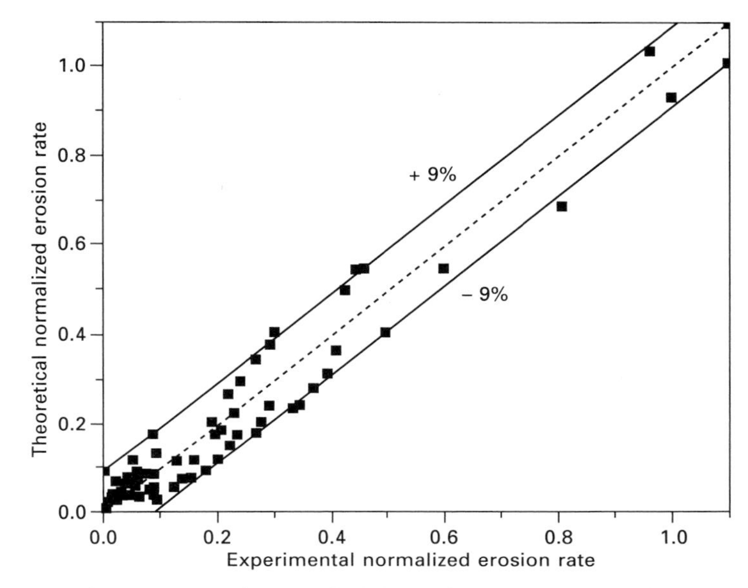

7.58 Comparison of theoretical (equation 7.4) and experimental normalized erosion rate for all target materials.

6 and Ni-based Colmonoy 88 coating powders. A new attempt has also been made to enhance the surface properties of coatings by the addition of various percentages of boron carbide (B_4C). The basic metallurgical aspects, such as microstructural details, hardness distribution across the cross-section and X-ray diffractogram (XRD) of laser surface modified coatings, were studied and reported. Slurry erosion tests have been performed on various laser surface alloyed coatings at room temperature and at various test conditions. Under identical test conditions, the slurry erosion performances of LSA coatings have been compared with substrate 13Cr–4Ni steel. The erosion mechanisms at various test parameters were identified from the extensive studies on eroded surfaces using scanning electron microscopy.

7.12.2 Conclusions

Metallurgical analysis

- Defect-free and about 300 μm thick newer coatings of Stellite 6 and Colmonoy 88 (with and without addition of various percentages of B_4C) have been obtained single-pass (without overlapping) by a two-stage process of laser surface alloying using continuous CO_2 laser system. The LSA parameters are: laser power 3 kW, laser scanning speed 3 mm/s and focusing distance 130 mm. It is understood from interface studies that coatings were continuous with good adhesion between substrate and alloyed region.

- Combined analysis by SEM and EDS on microstructure of Stellite 6 (with and without B_4C) coatings confirmed the fine (< 1 μm) carbides (richer with Cr- and W-phases) at dendrite regions, whereas Co-, Cr- and Fe-rich regions exist within the solid solution. Uniform distribution of coarse (2–10 μm) carbides/ borides (richer by Cr- and W-phases) and matrix regions richer in Ni-, Cr- and Fe-phases were observed in Colmonoy 88 (with and without B_4C) coatings. Volume fractions of secondary phases such as carbides and borides have been increased with the addition of various percentages of B_4C due to melting of B_4C during LSA process.
- XRD analysis of Stellite 6 (with and without B_4C) coatings confirmed the presence of solid solutions of Co, $Cr_{23}C_{26}$, Ni_3Si, Fe_3C and W_2C phases. Colmonoy 88 (with and without B_4C) coatings showed the presence of γ-nickel, Ni_3Si along with $Cr_{23}C_6$, WC, W_2C and (Ni, Fe)$_3$B, W_2B and Cr_5B_3 phases.
- The hardness of the substrate 13Cr–4Ni steel was about 250 HV_{100}. Stellite 6 coatings showed maximum hardness of 405 HV up to 0.3 mm from surface, whereas Stellite 6 + 5 wt% B_4C coatings showed 650 HV_{100} up to 0.2 mm from the top surface. Colmonoy 88, with addition of 5 and 10 wt% B_4C coatings, showed maximum hardness of 505 HV_{100} up to 0.3 mm, 790 HV up to 0.25 mm and 975 HV_{100} up to 0.25 mm from the top surface, respectively. The steep increment in hardness of LSA coatings is due to uniform distribution, presence of various phases and also increased volume fractions of carbides and borides.

Slurry erosion performances

- Characteristic curves of cumulative mass loss vs. time of exposure of all materials showed linear relations, indicating no significant change in erosion mechanisms. Erosion plots (erosion against each test parameter at different time intervals) indicated that erosion behaviour is independent of time, whereas only gradual increment in magnitude of erosion has been observed with time.
- When impacted with SiO_2 sand of size 375 and 100 μm, ductile and mixed (neither ductile nor brittle) mode of erosion behaviour was observed for substrate 13Cr–4Ni steel, mixed and brittle mode of erosion behaviour for Stellite 6 (with and without B_4C) coatings, and only brittle mode of erosion behaviour for the case of Colmonoy 88 (with and without B_4C) coatings with respect to impingement angles.
- Specific slurry erosion rates (SSER) were found to decrease by at least by 2 to 13 times when erodent size changed from 375 to 100 μm. Under

identical test conditions, basic LSA (without B_4C) coatings showed erosion resistance of about 1 to 2 times, whereas B_4C added LSA coatings showed significant enhancement (6 to 14 times) in erosion resistance compared to substrate 13Cr–4Ni steel. The steep increase in erosion resistance is attributed to the increase in hardness (in turn, volume fractions of carbide and boride phases) of B_4C added LSA coatings.

- Erosion rates of substrate and LSA (with and without B_4C) coatings increased due to increase in impact energy of erodents (as impact energy is directly proportional to erodent size[3]). However, the rate of erosion became constant above a certain erodent size for basic LSA (without B_4C) coatings due to shape factors of erodents and properties of substrate. Basic LSA (without B_4C) coatings showed erosion resistance of about 1 to 3 times, increasing up to 2 to 33 times for B_4C added LSA coatings. Again, this is due to increased volume fractions of carbides and borides and tougher matrix due to secondary phases.
- Power law relations were observed between SSER and slurry velocity. The velocity exponent varied between 1.04 to 2.25, depending on the properties of target materials and erosion test conditions. Due to cross-over of SSER values, it is concluded that B_4C added LSA coatings are suitable for low-head turbines, whereas B_4C addition is not necessary for high-head turbines. However, LSA (with and without B_4C) coatings showed better erosion resistance of at least 1.2 to 2.8 times compared with 13Cr–4Ni steel.
- Studies of effect of concentration on SSER of various target materials yielded a power law relationship (instead of linear relation) due to interference effects. The difference between SSER values of Colmonoy 88 (without B_4C) coatings and substrate 13Cr–4Ni steel was about 2.2 to 3.7 times.
- As in the case of commercial silica impacts, similar trends of erosion behaviour were observed for various target materials when impacted with river sand. Hence, mode of erosion behaviour is independent of erodent type, but the magnitude of SSER of target materials depends on it.
- Colmonoy 88 (with and without B_4C) coatings showed at least 1 to 2 times reduced erosion rates compared with Stellite 6 (with and without B_4C) coatings with respect to angle of impingement, erodent size, slurry velocity and erodent type. This is due to differences in type (carbides and borides), morphologies (coarser) and higher volume fractions of secondary phases in Colmonoy 88 coatings compared with the finer carbides in Stellite 6 coatings. The finer carbides in Stellite 6 coatings lead to strengthening of the Co-rich matrix rather than deflection of the erodent.
- 13Cr–4Ni steel and Stellite 6 (with and without B_4C) coatings showing erosion mechanisms such as formation of chips, fracturing of chips, deep

craters with lip formation, cutting and deep ploughing marks associated with excessive plastic deformations were observed under SEM. Progressive fracture of carbides/borides and other complex phases along with crater formations, intact nature and pullout of fractured secondary phases were main operating erosion mechanisms observed for the case of Colmonoy 88 coatings.

- A correlation for normalized erosion rate as a function of velocity, erodent size, slurry concentration, impingement angle and hardness of target materials has been developed based on power law relations by using erosion test data. The correlation coefficient between experimental and theoretical erosion rate was about 91%.

7.13 Acknowledgements

We gratefully acknowledge Shri. V. Ragupathy, Scientist–G and Dr T. R. Aggarwal, Scientist–E, DST, Govt. of India, New Delhi, for sanctioning funds to carry out the work. We are also grateful to Prof. M. S. Ananth, Director, IIT Madras, Chennai and Head of the Dept. of Metallurgical and Materials Engineering, IIT Madras, Chennai, for helping in utilizing the resources in IIT Madras, Chennai. We also acknowledge Dean (IC&SR), Registrar (Finance and Accounts) and the staff of Industrial Consultancy and Sponsored Research, IIT Madras, Chennai for helping in maintaining accounts of the project.

We express our sincere thanks to Dr B. S. Mann, Additional General Manager, Bharat Heavy Electricals Ltd (BHEL), Hyderabad, India, Shri. R. K. Girdonia, Additional General Manager, A. Mandal, Senior Technical Engineer, Hydroturbine Engineering Division, BHEL, Bhopal, India, and Mr P. Nath, Deputy General Manager, BHEL, Ranipur, Hardwar, India, for providing us with the required 13Cr–4Ni steel for carrying out laser surface alloying work. We also acknowledge to Dr Kulvir Singh, BHEL, R and D corporate, Hyderabad, for helping to acquire the 13Cr–4Ni steel.

We sincerely thank Dr G. Sundararajan, Director, ARCI, Hyderabad, for extending the laboratory facilities to us. We sincerely thank to Mr S. M. Shariff, Scientist-C, Mr Shyam Rao, Technician, Dr S. V. Joshi, Associate Director, Dr G. Padmanabham, Associate Director and Head, Center for Laser Processing of Materials (CLPM), Advanced Research Centre for Powder Metallurgy and Newer Materials (ARCI), Hyderabad, India, for their immense help during the processing of laser surface alloying for the research work.

We are grateful to Mr Kumar Ashwani, Construction Engineer, and Mr Rakesh Khatri, Engineer, BHEL, Allain Duhangan HEP, Manali, India, for helping us collect the river sand.

7.14 References

Arya V., K. Vidyasagar, P. Joshi and B. S. Mann (2003), HP HVAF and HVOF coatings to combat erosive and corrosive wear of Indian industry with a specific example of hydro industry, *Surface Engineering, Process Fundamentals and Applications, Vol. 2. Lecture Notes on Surface Engineering*, 23–1 to 23–30.

ASTM G 73 (2004), Standard Practice for Liquid Impingement Erosion Testing.

Bajracharya T. R., B. Acharya, C. B. Joshi, R. P. Saini and O. G. Dahlhaug (2008), Sand erosion of Pelton turbine nozzles and buckets: A case study of Chilime Hydropower Plant, *Wear* **264**, 177–184.

Barber J., B. G. Mellor and R. J. K. Wood (2005), The development of sub-surface damage during high energy solid particle erosion of a thermally sprayed WC–Co–Cr coating, *Wear* **259**, 125–134.

Bhushan B. *Introduction to Tribology*, John Wiley and Sons, New York, 2002.

Bitter J. G. A. (1963), A study of erosion phenomena, Part I and II, *Wear* **6**, 5–21 and 169–190.

Bukhaiti M. A., S. M. Ahmed, F. M. F. Badran and K.M. Emar (2007), Effect of impingement angle on slurry erosion behaviour and mechanisms of 1017 steel and high-chromium white cast iron, *Wear* **262**, 1187–1198.

Carpene E. (2002), Reactive surface processing by irradiation with excimer laser Nd : YAG laser, free electron laser and Ti : sapphire laser in nitrogen atmosphere, *Applied Surface Science* **186**, 195–199.

Chopra V. and S. Arya (1996), Silt – its effect on hydro power components and remedial measures, *Proc. Silt damages to equipment in hydro power stations and remedial measures*, New Delhi, 101–109.

Clark H. M. (1992), The influence of the flow field in slurry erosion, *Wear* **152**, 223–240.

Clark H. M. (2002), Particle velocity and size effects in laboratory slurry erosion measurements OR... do you know what your particles are doing?, *Tribology International* **35**, 617–624.

Clark H. M. and R. B. Hartwich (2001), A re-examination of the 'particle size effect' in slurry erosion, *Wear* **248**, 147–161.

Conde A., F. Zubiri and J. D. Damborenea (2002), Cladding of Ni–Cr–B–Si coatings with a high power diode laser, *Materials Science and Engineering A* **334**, 233–238.

Corengia P., G. Ybarra, C. Moina and A. Cabo (2005), Microstructural and topographical studies of DC-pulsed plasma nitrided AISI 4140 low-alloy steel, *Surface and Coatings Technology* **187**, 63–69.

Damborenea J. D., A. J. Vazquez and B. Fernandez (1994), Laser-clad 316 stainless steel with Ni-Cr powder mixtures, *Materials and Design* **15**, 41–44.

Draper C. W. and C. A. Ewing (1984), Review: laser surface alloying: a bibliography, *Journal of Materials Science* **19**, 3815–3825.

Finnie I. (1972), Some observations on the erosion of ductile metals, *Wear* **19**, 81–90.

Finnie I. (1995), Some reflections on the past and future of erosion, *Wear* **186-187**, 1–10.

Gandhi B. K. and S. V. Borse (2004), Nominal particle size of multi-sized particulate slurries for evaluation of erosion wear and effect of fine particles, *Wear* **257**, 73–79.

Gandhi B. K., S. N. Singh and V. Seshadri (1999), Study of the parametric dependence of erosion wear for the parallel flow of solid–liquid mixtures, *Tribology International* **32**, 275–282.

Goel D. B. and M. K. Sharma (1996), Present state of damages and their repair welding in Indian hydroelectric projects, *Proceedings of Silt Damages to Equipment in Hydro Power Stations and Remedial Measures*, New Delhi, 137–152.

Graham D. H. and A. Ball (1989), Particle erosion of candidate materials for hydraulic valves, *Wear* **133**, 125–132.

Gurumoorthy K. Wear behaviour of nickel-based hardfacing alloys deposited by plasma transferred arc welding, *Ph.D thesis*, IIT Madras, Chennai, 2006.

Haugen K., O. Kvernvold, A. Ronold and R. Sandberg (1995), Sand erosion of wear-resistant materials: erosion in choke valves, *Wear* **186-187**, 179–188.

Hawthorne H. M., B. Arsenault, J. P. Immarigeon, J. G. Legoux and V. R. Parameswaran (1999), Comparison of slurry and dry erosion behaviour of some HVOF thermal sprayed coatings, *Wear* **225–229**, 825–834.

Hotea V., I. Smical, E. Pop, I. Juhasz and G. Badescu (2008), Thermal spray coatings for modern technological applications, *Annals of the Oradea University, Fascicle of Management and Technological Engineering*, 1486–1492.

Hutchings I. M. (1993), Mechanisms of wear in powder technology, *Powder Technology* **76**, 3–13.

Iwai Y. and K. Nambu (1997), Slurry wear properties of pump lining materials, *Wear* **210**, 211–219.

Jiang, W. H. and R. Kovacevic (2004), Slurry erosion resistance of laser clad Fe–Cr–B–Si coatings, *Surface Engineering* **20**, 464–468.

Joshi S. V. and G. Sundararajan, Lasers for metallic and intermetallic coatings, 121–175. In N. B. Dahotre (Ed.), *Lasers in Surface Engineering, Surface Engineering Series* **1**, ASM, 1999.

Juliet P., A. Rouzaud, K. Aabadi, P. Monge-Cadet and Y. Pauleau (1996), Mechanical properties of hard W-C physically vapor deposited coatings in monolayer and multilayer configuration, *Thin Solid Films* **290–291**, 232–237.

Karimi A. and Ch. Verdon (1993), Hydroabrasive wear behaviour of high velocity oxyfuel thermally sprayed WC-M coatings, *Surface and Coatings Technology* **62**, 493–498.

Karimi A., Ch. Verdon, J. L. Martin and R. K. Schmid (1995), Slurry erosion behaviour of thermally sprayed WC-M coatings, *Wear* **186–187**, 480–486.

Kashani H., A. Amadeh and H. M. Ghasemi (2007), Room and high temperature wear behaviors of nickel and cobalt base weld overlay coatings on hot forging dies, *Wear* **262**, 800–806.

Kwok C. T., H. C. Man and F. T. Cheng (1998), Cavitation erosion and pitting corrosion of laser surface-melted stainless steels, *Surface and Coatings Technology* **99**, 295–304.

Kwok C. T., H. C. Man and F. T. Cheng (2000), Cavitation erosion and pitting corrosion behaviour of laser surface-melted martensitic stainless steel UNS S42000, *Surface and Coatings Technology* **126**, 238–255.

Lathabai S., M. Ottmuller and I. Fernandez (1998), Solid particle erosion behaviour of thermal sprayed ceramic, metallic and polymer coatings, *Wear* **221**, 93–108.

Lim L.C., Q. Ming and Z. D. Chen (1998), Microstructures of laser-clad nickel-based hardfacing alloys, *Surface and Coatings Technology* **106**, 183–192.

Lin H. C., S. K. Wu and C. H. Yeh (2001), A comparison of slurry erosion characteristics of TiNi shape memory alloys and SUS304 stainless steel, *Wear* **249**, 557–565.

Lin M. C., L. S. Chang, H. C. Lin, C. H. Yang and K. M. Lin (2006), A study of high-speed slurry erosion of NiCrBSi thermal-sprayed coating, *Surface and Coatings Technology* **201**, 3193–3198.

Lo K. H. (2003), Effects of laser treatment on cavitation erosion and corrosion of AISI 440 C martensitic stainless steel. *Materials Letter* **58**, 88–93.

Lynn R. S., K. K. Wong and H. M. Clark (1991), On the particle size effect in slurry erosion, *Wear* **149**, 55–71.

Magnee A. (1995), Generalized law of erosion: application to various alloys and intermetallics, *Wear* **181–183**, 500–510.

Majumdar J. D. and I. Manna (2003), Laser processing of materials, *Sadhana* **28**, 495–562.

Manisekaran T. (2005), Slurry erosion studies on surface modified 13Cr–4Ni hydroturbine steel, *M. S. Thesis*, IIT Madras, Chennai.

Manisekaran T., M. Kamaraj, S. M. Sharrif, and S. V. Joshi (2007), Slurry erosion studies on surface modified 13Cr–4Ni steel: effect of angle of impingement and particle size, *Journal of Materials Engineering and Performance* **16(5)**, 567–572.

Mann B. S. (1997), Boronizing of cast martensitic chromium nickel stainless and its abrasion and cavitation–erosion behaviour, *Wear* **2008**, 125–131.

Mann B. S. (1998), Erosion visualization and characteristics of a two dimensional diffusion treated martensitic stainless steel hydrofoil, *Wear* **217**, 56–61.

Mann B. S. (2000), High-energy particle impact wear resistance of hard coatings and their application in hydroturbines, *Wear* **237**, 140–146.

Mann B. S. and V. Arya (2001), Abrasive and erosive wear characteristics of plasma nitriding and HVOF coatings: their application in hydroturbines, *Wear* **249**, 354–360.

Mann B. S. and V. Arya (2002), An experimental study to correlate water jet impingement erosion resistance and properties of metallic materials and coatings, *Wear* **253**, 650–661.

Mann B. S. and V. Arya (2003), HVOF coating and surface treatment for enhancing droplet erosion resistance of steam turbine blades, *Wear* **254** (2003) 652–667.

Mann B. S., V. Arya, A. K. Maiti, M. U. B. Rao and P. Joshi (2006), Corrosion and erosion performance of HVOF/TiAlN PVD coatings and candidate materials for high pressure gate valve application, *Wear* **260**, 75–82.

Matsumura M. and B. E. Chen (2002), Erosion-resistant materials, In: Duan C. G. and Karelin V. Y. (eds), *Abrasive erosion and corrosion of hydraulic machinery*, Imperial College Press, London, 235–314

Mesa D. H., A. Toro, A. Sinator and A. P. Tschiptschin (2003), The effect of testing temperature on corrosion–erosion resistance of martensitic stainless steels, *Wear* **255**, 139–145.

Ming Q., L. C. Lim and Z. D. Chen (1998), Laser cladding of nickel-based hardfacing alloys, *Surface and Coatings Technology* **106**, 174–182.

Ocelík V., U. D. Oliveira, M. D. Boer and J. T. M. D. Hosson (2007), Thick Co-based coating on cast iron by side laser cladding: analysis of processing conditions and coating properties, *Surface & Coatings Technology* **201**, 5875–5883

Oliveira A. S. C. M. D., R. Vilar and C. G. Feder (2002), High temperature behaviour of plasma transferred arc and laser Co-based alloy coatings, *Applied Surface Science* **201**, 154–160.

Padhy M. K. and R. P. Saini (2008), A review on silt erosion in hydroturbines, *Renewable and Sustainable Energy Reviews* **12**, 1974–1987.

Pang W., H. C. Man and T. M. Yue (2005), Laser surface coating of Mo–WC metal matrix composite on Ti6Al4V alloy, *Materials Science and Engineering A*, **390**, 144–153.

Podgornik B. and J. Vizintin (2001), Wear resistance of pulse plasma nitrided AISI 4140 and A355 steels, *Materials Science and Engineering A* **315**, 28–34.

Przybylowicz J. and J. Kusinski (2000), Laser cladding and erosive wear of Co–Mo–Cr–Si coatings, *Surface and Coatings Technology* **125**, 13–18.

Qian M., L. C. Lim, Z. D. Chen and W. I. Chcn (1997), Parametric studies of laser cladding process, *Journal of Materials Science Technology* **63**, 590–593.

Rao D. R. K., B. Venkataraman, M. K. Asundi and G. Sundararajan (1993), The effect of laser surface melting on the erosion behavior of low alloy steel, *Surface Coating Technology* **58**, pp. 85–92.

Santa J. F., J. C. Baena and A. Toro (2007), Slurry erosion of thermal spray coatings and stainless steels for hydraulic machinery, *Wear* **263**, 258–264.

Santa J. F., L. A. Espitia, J. A. Blanco, S. A. Romo and A. Toro (2009), Slurry and cavitation erosion resistance of thermal spray coatings, *Wear* **267**, 160–167.

Schaff P. (1995), Laser nitriding of iron by excimer laser irradiation in air and N_2 gas *Materials science and Engineering A* **197**, 1–4.

Sharma M. K. (1996), Problems in repair welding of 13/4 stainless steel and their remedial measures, *Proceedings of Silt Damages to Equipment in Hydro Power Stations and Remedial Measures*, New Delhi, 166–181.

Singal S. K. and R. Singh (2006), Impact of silt on hydroturbines, *Proceedings of Himalayan Small Hydropower Summit*, Dehradun, 200–207.

Srinivasan K. (2008), Laser technology and materials processing, *IIM Metal News* **11**, 15–23.

Stachowiak G. W. (2000), Particle angularity and its relationship to abrasive and erosive wear, *Wear* **241**, 214–219.

Stachowiak G. W. and A. W. Batchelor (2001), *Engineering Tribology*, Elsevier, Amsterdam.

Steen W. M. (1991), *Laser Material Processing*, Springer.

Steen W. M. and K. G. Watkins (1993), Coating by laser surface treatment, *Journal De Physique IV* **3**, 581–590.

Stein K. J., B. S. Schorr and A. R. Marder (1999), Erosion of thermal spray MCr–Cr_3C_2 cermet coatings, *Wear* **224**, 153–159.

Thapa B. (2004), Sand erosion in hydraulic machinery, *Ph.D thesis*, University of Science and Technology (NTNU), Norway.

Tinamin S. (2003), Impact wear behaviour of laser hardened 2CrB martensitic stainless steel, Wear **255**, 444–455.

Tiziani A., L. Giordano, P. Matteazzi and B. Badan (1987), Laser Stellite coatings on austenitic stainless steels, *Materials Science and Engineering* **88**, 171–175.

Truscott G. F. (1972), Literature survey of abrasive wear in hydraulic machinery, *Wear* **20**, 29–50.

Tu J. P., J. Pan, H. X. Zhao and H. Fukunag (1999), Slurry erosion resistance of $Al_{18}B_4O_{33}$ whisker-reinforced AC4C Al alloy matrix composites, *Materials Science and Engineering A* **263**, 32–41.

Tucker T. R., A. H. Clauer, I. G. Wright and J. T. Stropki (1984), Laser-processed composite metal cladding for slurry erosion resistance, *Thin Solid Films* **118**, 73–84.

Turenne S., M. Fiset and J. Masounave (1989), The effect of sand concentration on the erosion of materials by a slurry jet, *Wear* **133**, 95–106.

Turenne S., Y. Chatigny, D. Simard, S. Caron and J. Masounave (1990), The effect of abrasive particle size on the slurry erosion resistance of particulate-reinforced aluminium alloy, *Wear* **141**, 147–158.

Vilar, R. (1999), Laser alloying and cladding, *Materials Science Forum* **301**, 229–252.

Wei R., E. Langa, C. Rincon and J. H. Arps (2006), Deposition of thick nitrides and carbonitrides for sand erosion protection, *Surface and Coatings Technology* **201**, 4453–4459.

Wood R. J. K. (1999), The sand erosion performance of coatings, *Materials and Design* **20**, 179–191.

Wood R. J. K., B. G. Mellor and M. L. Binfield (1997), Sand erosion performance of detonation gun applied tungsten carbide/cobalt-chromium coatings, *Wear* **211**, 70–83.

Wu P., C. Z. Zhou and X. N. Tang (2003), Microstructural characterization and wear behavior of laser cladded nickel-based and tungsten carbide composite coatings, *Surface and Coatings Technology* **166**, 84–88.

Xu G. J. and M. Kutsuna (2006), Cladding with Stellite 6-WC using a YAG laser robot system, *Surface Engineering* **22**, 345–352.

Xu G., M. Kutsuna, Z. Liu and H. Zhang (2006), Characteristics of Ni-based coating layer formed by laser and plasma cladding processes, *Materials Science and Engineering A* **417**, 63–72.

Yarrapareddy E., S. Zekovic, S. Hamid and R. Kovacevic (2006), The development of nickel-tungsten carbide functionally graded materials by a laser-based direct metal deposition process for industrial slurry erosion applications, *Proceedings of Institute of Mechanical Engineer, Part B: Journal of Engineering Manufacture*, **220**, 1923–1936.

Yellup J. M. (1995), Laser cladding using the powder blowing technique, *Surface and Coatings Technology* **71**, 121–128.

Zhang D. and X. Zhang (2005), Laser cladding of stainless steel with Ni–Cr_3C_2 and Ni–WC for improving erosive–corrosive wear performance, *Surface and Coatings Technology* **190**, 212–217.

Zhang J., M. O. W. Richardson, G. D. Wilcox, J. Min and X. Wang (1996), Assessment of resistance of non-metallic coatings to silt abrasion and cavitation erosion in a rotating disk test rig, *Wear* **194**, 149–155.

Zhong M., W.-J. Liu, K. Yao, J.-C. Goussain, C. Mayer and A. Becker (2002), Microstructural evolution in high power laser cladding of Stellite 6+WC layers, *Surface and Coatings Technology* **157**, 128–137.

Zielinski A., H. Smolenska, W. Serbinski, W. Konczewicz and A. Klimpel (2005), Characterization of the Co-base layers obtained by laser cladding technique, *13th International Scientific Conference on Achievements in Mechanical and Materials Engineering*, 723–726.

Zu J. B., G. T. Burstein and I. M. Hutchings (1991), A comparative study of the slurry erosion and free-fall particle erosion of aluminium, *Wear* **149**, 73–84.

8
Laser surface alloying (LSA) of copper for electrical erosion resistance

P. K. WONG and C. T. KWOK, University of Macau, China and H. C. MAN and F. T. CHENG, The Hong Kong Polytechnic University, China

Abstract: In order to enhance the electrical erosion and corrosion resistance and hence the lifespan of copper for applying as the electrical contacts, surface modification employing various metals such as Cr, W, Ni, Al and Ti on copper was attempted using laser surface alloying with a 2.5 kW continuous wave (CW) Nd : YAG (yttrium aluminum garnet) laser. The microstructure, hardness, electrical erosion and corrosion resistance and interfacial contact resistance (ICR) of the laser-alloyed copper were compared with those of cold-drawn pure copper. Among all specimens, the laser-alloyed copper with Ti shows the most significant improvement in electrical erosion while the electrical conductivity can be preserved.

Key words: laser surface alloying, electrical erosion, corrosion, electrical contact materials.

8.1 Introduction

Electrical contacts are conductive devices that make or break electrical circuits. Electrical contacts can be made of metals, alloys or composites, depending on the application, and should possess the following characteristics:

- high electrical conductivity to minimize the heat generated during the passage of electricity;
- high thermal conductivity to dissipate both resistive and arc heat;
- high corrosion and chemical resistance to prevent degradation;
- high melting temperature to avoid arc erosion and metal transfer;
- high hardness to bear wear applications;
- high environmental friendliness;
- high abundance and low cost.

Copper and its alloys are often widely used as critical electrical contact components such as overhead catenaries and current collectors for electric railway systems and commutators of electric motors and generators. Overhead catenaries constitute an important traction system of overhead contact wires used in conjunction with sliding pantographs (current collectors) to supply electrical energy to moving trains. The increasing speed of modern high-speed trains has become a challenging problem in the design of electric railway systems over the world. Particular attention has been paid to the contact between the catenaries and pantographs because electric contact is indispensable to reliable operation of trains. High-speed trains running on the electric railways will result in severe erosion or wear of the catenaries and collector strips and affect their service life. In recent technological developments of electric railway systems, major efforts have been devoted to speeding up trains, prolonging lifetime and reducing maintenance costs of the overhead catenaries and collector strips.

To withstand the harsh service conditions in real-life applications, electrical contact materials for the catenaries and collector strips should possess high hardness and mechanical strength, high electrical erosion and corrosion resistance, and excellent electrical conductivity. The service lifespan of the contact members generally depends on:

- types of material at contact;
- operating conditions such as sliding speed, contact force and current intensity;
- level of arcing and/or sparking [1]; and
- environmental factors such as applications in tunnels or in open space [2].

Conventionally, hard-drawn copper is often used as overhead catenary for railway systems because of its excellent electrical and thermal conductivities [3] and moderately low price but its limitations are low hardness, susceptibility to electrical erosion and atmospheric corrosion, which lead to shortening of its service life [3–5], interruption of the system and safety problems. In the literature, several studies have reported the electrical erosion behavior of the precipitation hardenable copper alloys (CuCr, CuZr, CuCrZr, CuNiSiCr, CuBe, CuBeNi, CuAg and CuAgCr) and copper matrix composite (Cu–La_2O_3) for use as electrical contact materials [6–10]. Some of them were reported to have improved resistance to electrical erosion due to their high hardness; however, gradual damage of the electric contact materials is inevitable for long-term service and results in regular replacement of the parts.

Remanufacturing is the ultimate form of recycling. It conserves not only the raw materials but also much of the value added during the processes required to manufacture new products [11,12]. Laser remanufacturing technology is based on the application of lasers and can be classified into two major types: laser surface treating and laser forming (Fig. 8.1). In recent years, laser surface treating has been widely used for remanufacturing the engineering components with the merits of refined microstructure, strong metallurgical bond, minimum thermal distortion and high processing speed, precision and versatility. It brings great economic and social benefits including lower energy consumption, reduction of pollution, saving of precious materials

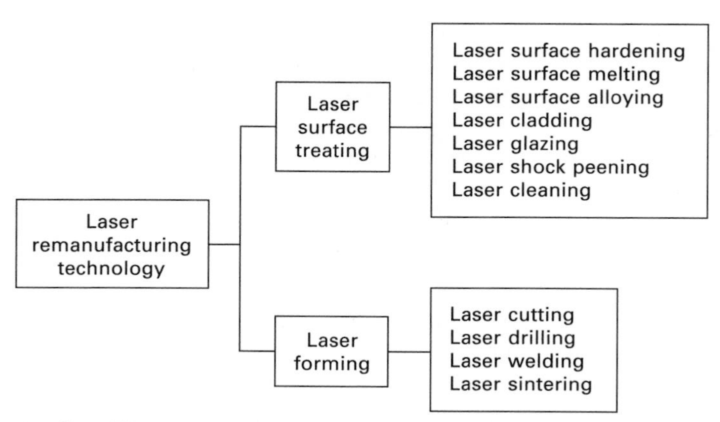

8.1 Classification of laser remanufacturing technology.

and high value added to the components. In many circumstances, high resistance to erosion and corrosion are only required on an applied surface. Laser surface treating can enhance the surface properties of the components against corrosion, erosion, wear, oxidation or fatigue while the mechanical, electrical and thermal properties of the bulk can be preserved. Laser surface alloying (LSA) is a technique that involves melting of a pre-/co-deposited layer of alloying material(s) along with a part of the underlying substrate to form an alloyed zone (AZ) confined to a shallow depth from the surface within a very short interaction time. With the addition of alloying element(s), LSA can produce tailor-made surface compositions which are not attainable by other methods. LSA is a feasible technique for fabricating coatings and repairing undersized and worn electric contact components, especially at localized regions suffering from electrical erosion or corrosion. It can be used to substitute expensive powder metallurgical process, for which it is not easy to control the degree of compaction and the deterioration of electrical conductivity of the contact materials.

In fact, laser surface treating of copper is extremely challenging due to its high reflectivity to infra-red wavelength and high thermal conductivity. There are very few studies on laser surface treatment of copper for electrical contact applications reported in the literature. Hirose and Kobayashi firstly reported that Cu–Cr alloyed layers of thickness 100 to 200 μm were produced by LSA of Cr powder beds or a plasma spray-coating on the Cu substrate with a 2.5 kW CO_2 laser (beam diameter of 1.2 mm, transverse speed of 0.33 mm/s, He shielding) for improving hardness and wear resistance at elevated temperatures without sacrificing its electrical properties [13]. The Cr concentration of the alloyed layers is approximately 20 wt%. Owing to the limited solid solubility between Cr and Cu, LSA causes pure Cr particles to precipitate and be finely dispersed upon rapid cooling. The alloyed layers had higher hardness (120 HV) than pure Cu (60–70 HV) due to the dispersion of the Cr particles. Moreover, the hardness of the alloyed layers was 3–4 times that of pure Cu at 873 K. Later, similar work was investigated by Majumdar and Manna who attempted LSA of Cu substrate with Cr for improving abrasion and erosion resistance using a continuous wave (CW) CO_2 laser [14,15]. It has been reported that rapid quenching in the LSA process has extended the solid solubility of Cr in Cu to about 4 at% [14]. The AZ consists of dispersed Cr-rich particles in the Cr-alloyed Cu-rich solid solution and is consistent with the findings of Hirose and Kobayashi. The laser alloyed Cu–Cr possesses 2–3 times increase in hardness and the wear resistance was significantly improved. Moreover, laser surface melting/remelting (LSM) of a powder metallurgically manufactured 48Cu–48Cr–4Fe (wt%) contact material was studied by Geng and co-workers [16]. After LSM, the refined spherical Cr phase was uniformly dispersed in the Cu-rich matrix. Increased compactness and refined microstructure of the remelted alloy yielded increased hardness (by 80%), wear resistance, and a reduced friction coefficient as compared with the base material.

In order to enhance the hardness, wear resistance and hence the service life of copper as electrical contacts, Ng and co-workers reported that laser cladding of Ni (intermediate layer) and Mo (top layer) on Cu as a sandwich layer of Mo/Ni/Cu using 2.5 kW CW Nd : YAG (yttrium aluminum garnet) laser was successfully fabricated for overcoming the difficulties arising from the large difference in thermal properties and the low mutual solubility between Cu and Mo [17]. Excellent bonding between the clad layer and the Cu substrate was ensured by strong metallurgical bonding. The hardness of the surface of the clad layer is 7 times higher than that of the Cu substrate. Pin-on-disc wear tests also showed that the abrasive wear resistance of the clad layer was also improved by a factor of 7 as compared with the Cu substrate. The specific electrical contact resistance of the clad surface was about $5.63 \times 10^{-7}\,\Omega\,cm^2$ as compared to the negligible resistance of the pure Cu.

W–Cu alloys are widely used for heavy-duty electrical contacts and arc-resistant electrodes. Recently, studies on selective laser melting (SLM) and direct metal laser sintering (DMLS) of W–Cu alloys were reported [18,19]. Li and co-workers reported that SLM of W–10 wt%Cu alloy with a 100 W CW/modulated fiber laser resulted in complete melting of Cu but without melting of W [18]. The binder (Cu) was molten and infiltrated the W particles, then solidified as a continuous solid phase with uniformly distributed W particles in the Cu matrix. On other hand, Gu and co-workers reported that DMLS of a composite system consisting of 40 wt% submicron W–20Cu powder and 60 wt% micron Cu powder was performed with a 2 kW CW CO_2 laser to produce a series of regularly shaped W-rim/Cu-core structures [19]. The formation of the W-rim was attributed to the combined action of the clockwise thermal Marangoni flow and the counterclockwise solutal one, which is induced by temperature gradient or chemical concentration at a solid/liquid interface. The repulsion forces between the W particles in the Cu liquid prevent the W-rim from merging, thereby forming the Cu–core after solidification. The difference in resultant microstructure by SLM and DMLS is ascribed to different processing routes. In both studies, only the microstructural evolution of the W–Cu systems was addressed. However, fabrication of W–Cu surface coatings and wear testing were not attempted.

So far no systematic research has been carried out on LSA of Cu to improve its electrical erosion and corrosion resistance. In the present work, LSA of Cu with various metals including Ni, Al, Cr, W and Ti using a 2.5 kW Nd : YAG laser is attempted to improve its electrical erosion and corrosion resistance without sacrificing its electrical properties.

8.2 Experimental details

Hard-drawn pure copper (Cu) in the form of plate was cut into specimens with dimensions of 25.5 mm × 15 mm × 6.5 mm. The specimens were first ground with 220-grit SiC paper prior to LSA for removing surface oxide and enhancing adhesion to preplaced powder. The specimens were ultrasonically cleaned in ethanol and then in distilled water and dried. Reagent grade pure Cr, W, Ni, Al and Ti powders were separately mixed with 4 wt% polyvinyl alcohol as a binder and then pasted on the Cu substrate by a paintbrush. The thickness of the preplaced layer was controlled to about 0.2 mm because coatings thicker than 0.3 mm will absorb excessive laser energy and burn down [16]. The physical, mechanical and electrical properties of various metals are listed in Table 8.1.

LSA of precoated specimens was carried out using a 2.5 kW CW Nd : YAG laser system (Lumonics MW2000) at a power of 1.8 kW, a beam diameter of 2 mm and a scanning speed of 15 mm/s. Argon flowing at 20 l/min was used as the shielding gas to prevent oxidation during laser surfacing. The alloyed surfaces were obtained by running parallel tracks with 50% overlap between adjacent tracks.

Table 8.1 Physical, mechanical and electrical properties of various metals [9]

Metal	Density (g/cm^3)	Softening temperature (°C)	Melting temperature (°C)	Thermal conductivity (W/m K)	Resistivity (nΩ m)	Brinell hardness (H_B)
Cu	8.9	190	1083	397	17	35
Cr	7.0	600 [7]	1615	91.3	125	90
W	19.3	1000	3390	147	52.8	350
Ni	8.8	520	1452	88.5	70	70
Al	2.7	150	660	238	28	27
Ti	4.5	500	1668	21.6	420	110 [8]

After LSA, the specimens were sectioned and etched with acidified ferric chloride solution (25 g $FeCl_3$, 25 ml HCl and 100 ml H_2O) for microstructural examination by means of scanning electron microscopy (SEM, Hitachi S-3400N). Compositional analysis along the depth of the alloyed layers and at specific microstructural locations was carried out by energy-dispersive X-ray spectrometry (EDS, Horiba EX-250). The phases present in the surface layer were identified by X-ray diffractometry (XRD, Bruker D8 Advance). The hardness profiles along the depth of the surface layer were measured with a microhardness tester at a load of 9.8 N and a loading time of 10 s. At least three sets of indentations were taken at each location, and the average hardness values were reported with errors of ±5% approximately. In addition, the interfacial contact resistance (ICR) was measured by the modified Davies method, described elsewhere [20].

Electrical erosion (sliding wear) testing for the laser-alloyed and untreated Cu was conducted using a pin-on-disc tribometer (Fig. 8.2). The specimens were forced to slide against a counterface steel rotating disc in air under unlubricated condition at a sliding speed of 10 km/h, a load of 20 N and a d.c. current intensity of 10 A. Owing to the presence of electric current, an electric arc will be produced and arc erosion was observed as shown in Fig. 8.3. The weight loss of the specimens was measured using an electronic balance with an accuracy of ±0.1 mg. The weight loss ΔW (in g) of the specimens in time t (in hour) was recorded. The volume loss ΔV (in mm^3) and the erosion rate E (mm^3/h) were calculated by:

$$\Delta V = 1000 \frac{\Delta W}{\rho} \quad [8.1]$$

and

$$E = \frac{\Delta V}{t} \quad [8.2]$$

where ρ is the density of Cu or laser-alloyed layers in g/cm^3. The densities of the laser-alloyed layers were estimated by the rule of mixture. The wear resistance R (in h/mm^3) is defined as:

$$R = \frac{1}{E} \quad [8.3]$$

In order to study the electrochemical corrosion behavior of the laser-alloyed specimens, open-circuit potential (OCP) measurement and potentiodynamic polarization test were performed in 3.5 wt% NaCl solution (open to air) at 25±1 °C using a PAR VersastatII potentiostat conforming to ASTM Standard G5-92 [21]. All potentials were measured with respect to a saturated calomel electrode (0.244 V versus SHE at 25 °C) as the reference electrode. Two parallel graphite rods served as the counter electrode for current measurement. After the OCP measurement for 2 h, the potential was increased at a rate of 0.167 mV/s, starting from 0.2 V below the OCP and stopped at 0.8 V. From

(a)

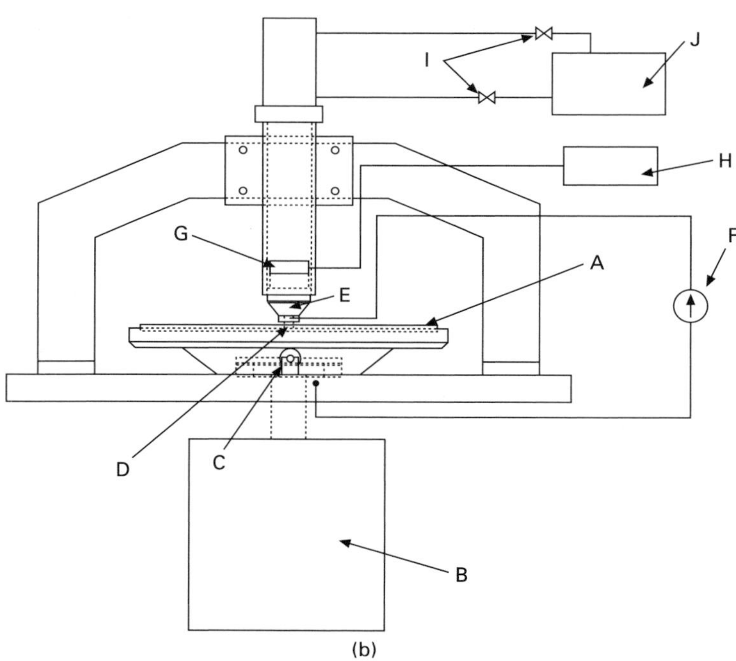

(b)

8.2 (a) Overall view of pin-on-disc tribometer and (b) schematic diagram: (A) rotating disc; (B) high speed motor; (C) bearing support; (D) specimen; (E) specimen holder; (F) power supply; (G) force sensor; (H) manometer; (I) oil control valves; (J) oil pressure pump and tank.

8.3 Electrical erosion testing of Cu showing arc erosion.

the polarization curve, the corrosion current density (I_{corr}) were extracted by Tafel extrapolation. The corrosion morphologies of the specimens after polarization test were studied by SEM.

8.3 Microstructural analysis

Owing to the high reflectivity of Cu to Nd : YAG laser (90%) and high thermal conductivity (395 W mK^{-1}), it is difficult to melt the surface by direct irradiation using the Nd : YAG laser. In the present work, LSA of Cu was achieved by preplacing the various kinds of metal powder on its surface which was easier to melt owing to its higher absorptivity to the laser beam, and the subsequently molten metal absorbed further laser energy. Heat absorbed by the melting metal layers was then effectively transferred to the Cu substrate. Melting, intermixing and rapid solidification then resulted in the formation of an alloyed zone with the desired properties.

The specimens of Cu laser-alloyed with Cr, W, Ni, Al, Ti, are designated as LA–Cr–Cu, LA–W–Cu, LA–Ni–Cu, LA–Al–Cu and LA–Ti–Cu respectively. The transverse cross-section and microstructure of the laser-alloyed specimens are shown in Figs 8.4 to 8.8 and the compositional profiles and XRD patterns in Figs 8.9 and 8.10, respectively. The thicknesses of the alloyed layers are given in Table 8.2. A strong metallurgical bond was formed between the laser-alloyed layers and the Cu substrate. Cr and W are almost undiluted except at the interface because of the large difference in melting temperature and the low solubility of Cr and W in Cu. Except for LA–W–Cu, the surface layers were free of cracks and porosity. This shows that the convection flows are sufficient to eliminate the porosities in the preplaced layer [22]. The

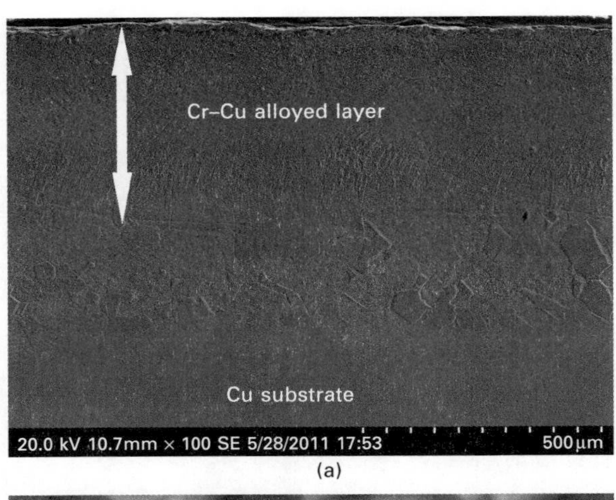

(a)

(b)

(c)

8.4 Laser-alloyed specimen LA–Cr–Cu showing (a) cross-sectional view; (b) alloyed zone; and (c) boundary between alloyed zone and substrate.

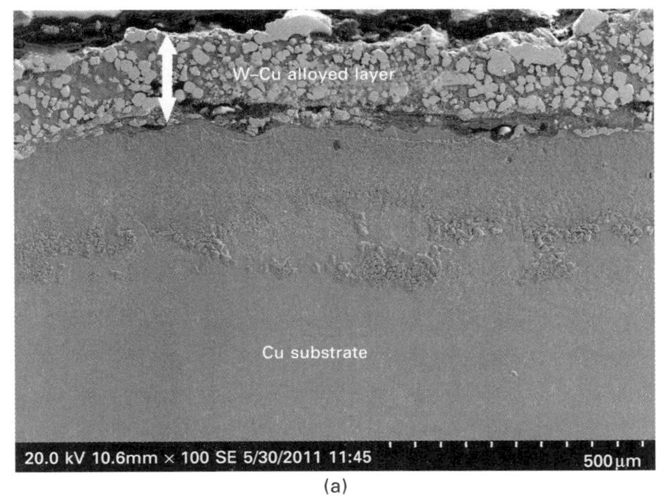

(a)

(b)

8.5 Laser-alloyed specimen LA–W–Cu showing (a) cross-sectional view; (b) alloyed zone; (c) boundary between alloyed zone and substrate; and (d) EDS line scan across the W and Cu phases.

composition and resultant microstructure of the laser-alloyed layers depend on the thermal properties, mutual solubility of the alloy systems according to the relevant binary phase diagrams (Cu–X, where X represents the alloying element) [23] and laser processing parameters.

8.3.1 LSA of Cu with Cr

From the equilibrium binary phase diagram of Cu–Cr system [23], solid solutions in both Cr and Cu have very limited solid solubility. In laser-alloyed

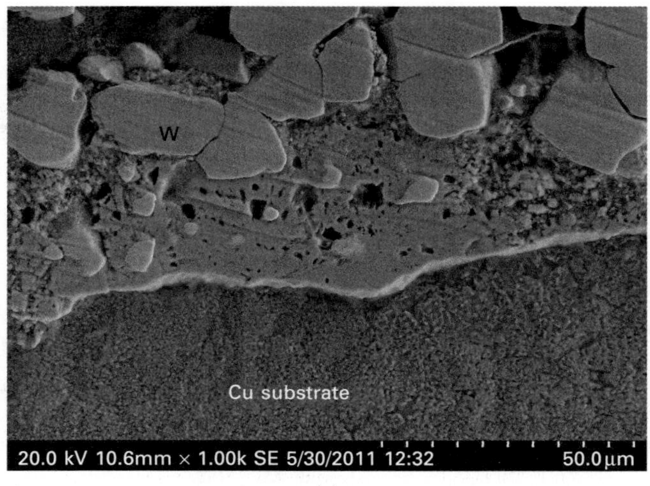

(c)

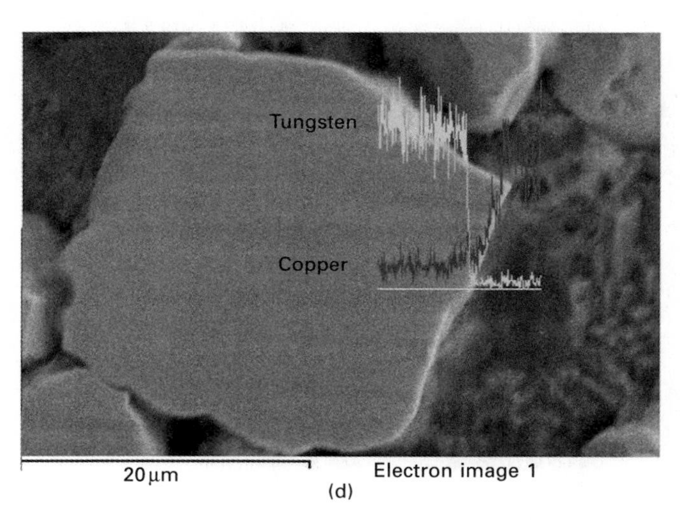

(d)

8.5 Continued

specimen LA–Cr–Cu with Cr content exceeding the eutectic composition (1.5 wt%), pure Cr particles precipitated and were finely dispersed. The Cu-rich matrix with 5.9 wt% Cr is a supersaturated solid solution caused by rapid cooling as shown in the EDS results (Fig. 8.4b). In the AZ of LA–Cr–Cu, as shown in Fig. 8.4a, numerous fine spherical Cr particles (0.5–1 μm) were dispersed in the Cu-rich matrix whereas fewer Cr particles were observed near the boundary of the AZ and Cu substrate. From the compositional profile of LA–Cr–Cu, the average Cr content in the overall specimen was

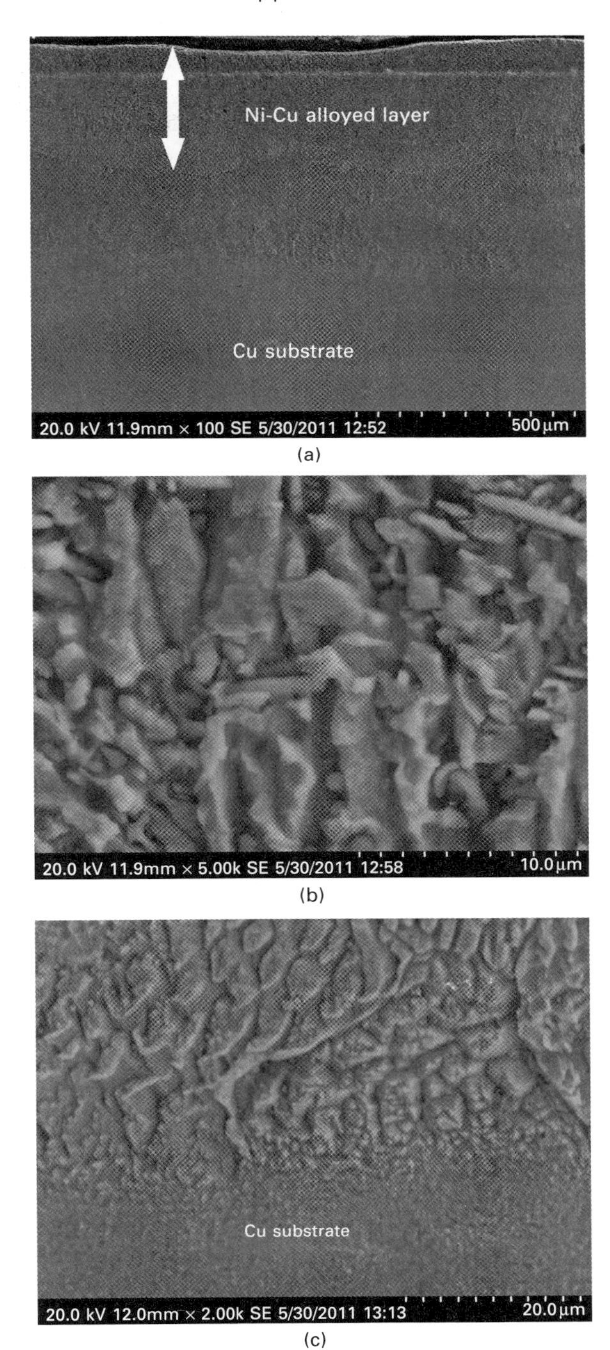

(a)

(b)

(c)

8.6 Laser-alloyed specimen LA–Ni–Cu showing (a) cross-sectional view; (b) alloyed zone; and (c) boundary between alloyed zone and substrate.

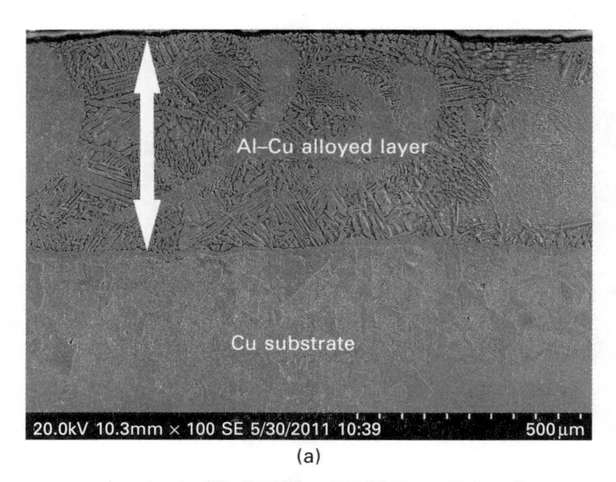

(a)

(b)

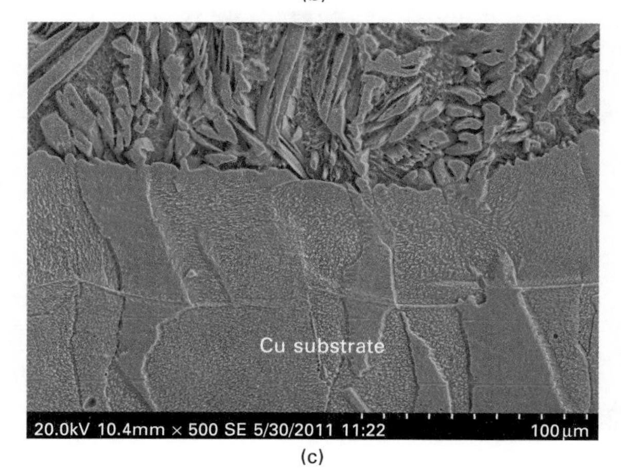

(c)

8.7 Laser-alloyed specimen LA–Al–Cu showing (a) cross-sectional view; (b) alloyed zone; and (c) boundary between alloyed zone and substrate.

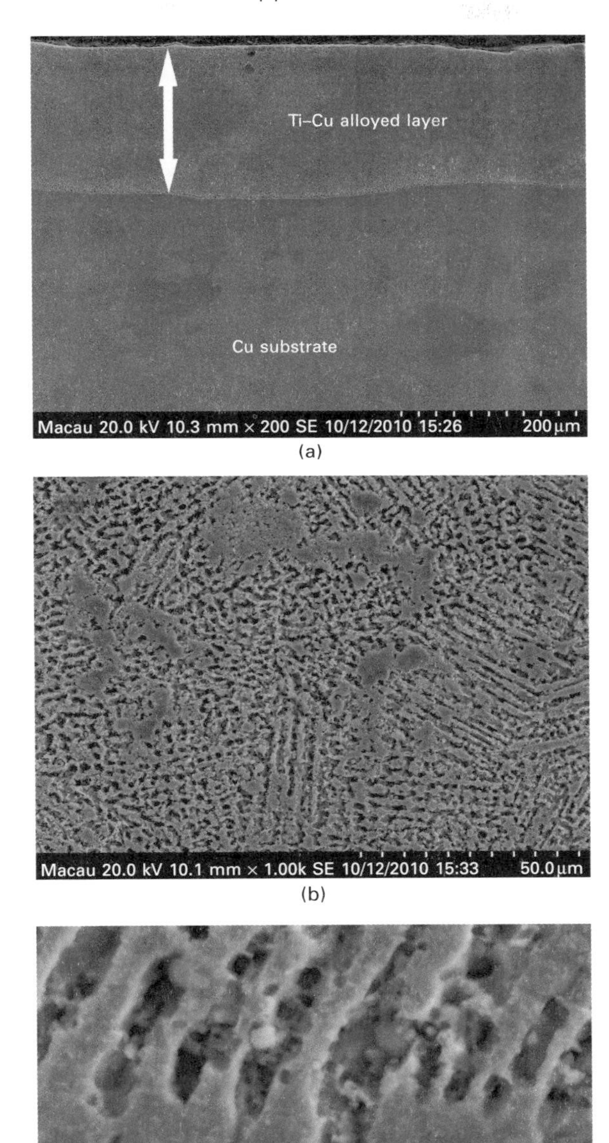

8.8 Laser-alloyed specimen LA–Ti–Cu showing (a) cross-sectional view; (b) alloyed zone; and (c) boundary between alloyed zone and substrate.

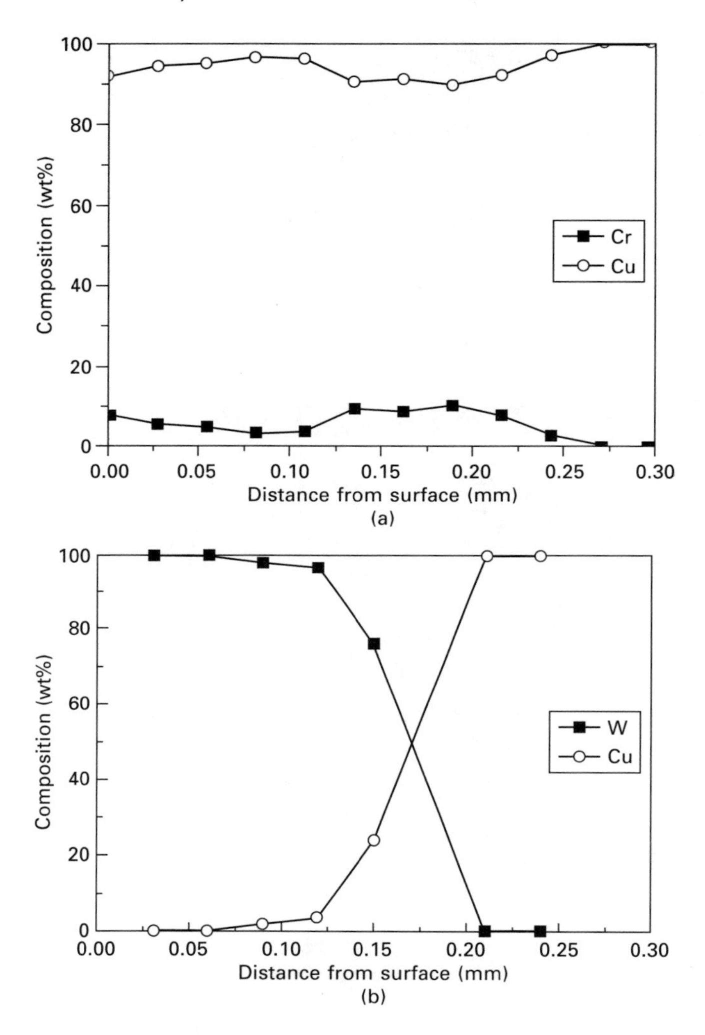

8.9 Compositional profiles of (a) LA–Cr–Cu; (b) LA–W–Cu; (c) LA–Ni–Cu; (d) LA–Al–Cu; and (e) LA–Ti–Cu.

approximately 10 wt%. From the XRD patterns of LA–Cr–Cu, distinct major Cu peaks and a Cr minor peak are observed (Fig. 8.10a).

8.3.2 LSA of Cu with W

During LSA, laser energy is directly absorbed by the melting W layer and the heat is transferred to the underlying Cu substrate. W and Cu were molten and resolidified. Owing to the mutual insolubility, low wettability and large

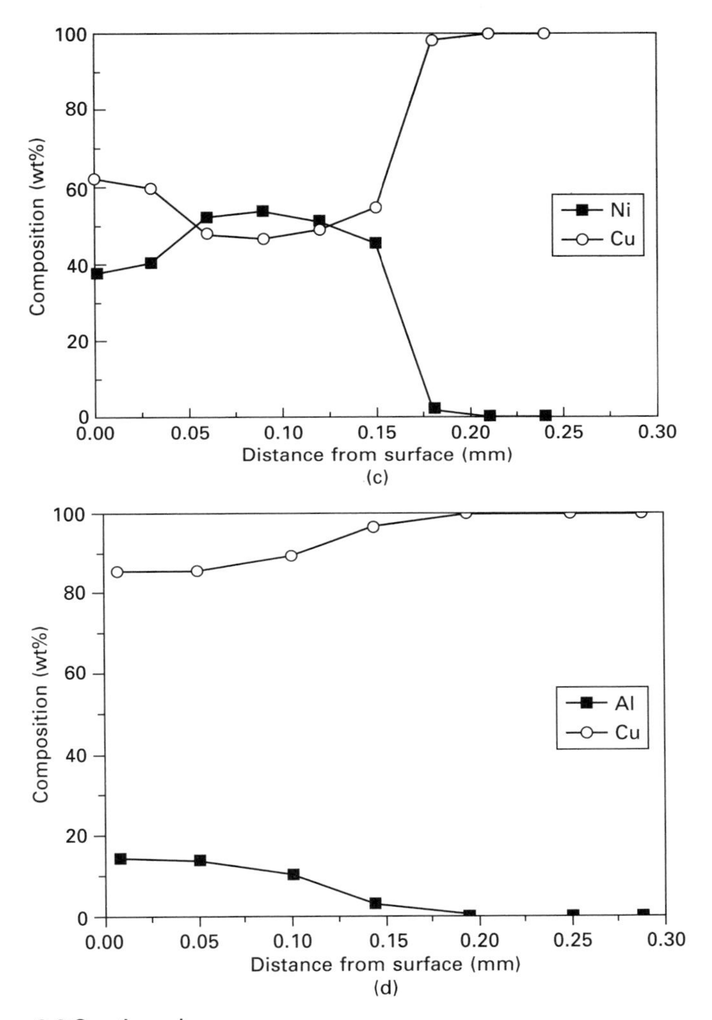

8.9 Continued

difference in melting temperature, intermixing of W and Cu is difficult. A relatively rough surface can be seen from the cross-sectional view of the alloyed layer (Fig. 8.5a) and some pores are present. For LA–W–Cu, large W particles with average size of 30 μm are uniformly dispersed in a Cu matrix as shown in Fig. 8.5b. From the XRD patterns shown in Fig. 8.10b, it is found that the individual peaks of Cu and W are detected in the laser-alloyed specimen LA–W–Cu. In the AZ, the volume fraction of W (90%) is much higher than that of Cu. The compositional profile of LA–W–Cu shows high W content (Fig. 8.9b). From Fig. 8.5d, EDS line scan across the two phases reveals that the brighter contrast region is the W-rich phase

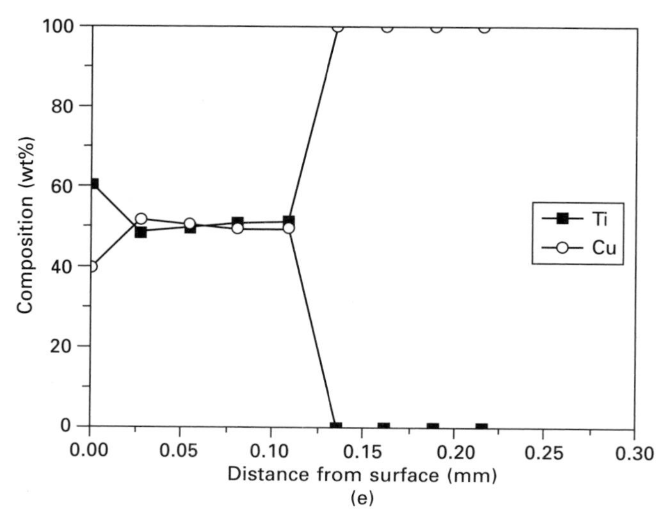

8.9 Continued

while the darker phase is Cu-rich. This microstructure is quite different from the W-rim/Cu-core structure fabricated by DMLS because of the different processing mode [19].

8.3.3 LSA of Cu with Ni

Since Cu–Ni (both are face centered cubic (fcc) structure) is a binary isomorphous system [23], Cu and Ni have no solid solubility limit. The average Ni content in the AZ of LA–Ni–Cu is about 50%. Owing to rapid solidification after LSA process, a very fine dendritic structure is observed, as shown in Fig. 8.6b. From the XRD patterns of LA–Ni–Cu (Fig. 8.10c), peaks of solid solution of Cu–Ni were detected, indicating that homogeneous solid solution was formed in the laser-alloyed layer.

8.3.4 LSA of Cu with Al and Ti

Referring to the phase diagram of the Cu–Al system [23], the maximum solid solubility of Al in Cu was about 8 wt%. This solubility caused the Cu–Al peaks (compared with Cu) to shift to smaller Bragg angles. From the XRD patterns of LA–Al–Cu, peaks of solid solution of Cu–Al and intermetallic Al_2Cu were detected in the laser-alloyed layer, probably due to the solid solution supersaturated with 17 wt% Al (Fig. 8.9d). Manna *et al.* reported the presence of precipitates such as Al_2Cu, in addition to the formation of solid solution of Al in Cu, in the laser-alloyed specimen fabricated by

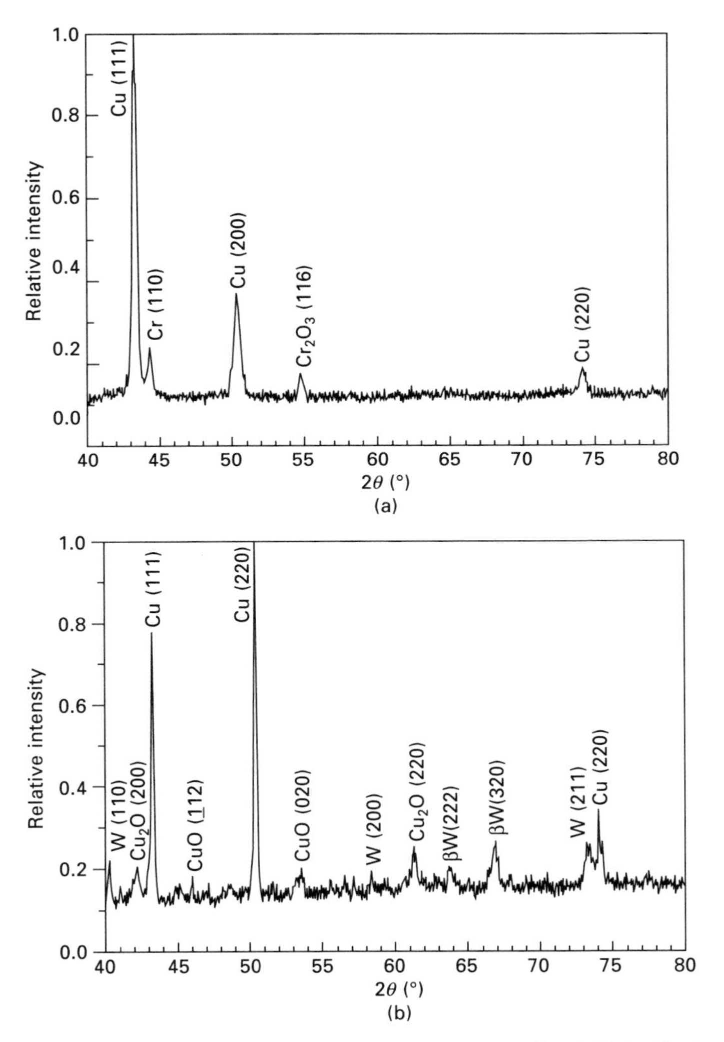

8.10 XRD patterns of (a) LA–Cr–Cu; (b) LA–W–Cu; (c) LA–Ni–Cu; (d) LA–Al–Cu; and (e) LA–Ti–Cu.

higher-energy density laser pulse irradiation [24]. This leads to the formation of novel microstructures, including extended (metastable) solid solution through melting, intermixing and rapid diffusion within an extremely short period of time. The present microstructure is consistent with that obtained by Manna *et al.*

In an attempt reported by Bateni *et al.* [25], Cu–Ti intermetallic phases were formed by physical vapor deposition of Ti on Cu followed by diffusion annealing. Microstructural examination revealed that the coatings consisted

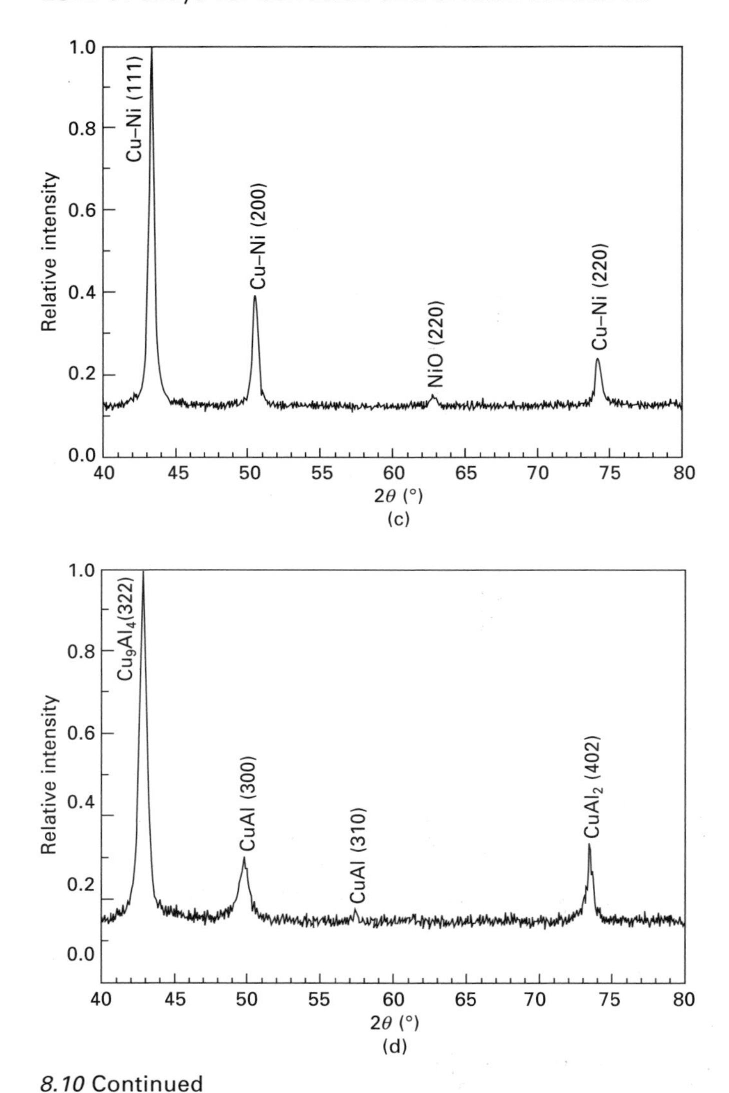

8.10 Continued

of α-Ti and $CuTi_2$, Cu_4Ti phases. Intermetallics such as CuTi and $CuTi_2$ were also obtained by rapid solidification during micropyretic synthesis [26]. For LA–Ti–Cu, peaks of solid solution of Cu–Ti and intermetallic phases (Cu_2Ti and Cu_4Ti) were detected in the XRD patterns of LA–Ti–Cu. As Ti has an extremely high affinity for oxygen, a very stable and highly adherent protective oxide film (rutile, TiO_2) was formed on its surface after LSA, as depicted in Fig. 8.10e.

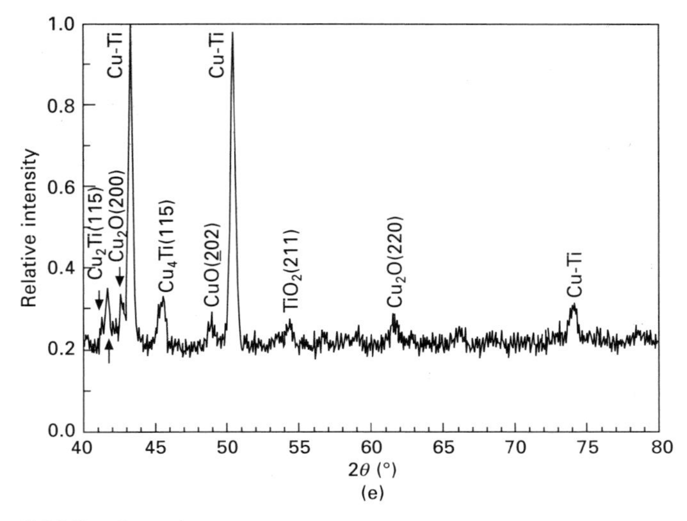

8.10 Continued

Table 8.2 Thickness, hardness and wear resistances of Cu and laser-alloyed specimens

Specimen	Thickness (mm)	Average hardness (HV)	Erosion resistance R (h/mm^3)	Normalized erosion resistance R_N (with respect to Cu) (h/mm^3)
Cu	–	82	0.065	1.00
LA–Cr–Cu	0.45	124	0.092	1.42
LA–W–Cu	0.18	152	0.280	4.31
LA–Ni–Cu	0.50	140	0.073	1.12
LA–Al–Cu	0.50	120	0.067	1.03
LA–Ti–Cu	0.18	619	0.490	7.53

8.4 Hardness and strengthening mechanisms

Figure 8.11 shows the hardness profiles along the cross-section of the laser-alloyed specimens and the average hardness is summarized in Table 8.2. All alloyed layers show increased hardness, in the range of 124 to 619 HV as compared with 82 HV for pure Cu. Compared with Cu, the hardness of LA–Cr–Cu, LA–W–Cu, LA–Ni–Cu, LA–Al–Cu and LA–Ti–Cu has increased by factors of 1.5, 1.9, 1.7, 1.5 and 7.5 respectively. The laser-alloyed layers were strengthened by the following mechanisms as proposed by Geng *et al.* [16]:

- *Solid solution strengthening*. For the laser-alloyed specimen, the solubility limit of solute (X) in solvent (Cu) is extended. The supersaturating of solute X in Cu distorts the crystal lattice, resulting in an increase in hardness.

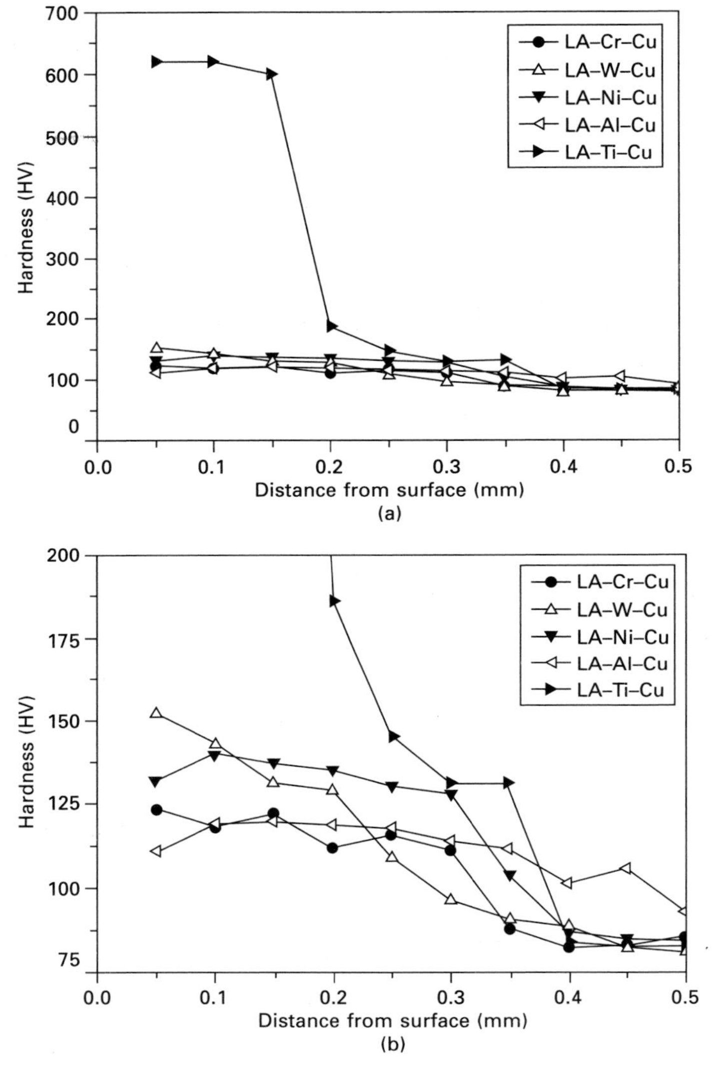

8.11 Hardness profiles of (a) all laser-alloyed specimens and (b) enlargement in scale for specimens with lower hardness.

- *Fine grain strengthening*. Rapid solidification of the melt pool in LSA results in much finer grains. This gives rise to the strengthening effects by the fine grain or phase boundary.
- *Dispersion strengthening*. LSA allows the hard intermetallic and other secondary phases to refine greatly and disperse well throughout the Cu matrix as a result in strengthening.
- *Dislocation strengthening*. Rapid melting and resolidification of the

specimens increase the dislocation density in the AZ and hence increase the hardness.

All laser-alloyed specimens (e.g. LA–Ni–Cu) are strengthened by fine grain and dislocation strengthening due to rapid solidification after LSA. In addition to these two mechanisms, the increase in hardness of LA–Cr–Cu has also been attributed to solid solution and dispersion strengthening (due to the presence of individual fine Cr particles). LA–W–Cu was strengthened by solid solution and dispersion strengthening by large volume fraction of large W particles.

Owing to the hard intermetallic phases (Cu_2Ti and Cu_4Ti) for LA–Ti–Cu and Al_2Cu for LA–Al–Cu, these two laser-alloyed specimens were mainly strengthened by dispersion and solid solution strengthening. Kac *et al.* reported that LSA of Cu–10%Al–4%Fe–2%Mn bronze with Ti exhibited an increase in hardness from 230 V to 467 HV and improved wear resistance [27]; however, the strengthening and wear mechanisms are not known.

8.5 Electrical erosion behavior and damage mechanism

The plots of cumulative volume loss and erosion rate versus time are shown in Fig. 8.12. Accompanying the increase in hardness of the laser-alloyed specimens, the electrical erosion resistance increased as compared with the untreated Cu. The ranking of electrical erosion resistance R is:

$$\text{LA–Ti–Cu} > \text{LA–W–Cu} > \text{LA–Cr–Cu} > \text{LA–Ni–Cu} > \text{LA–Al–Cu} \sim \text{Cu}$$

The normalized electrical erosion resistances (R_N) of the laser-alloyed specimens (with respect to Cu) are shown in Table 8.2. The laser-alloyed specimen LA–Ti–Cu show the most significant improvement in R, followed by LA–W–Cu. LA-Ti-Cu possesses the highest erosion resistance due to the dispersion of hard intermetallic phases (Cu_2Ti and Cu_4Ti) and solid solution strengthened Cu–Ti alloy. In addition, the titanium oxide film (TiO_2) formed on the surface of the as-alloyed Cu–Ti layer can prevent adhesion from the steel disc, thus reducing adhesive wear [28]. Moreover, the presence of TiO_2 on the surface of the specimen as a result of oxidation during the electrical erosion process could efficiently reduce the erosion loss [29].

Compared with the worn surface of Cu (Fig. 8.13a), a much lower degree of damage is observed in LA–Ti–Cu (Fig. 8.13c). Moreover, LA–W–Cu also possesses a high erosion resistance due to the hard W phase (Fig. 8.13b) with the highest softening temperature. It was reported that hard W particles were pulled out of the Cu matrix during the sliding wear test [5, 30]. However, a strong metallurgical bond was formed between the Cu and W phases in the laser-alloyed layer so such a pullout is not observed. Both LA–Ni–Cu and

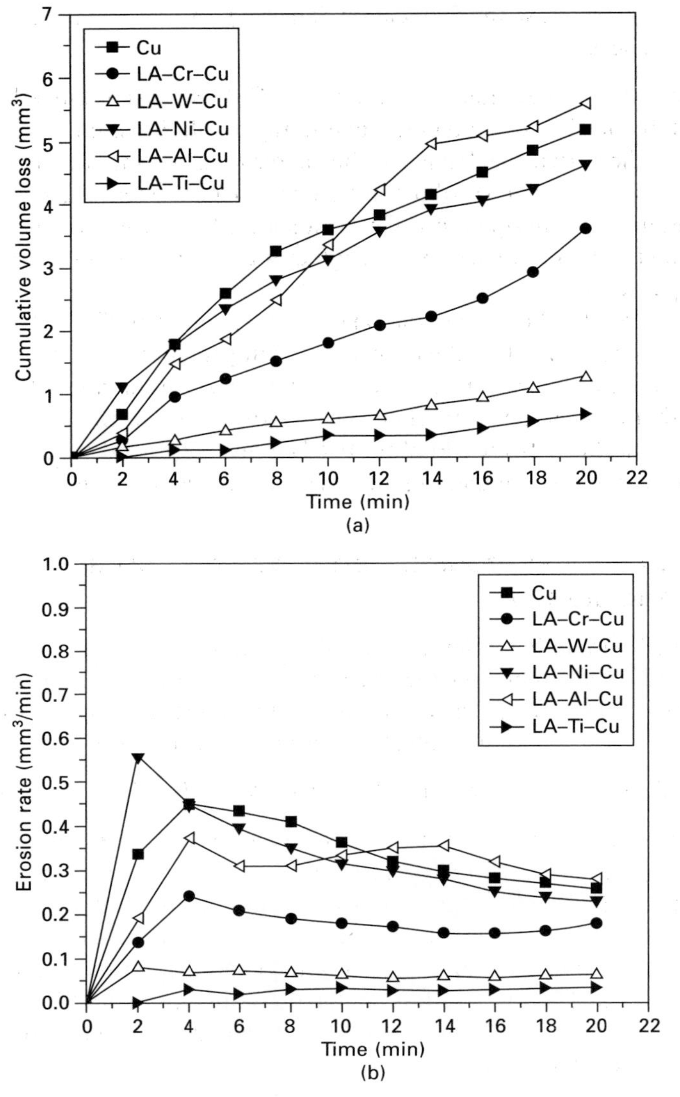

8.12 Plots of (a) cumulative volume loss and (b) erosion rate against time for various specimens.

LA–Al–Cu have a slight improvement in the electrical erosion resistance due to the slight increment in hardness of the laser-alloyed layers. It was reported when 60 wt% of Al is mixed with Cu substrate, hard Al_2Cu phases can be formed, the hardness can be up to 250 $HV_{0.2}$ and the wear loss can decrease by a factor of 15 compared with untreated copper [31]. As LA–Al–Cu possesses around 18 wt% Al, only a small amount of hard Al_2Cu is observed,

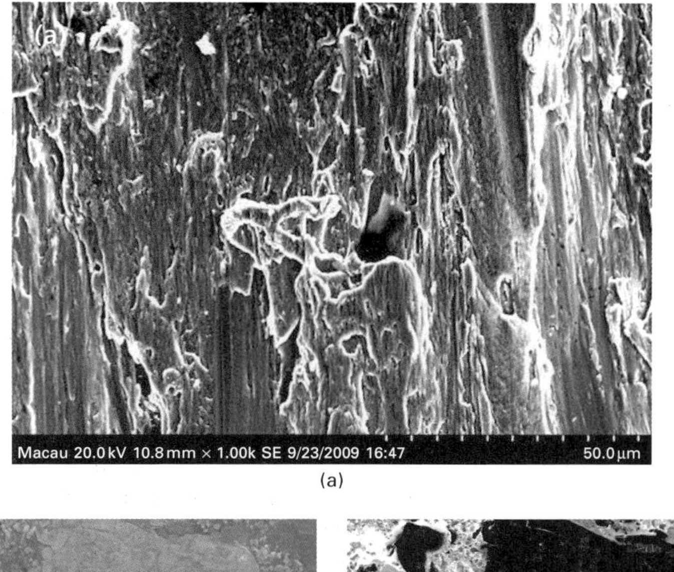

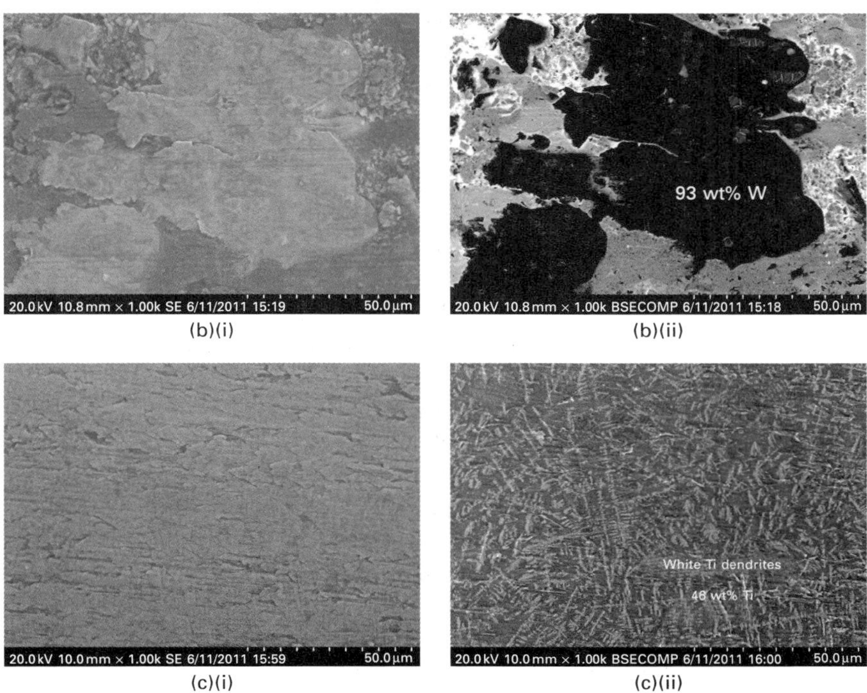

8.13 Worn surfaces of (a) Cu; (b) LA–W–Cu: (i) scanning electron (SE) micrograph, (ii) backscattered electron (BSE) micrograph; and (c) LA–Ti–Cu: (i) SE micrograph, (ii) BSE micrograph.

as shown in Fig. 8.10d. It is noticed that LA–Cr–Cu has a lower hardness than LA–Ni–Cu, but LA–Cr–Cu has a higher electrical erosion resistance than LA–Ni–Cu. It was reported that the alloyed layers had 3–4 times the hardness of pure Cu at 600 °C [13]. Thus the dispersion of Cr particles strengthened the structure effectively at elevated temperatures. Because of friction and Joule heating, the laser-alloyed layers were heated to an elevated temperature, resulting in softening. The higher erosion resistance of LA–Cr–Cu than that of LA–Ni–Cu is due to the higher softening temperature of the former.

The combined effects of friction and electric current lead to electromechanical damage, including arc erosion, abrasive and adhesive wear. Total material loss is the sum of loss in electric arc erosion and mechanical erosion during electrical contact sliding. Typical continuous parallel scratches and ploughing grooves in the direction of motion are observed on the eroded surface of the specimens (Fig. 8.13) due to the action of eroded particles of metal and its oxide. Friction between the specimens and the disc during the passage of an electric current favors oxidation with an increase in surface temperature by Joule heating, and causes thermal softening of the surface [32,33]. In the ambient condition, metal oxide films might be formed, which then behave like a lubricant to protect the surface [34]. As the sliding contacts repeat, the oxide film will fracture. Joule heating from electric current and arc discharge, which is generated between the specimens and the rotating disc when the electric contact break at current up to 10 A, also make a contribution to the erosion loss of the specimens. The electrical erosion leads to oxide film formation and removal, metal transfer, debris generation and cyclic surface deterioration of the specimens.

Dong and his co-workers reported that if normal stress is larger than or equal to the threshold value (0.64 MPa), friction coefficients increase with electric current, wear volume losses of the copper-impregnated metallized carbon material increase slightly with the increase of normal force, and contact resistance heat enhances mechanical wear [35]. On the other hand, if the normal stress is smaller than the threshold value, friction coefficients decrease with increasing electric current, wear volume losses increase intensely with the decreasing of normal stress, and arc heat induces arc erosion. Sliding wear with adhesion and abrasion are the major wear mechanisms of the frictional pair, but the material volume loss caused by arc erosion is much larger than those caused by the two kinds of wear mechanisms. In the present study, the arc erosion is the dominant damage mechanism for the laser-alloyed specimens.

8.6 Corrosion behavior

The plot of OCP against time for untreated Cu and laser-alloyed specimens in 3.5% NaCl solution (open to air) at 25 °C is shown in Fig. 8.14a. After

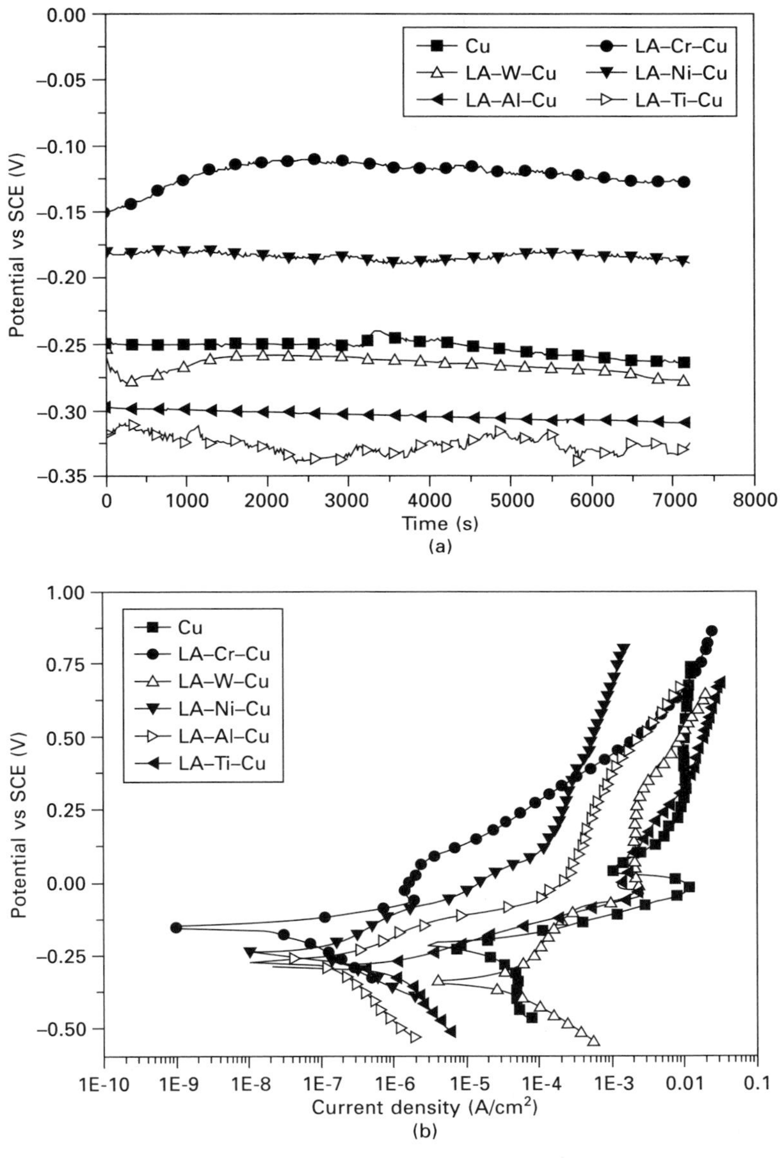

8.14 (a) Plot of OCP versus time and (b) potentiodynamic polarization curves of the laser-alloyed specimens in 3.5% NaCl solution at 25 °C.

2 h, the OCP became stable and the steady values are summarized in Table 8.3. Compared with Cu, the OCP of LA–Cr–Cu and LA–Ni–Cu shifted in the noble direction whereas that of LA–W–Cu, LA–Al–Cu and LA–Ti–Cu shifted in the active direction. The OCP of LA–Cr–Cu in 3.5% NaCl solution

Table 8.3 Corrosion parameters and ICR values of Cu and laser-alloyed specimens

Specimen	OCP (V)	I_{corr} (μA/cm^2)	ICR (mW cm^2)
Cu	–0.26	3.330	10.8
LA–Cr–Cu	–0.13	0.002	10.9
LA–W–Cu	–0.28	5.220	20.0
LA–Ni–Cu	–0.19	0.016	14.0
LA–Al–Cu	–0.31	0.024	12.3
LA–Ti–Cu	–0.33	0.014	11.0

is the noblest among the specimens, indicating the highest thermodynamic stability. The ranking of OCP of various specimens is:

LA–Cr–Cu > LA–Ni–Cu > Cu > LA–W–Cu > LA–Al–Cu > LA–Ti–Cu

The potentiodynamic polarization curves for the various specimens in 3.5% NaCl solution at 25 °C are shown in Fig. 8.14b. All laser-alloyed specimens showed no passivation in 3.5% NaCl solution. The I_{corr} values are small (Table 8.3). Based on I_{corr}, the ranking of the corrosion resistance of the specimens is:

LA–Cr–Cu > LA–Ti–Cu > LA–Ni–Cu > LA–Al–Cu > Cu > LA–W–Cu

Among the laser-alloyed specimens, LA–Cr–Cu possesses the highest corrosion resistance as reflected by the noblest OCP and the lowest I_{corr}, indicating that it is a good candidate to be used in chloride-containing environments. Moreover, LA–Ti–Cu has lower OCP but lower I_{corr} than Cu, indicating that it can also be considered for application as the contact material in a corrosive environment because of its excellent electrical erosion resistance. LA–W–Cu is the least corrosion resistant as reflected by the highest I_{corr} and low OCP value. W is more active than Cu according to the galvanic series. As high volume fraction of W particles are dispersed in a small volume fraction of Cu matrix, corrosion attack initiates at their interface and then propagates into W.

Owing to the existence of separated metallic phases (Cr and W particles) and intermetallic phases (Al_2Cu, Cu_2Ti and Cu_4Ti) in the Cu-rich matrix, the phase boundaries probably are the active sites for corrosion attack. From the corrosion morpologies of the laser-alloyed specimens shown in Fig. 8.15, corrosion attack generally initiates at the phase boundaries or interdendritic boundaries.

8.7 Interfacial contact resistance (ICR)

The plot of ICR of Cu and the laser-alloyed specimens versus applied compaction force is shown in Fig. 8.16 and the steady values of ICR at 25 N/cm^2 are listed in Table 8.3. The ranking of ICR values is:

(a)

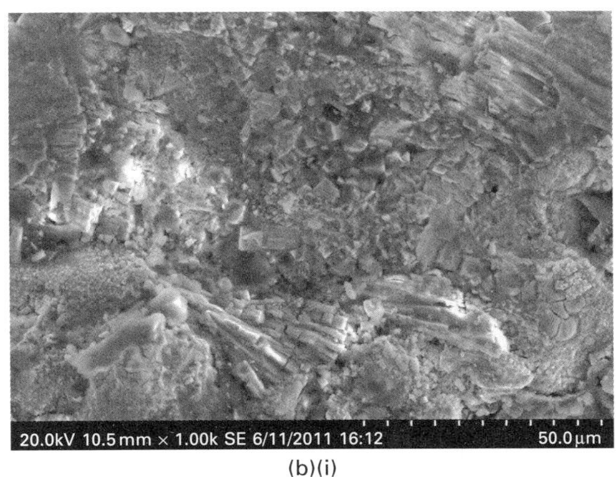

(b)(i)

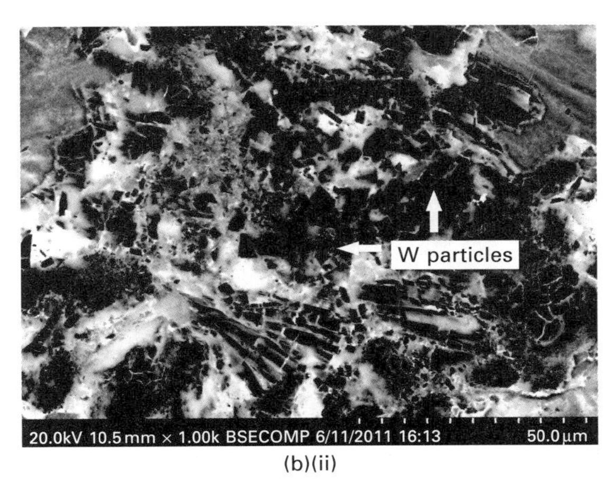

(b)(ii)

8.15 Morphologies of corroded surface of (a) LA–Cr–Cu; (b) LA–W–Cu, (i) SE micrograph; (ii) BSE micrograph; (c) LA–Ni–Cu; (d) LA–Al–Cu; and (e) LA–Ti–Cu.

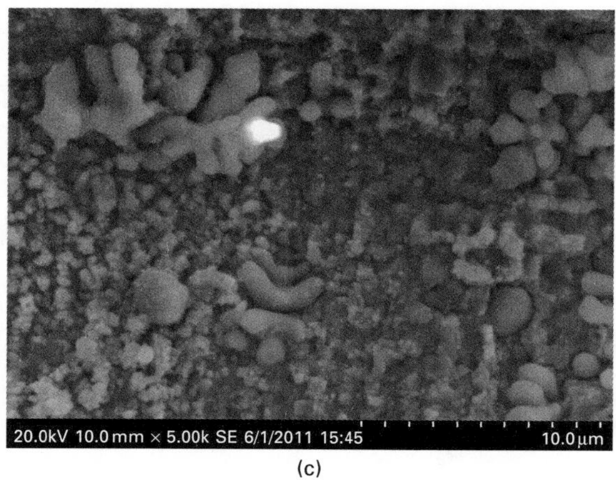

(c)

(d)

(e)

8.15 Continued

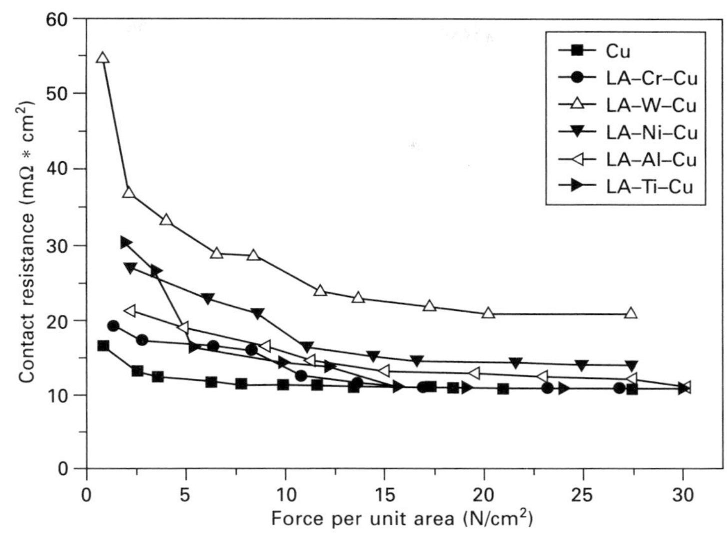

8.16 Plot of ICR versus applied compaction force for various specimens.

Cu ~ LA–Cr–Cu < LA–Ti–Cu < LA–Al–Cu < LA–Ni–Cu < LA–W–Cu

The ICR of LA–Cr–Cu is closest to that of Cu, followed by LA–Ti–Cu. This indicates that LA–Cr–Cu has some enhancement in electrical erosion resistance while the electrical conductivity can be preserved. In contrast, the ICR value of LA–W–Cu is the highest because of the large volume fraction of W phase, which has higher resistivity. According to reference data [36], the resistivities of Cu, Cr, Ni, W, Al and Ti are 17, 125, 70, 52.8, 420 and 28 nΩ m, respectively (Table 8.1). Since all elements possess higher resistivity than Cu, it is expected that the resistance of the laser-alloyed specimens will increase as compared with the untreated copper.

Among the laser-alloyed specimens, LA–Ti–Cu has the highest electrical erosion resistance whereas LA–Cr–Cu has the highest corrosion resistance. By considering the electrical erosion resistance, LA–Ti–Cu seems to be a promising protective laser-alloyed layer for electrical contact applications.

8.8 Conclusions

- Laser surface alloying of Cu with Cr, W, Ni, Al and Ti was successfully achieved by Nd : YAG laser. Ni forms a solid solution, while Cr and W are precipitated as separate phases in the Cu matrix; Al and Ti form intermetallics in the composite layer.
- The hardness of the LA–Cr–Cu, LA–W–Cu, LA–Ni–Cu, LA–Al–Cu and LA–Ti–Cu has been improved by 1.5, 1.85, 1.7, 1.5 and 7.5 times, respectively.

- The electrical erosion resistance of LA–Ti–Cu has been significantly increased by a factor of 7.5 times, which corresponds to the high hardness of the intermetallic phases (Cu_2Ti and Cu_4Ti). The electrical erosion resistance ranking of the laser-alloyed specimens is:

  LA–Ti–Cu > LA–W–Cu > LA–Cr–Cu > LA–Ni–Cu > LA–Al–Cu ~ Cu

- The ranking of corrosion resistance of the laser-alloyed specimens in 3.5% NaCl solution at 25 °C is:

  LA–Cr–Cu > LA–Ti–Cu > LA–Ni–Cu > LA–Al–Cu > Cu > LA–W–Cu

  The corrosion resistance of all the laser-alloyed specimens has increased as evidenced by the reduction in corrosion current density. For LA–W–Cu, the corrosion resistance is reduced due to the high volume fraction of W particles which act as active sites for corrosion attack.
- The ICR values of all the laser-alloyed specimens are close to that of untreated copper, except for LA–W–Cu. This shows that the electrical erosion and corrosion resistances of the laser treated specimens have been enhanced while the high electrical conductivity can be preserved. For LA–W–Cu, the highest ICR value is due to the high resistivity of W.
- Among the various specimens, LA–Ti–Cu possesses the highest wear resistance whereas LA–Cr–Cu has the highest corrosion resistance and the lowest ICR.

8.9 Acknowledgments

The work described in this chapter was fully supported by research grants from the Science and Technology Development Fund (FDCT) of Macau SAR (Grant no. 018/2008/A) and the Research Committee of University of Macau (Project no. RG068/07-08S/KCT/FST).

8.10 References

[1] B. Pizzigoni, A. Collina, B. Flapp, S. Melzi, *Tribo Test*, 2007, **13** (1), 35–47.

[2] F. Kieβling, R. Puschmann, A. Schmieder, *Contact Lines for Electric Railways Planning Design Implementation*, Publicis, 2001, 504.

[3] Y. Oura, Y. Mochinaga, H. Nagasawam, *Japan Railway & Transport Review*, 1998, **16**, 48–58.

[4] A. Senouci, H. Zaidi, J. Frene, A. Bouchoucha, D. Paulmier, *Applied Surface Science*, 1999, **144–145**, 287–291.

[5] C.R.F. Azevedo, A. Sinatora, *Engineering Failure Analysis*, 2004, **11** (6), 829–841.

[6] C.T. Kwok, P.K. Wong, H.C. Man, F.T. Cheng, *International Journal of Railways*, 2010, **3**, 19–27.

[7] J.P. Tu, W.X. Qi, Y.Z. Yang, F. Liua, J.T. Zhang, G.Y. Gan, N.Y. Wang, X.B. Zhang M.S. Liu, *Wear*, 2002, **249**, 1021–1027.

[8] S.G. Jia, P. Liu, F.Z. Ren, B.H. Tian, M.S. Zheng, G.S. Zhou, *Wear*, 2007, **262**, 772–777.
[9] R.G. Zheng, Z.J. Zhan, W.K.Wang, *Wear*, 2010, **268**, 72–76.
[10] R.G. Zheng, Z.J. Zhan, W.K. Wang, *Journal of Rare Earths*, 2011, **29**, 247–252.
[11] R. Giuntini, K. Gaudette, *Business Horizons*, 2003, **46** (6), 41–48.
[12] P.K Wong, W.K. Chan, C.T. Kwok, K.H. Lo, Proceedings of 6th International Symposium on Environmentally Conscious Design and Inverse Manufacturing (EcoDesign 2009), December 7–9, Sapporo, Japan, 2009, 303–307.
[13] A. Hirose, K.F. Kobayashi, *Materials Science and Engineering A*, 1994, **174**, 199–206.
[14] J. Dutta Majumdar, I. Manna, *Materials Science and Engineering*, 1999, **A268**, 216–226.
[15] J. Dutta Majumdar, I. Manna, *Materials Science and Engineering*, 1999, **A268**, 227–235.
[16] H.R. Geng, Y. Liu, C.Z. Chen, M.H. Sun, Y.Q. Ga, *Materials Science and Technology*, 2000, **16**, 564–567.
[17] K.W. Ng, H.C. Man, F.T. Cheng, T.M. Yue, *Applied Surface Science*, 2007, **253**, 6236–6241.
[18] R. Li, Y. Shi, J. Liu, Z. Xie, Z. Wang, *Int J Adv Manuf Technol*, 2010, **48**, 597–605.
[19] D. Gu, Y. Shen, X. Wu, *Materials Letters*, 2008, **62**, 1765–1768.
[20] H.L. Wang, M.A. Sweikart, J.A. Turner, *J. Power Sources*, 2003, **115**, 243.
[21] ASTM Standard G5-92, ASTM Standards, Philadelphia, USA.
[22] L. Dubourg, H. Pelletier, D. Vaissiere, F. Hlawka, A. Cornet, *Wear*, 2002, **253**, 1077–1085.
[23] T.B. Massalskim (Ed.), *Binary Alloy Phase Diagrams*, ASM, Metals Park, OH, 1990.
[24] I. Manna, S. Abraham, G. Reddy, D.N. Bose, T.B. Ghosh, S.K. Pabi, *Scripta Metallurgica et Materials*, 1994, **31**, 713–718.
[25] M.R. Bateni, S. Mirdamadi, F. Ashrafizadeh, J.A. Szpunar, R.A.L. Drew, *Materials and Manufacturing Processes*, 2001, **16 (2)**, 219–228.
[26] G.K. Dey, A. Biswas, S.K. Roy, S. Banerjee, *Materials Science and Engineering*, 2001, 304–306, 641–645.
[27] S. Kac, A. Radziszewska, J. Kusinski, *Applied Surface Science*, 2007, **253**, 7895–7898.
[28] M.R. Bateni, F. Ashrafizadeh, J.A. Szpunar, R.A.L. Drew, *Wear*, 2002, **253**, 629–639.
[29] T. Sasada, S. Ban, S. Norose, T. Nakano, *Wear*, 1992, **159**, 191–199.
[30] S.V. Prasad, P.K. Rohatgi, T.H. Kosel, *Materials Science and Engineering*, 1986, **80** (2), 213–220.
[31] G.W. Stachowiak, A.W. Batchelor, *Engineering Tribology*, 2nd edn, Butterworth-Heinemann Pub., 2005.
[32] M. Hashempour, H. Razavizadeh, H. Rezaie, *Wear*, 2010, **269**, 405–415.
[33] A. Bouchoucha, S. Chekroud, D. Paulmier, *Applied Surface Science*, 2004, **223**, 330.
[34] D. Paulmier, A. Bouchoucha, H. Zaidi, *Vacuum*, 1990, **41** (7–9), 2230–2232.
[35] L. Dong, G.X. Chen, M.H. Zhu, Z.R. Zhou, *Wear*, 2007, **263**, 598–603.
[36] P.L. Rossitor, *The Electrical Resistivity of Metals and Alloys*, Cambridge University Press, 1987, 290.

9 Laser remanufacturing to improve the erosion and corrosion resistance of metal components

J. H. YAO, Q. L. ZHANG and F. Z. KONG,
Zhejiang University of Technology, China

Abstract: This chapter looks at laser remanufacturing processes and their application to typical failed components. The failure of four typical components is analyzed and then laser solutions are described, including special material design, technique optimization, field testing and application. The results show that laser remanufacturing is an effective method for hardening and repairing failed components which can be reused with high corrosion and erosion resistance.

Key words: corrosion resistance, erosion resistance, laser remanufacturing.

9.1 Introduction

Remanufacturing technology is an effective way to reduce waste and environmental pollution. This area is a developing new research field, and a growing and developing advanced manufacturing technology, offering an extension to the whole life cycle of many manufacturing processes. Remanufacturing technology provides an important technical support to industrial sustainable development, bringing great benefits to the development of the national economy, and becoming a new point of economic growth [1–5].

Laser remanufacturing uses laser beams as a heat source to renew and improve failed metal components, whilst simultaneously improving their performance. Based on laser surface modification technology, specially developed alloy materials with high erosion and corrosion resistance (as required) are added to the metal components, and optimum processing parameters are chosen.

9.2 Laser remanufacturing technology

9.2.1 Characteristics of laser remanufacturing

Laser surface modification technology includes laser transformation hardening, laser solution strengthening, laser remelting strengthening, laser shock peening (LSP), laser glazing, laser surface alloying, laser cladding and composite strengthening combined with other surface treatments (such as spraying or electroless plating) [6]. The three basic techniques (laser solution strengthening, laser surface alloying and laser cladding) have a close relationship with corrosion and erosion resistance; the first two are mainly used to improve corrosion and erosion resistance of old products without increasing their sizes, while the third (laser cladding) is used to repair wasted products by restoring their size.

During the laser remanufacturing process, a laser beam is used to heat injected alloy materials on a surface to generate rapid melting and solidification, or to irradiate directly onto the surface of a precipitation-hardenable alloy to achieve rapid solidification. Because of the fast heating and cooling speeds involved, laser remanufacturing features a low dilution ratio, small distortion, high production efficiency and no environmental pollution, etc. As a result, laser remanufacturing can renew failed and abandoned products caused by corrosion, erosion and processing errors, and then repair them so that they are as good, if not better, than new.

9.2.2 Laser remanufacturing equipment

As shown in Figs 9.1 and 9.2, laser remanufacturing equipment usually consists of a laser system (a CO_2 laser, Nd : YAG (yttrium aluminium garnet) laser, diode laser, fiber laser, etc.), a light guiding and focusing system, a computer numerical control (CNC) machine or robot, powder feeding system,

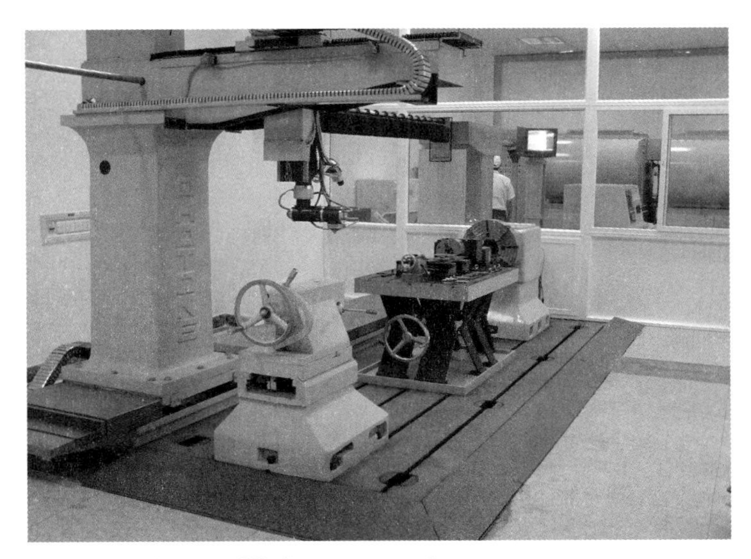

9.1 Seven kilowatt CO_2 laser processing system.

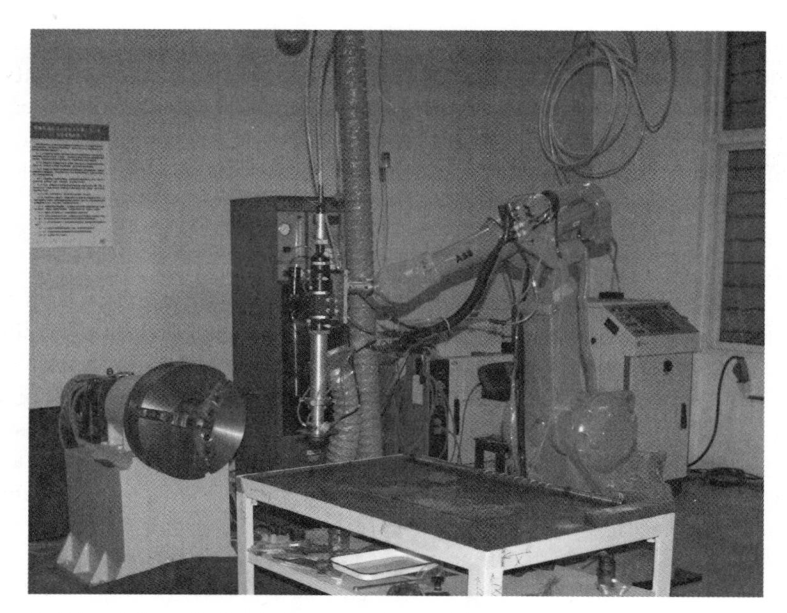

9.2 Two kilowatt diode laser system with robot.

online measuring and monitoring system and auxiliary devices (shielding gases, chillers, etc.).

The laser system generates a laser beam, whilst the light guiding and focusing system guides and focuses the laser beam onto the desired spot before irradiating the component's surface. A CNC machine and robot are used to control the movement of the laser beam, or the movement of the part. A powder feeder is used to deliver alloy powder to the component surface by means of coaxial or lateral powder delivery. Online measuring and monitoring can detect the height and width of the cladding region and the temperature during the laser cladding process. A control system then adjusts the parameters of the laser cladding according to the monitored results, with the aim of stabilizing the cladding process and achieving better cladding quality.

9.2.3 Laser remanufacturing techniques

Laser surface alloying

In laser surface alloying, alloy materials in powder form are deposited by feeding, or by brushing or painting, onto the surface of a component, before irradiating with a laser beam of high power density. The alloy elements quickly fuse into the surface and the melted surface then solidifies rapidly to form an alloy coating (by heat transfer) of the component itself. After laser surface alloying, components can again achieve the performances required, such as wear resistance, corrosion resistance, high temperature resistance and anti-oxidation [7].

Laser surface solution strengthening

Laser surface solution strengthening is used on precipitation-hardening stainless steel with low carbon content (about 0.05 wt%) to improve its corrosion and erosion resistance. A high power density laser beam is used as the heat source. The soluble solid elements in the substrate diffuse during the laser process and, after aging, the ε-Cu second-phase of 0Cr17Ni4Cu4Nb (17-4PH) precipitates to strengthen the matrix. The precipitation can lead to a distortion of the crystal lattice, which increases the corrosion resistance. If the erosion is serious, both laser surface alloying and the laser surface solution composite strengthening process can be used together to improve surface hardness and enhance its erosion resistance [8,9].

Laser cladding

With laser cladding, a high-power density laser beam is used to apply (by rapidly melting and solidifying injected alloy materials, applied by powder or a wire feeding system) a thin layer onto the surface, achieving a cladding layer with metallurgical bonding, with low rates of dilution and some special performance characteristics [10].

9.2.4 Materials for laser remanufacturing

Currently, most research into special materials for laser remanufacturing considers two aspects: improving the chemical composition of the powder materials used and choosing chemical compositions with eutectic high carbon or peritectic low carbon, based on the phase transformation characteristics of steels.

These special materials have high requirements in terms of their thermal expansion coefficients and thermal stress matching, so that it has been necessary to develop special powder materials. Researchers at the Zhejiang University of Technology have developed two series of special alloy powders [11,12]: Type H and Type F. Type H is used to fabricate a hardened layer with high erosion resistance and with a high hardness (descending in hardness from the surface to the substrate; as shown in Table 9.1). Type H is a special material for laser remanufacturing used to get a hardened layer with high corrosion resistance as shown in Table 9.2.

Table 9.1 Type H for laser remanufacturing to increase erosion resistance

No.	Objects	Substrate materials	Hardness ($HV_{0.2}$)	Adding methods of powder	Applications
H_1-1	Flake parts easily deforming and needing wear-resistance locally	SUS 420J2 GB 4Cr13 GB 5Cr15	1000–1100	Spraying	Blades of small recuperation unit, kitchen knife with thickness ≤ 2 mm
H_1-2	Parts with large deviation of section size and easy deformation, needing wear and erosion resistance	AISI 420 mild carbon steel	620	Brushing	300 000 kW pump blades, cold die blades
H_1-3	Parts requiring high wear and erosion resistance	17-4PH 316 AISI 4140	700	Spraying	Blades of large capacity unit, pump impeller in nuclear power plant, mouth rings, screws, check inverse rings, etc.
H_2-2	Parts requiring high hardness, wear-resisting and erosion-resisting with work temperature below 550 °C	Fe-based materials	900–1000	Feeding	Fan blades, high temperature and high-pressure valves, helical conveyors, scraper blades

Table 9.2 Type F for laser remanufacturing to increase corrosion resistance

No.	Objects	Substrate materials	Hardness ($HV_{0.2}$)	Adding methods of powder	Applications
F_1	Parts needing good wear resistance with any thickness	AISI 1045 AISI 5140 AISI 4140	800–900	Feeding or spraying	Screws of rubber and plastic machines
F_2	Parts needing corrosion resistance	Mild carbon steel AISI 410 AISI 304	160–240	Feeding	Stirring shafts of polymerizer, blades of lye pump
F_3	Parts needing corrosion and wear resistance	SUS 420J2	800	Feeding	Dental surgeon pincers
F_5	Parts requiring high corrosion resistance with work temperature below 550 °C	Fe-based materials	200	Feeding	Used for cladding or rendering

9.3 Application of laser remanufacturing for corrosion and erosion resistance of turbine blades

9.3.1 Turbine blade failure

As key components of any power generation equipment, turbine blades are the typical easily damaged consumable parts, of which about one-third must be changed every year because of failure, mainly caused by the severe working conditions in which they operate. During different stages, blades work at different working temperatures, ranging from 500 °C to about 100 °C. During their running, the circumference speed of blades is over 300 m/s. Backflow steam with water drops causes erosion on the leading edge of the blades, which has a honeycomb appearance on the surface and a zigzag on the edge, even forming notches if the erosion is serious, as shown in Figs 9.3 and 9.4. These faults affect the vibration characteristics of blades, reducing blade intensity and making the aerodynamic performance of the blade worse with grade efficiency decrease, and even increasing the risk of fracture. With the newly developed supercritical and ultra-supercritical power units, which have obvious energy saving and environment-improving advantages, the working conditions are even worse: the steam working temperature of ultra-supercritical power units is not less than 593 °C, whilst the pressure is not less than 31 MPa. However, conventional surface hardening methods such as flame and induction hardening cannot meet the requirements, leading

9.3 Zigzag on the edge of blades.

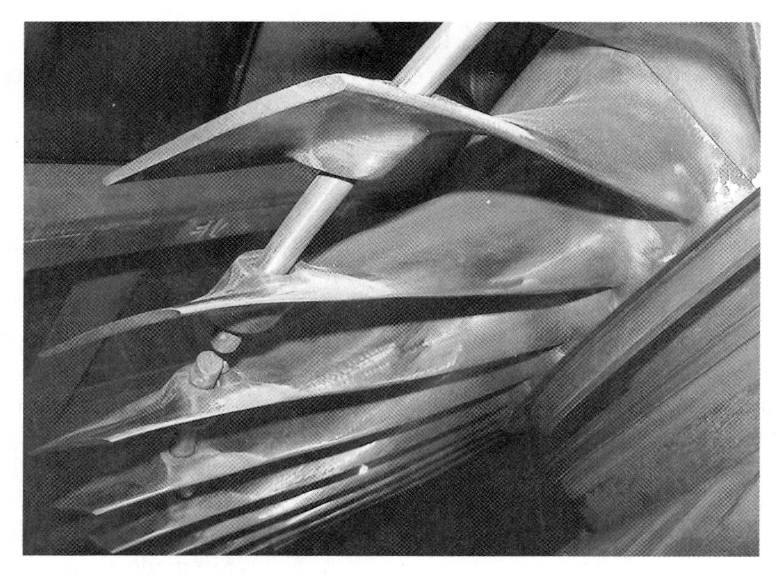

9.4 Failed blades by erosion.

to several blades being discarded as useless. However, no other effective ways have been proposed to renew the wasted turbine parts with high added value [13–16].

9.3.2 Laser solutions for turbine blades

At present, the utilization of high alloy materials as the matrix, plating of hard chromium, flame hardening, induction hardening, overlaying, inserting Stellite alloy and laser remanufacturing are the methods usually used to solve turbine blade failure problems. The costs and performance of these solutions are shown in Table 9.3.

9.3.3 Laser remanufacturing techniques used on turbine blades

Depending to the working conditions and performance requirements, five main techniques of laser remanufacturing are proposed for the hundreds of types of turbine blades:

(1) For AISI 420 steel blades in mini-type turbine units, flame hardening is being substituted by laser transformation hardening, with a laser power density of 18–22 W/mm^2, a scanning speed of 0.5–1 m/min, and beam spot size of 10 × 2 mm^2. As a result, a hardened layer with hardened depth of 0.25–0.45 mm can be obtained, the hardness of the hardened

Table 9.3 Cost and performances of the solutions to failure of blades

Solutions	Cost	Operating ability	Percent of pass	Corrosion resistance	Distortion	Bonding
Utilization of high-alloy materials as matrix	High	Easy		Bad		
Plating hard chromium	Low	Easy	High	Bad	No	Easy to fall off
Flame hardening	Low	Easy	Low	Below normal	Large	Non-uniformity of hardened layer
Induction hardening	Middle	Hard to make induction coil	Low	Middle	Large	Non-uniformity of hardened layer
Overlaying	Middle	Complex(preheat)	Low	Above normal	Large	Metallurgical
Inserting stellite alloy	Above normal	Middle	Above normal	Above normal	Small	Easy to fall off
Laser remanufacturing	Above normal (1/7~1/8 of the new one)	Automatic	High	Good	No	Metallurgical

layer decreasing from the surface in a gradient of maximum 690 $HV_{0.2}$ and an average of 588 $HV_{0.2}$ [13]. The substrate (AISI 420 steel) used in this experiment is processed by hardening and tempering. After laser quenching, fine martensites with a staggered distribution are found in the strengthened layer (Fig. 9.5). During the rapid heating and cooling process of laser strengthening, the growth of austenite is restrained, forming a highly refined microstructure, which can result in an improvement of the surface hardness and corrosion resistance. The yield rate of production after laser transformation hardening can reach 100% and this technique has now been listed as a business standard (instead of flame hardening) in some Chinese businesses.

(2) For AISI 420 steel blades in large capacity turbine units, induction hardening is now being replaced by laser surface alloying. The composition of alloy powder is shown in Table 9.4, and the parameters chosen are of laser power density of 22–44 W/mm^2 and a scanning speed of 0.2–0.4 m/min.

The microstructure of an AISI 420 substrate, which includes ferrite and

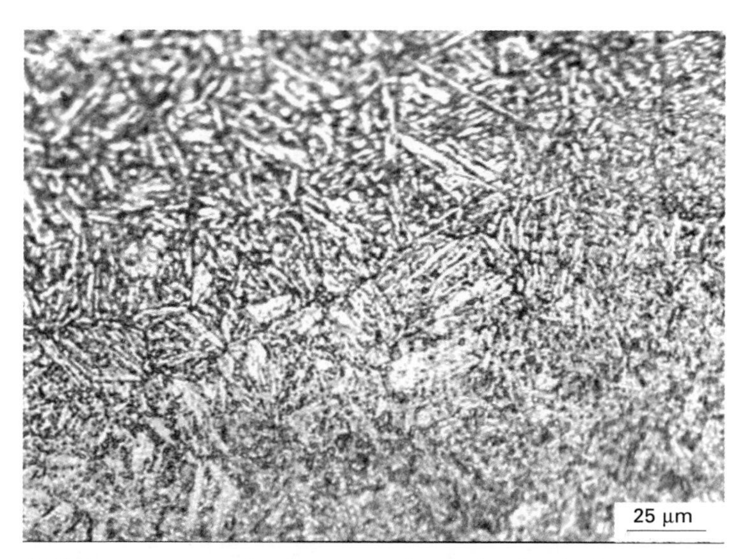

9.5 Microstructure after laser transformation hardening.

Table 9.4 Chemical composition of alloying powder and laser surface alloying layer (wt%)

Elements	Si	Cr	W	Ni	Mo	Co	C	Fe
Alloying powder	1.30	2.86	40.24	3.29	–	51.33	–	0.98
Laser surface alloying layer	0.29	13.78	10.59	0.47	1.32	–	0.04	Balance

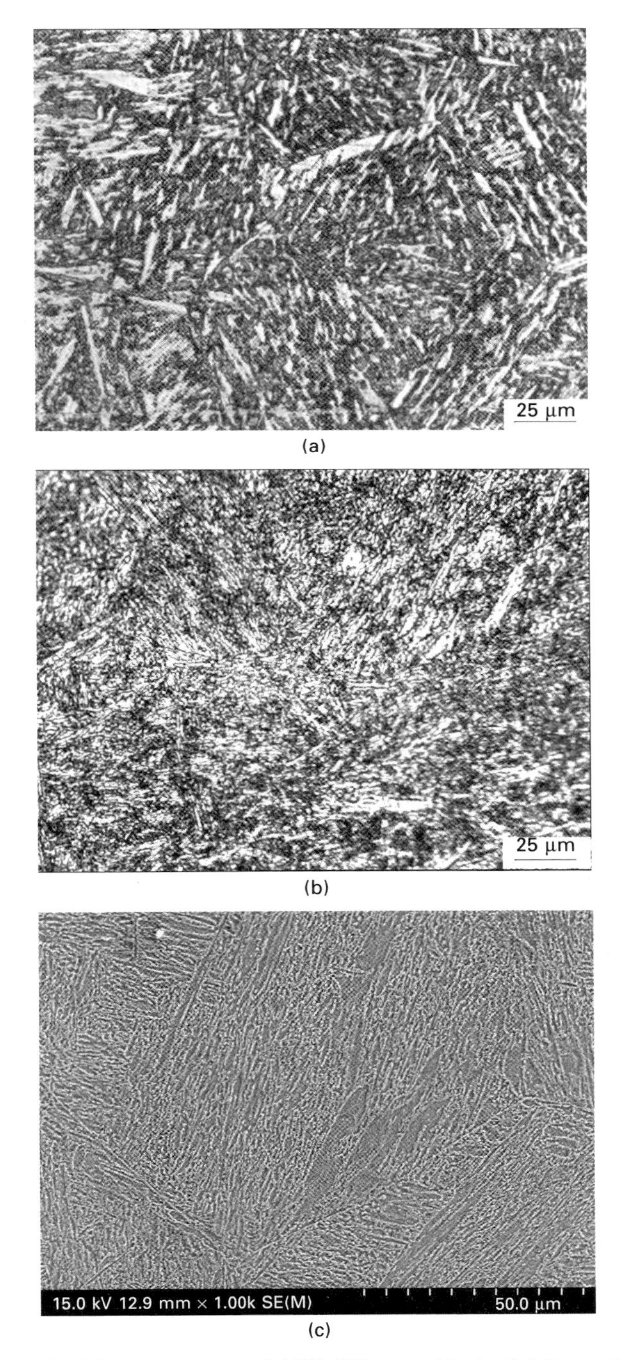

9.6 Microstructure of AISI 420 steel blade: (a) the substrate; (b) the optical microscopy morphology of laser alloyed layer; (c) the scanning electron microscopy (SEM) graph of laser alloyed layer.

sorbite composed of granular carbide, is shown in Fig. 9.6(a). Figure 9.6(b) shows the morphology of the treated layer after the laser alloying process, and Fig. 9.6(c) is a scanning electron microscopy (SEM) image of the alloyed layer. It can be seen that a dense alloyed layer with fine grains was generated. During the rapid heating and cooling process of laser alloying, the growth of austenite is restrained, forming a highly refined microstructure, which is one reason why the hardness increases. The alloyed layer and the substrate are metallurgically bonded, without crack and porosity. During the laser alloying process, under high laser power irradiation, the surface of the sample is molten and the alloying materials fuse into the melting layer, which can be proved by the energy-dispersive spectrometry X-ray (EDS) results shown in Table 9.4. The added alloying powders (mainly W, Cr, etc.) are melted into the molten pool in a very short time and carbonization occurs. Figure 9.7 shows that the phases of the alloyed layer are WC, Fe_2C and Cr_7C_3. The highly dispersed WC hard phases are the main reason for the improvement in microhardness. The hardenability of the materials is enhanced because of the existence of Cr. Meanwhile, the hardness is increased due to the Cr_7C_3 formed. The added Ni as an element to expand the austenite region prevents the formation of the second-phase particles and improves the surface erosion resistance performance.

The highest microhardness of the laser alloying layer is 785 $HV_{0.2}$, whilst the average is 701 $HV_{0.2}$, which is 1.8 times higher than that of the

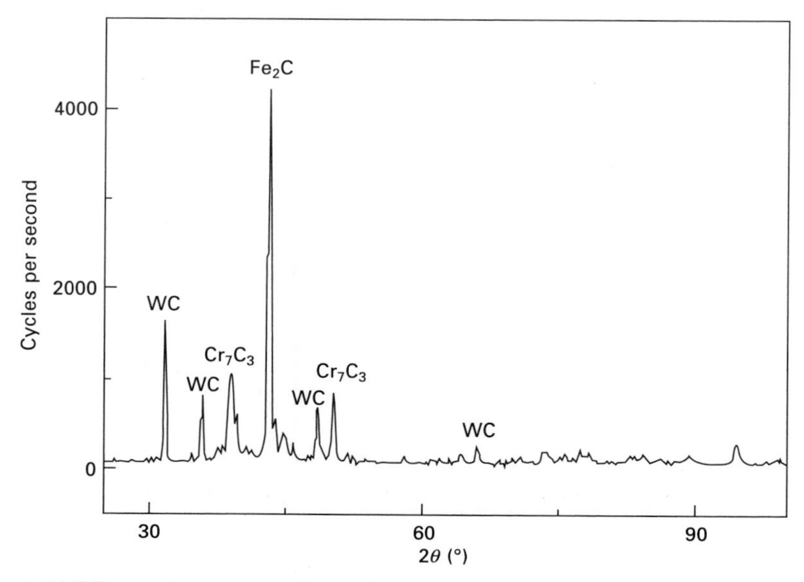

9.7 XRD result on the surface of alloyed layer.

substrate. The thickness of the hardened layer is around 0.4 mm and the hardness decreases gradually from the alloyed layer to the substrate.

The AISI 420 steel blade samples before and after laser alloying were included in experiments to measure cavitation erosion. The results of weight loss vs. time are shown in Fig. 9.8. The weight loss of the substrate is significantly higher than that of the alloyed sample, which shows that the material after laser alloying has higher cavitation resistance under the same conditions. Figure 9.9 shows SEM images of the cavitation erosion of the surfaces. For the substrate sample, deep and nubby erosion holes are uniformly distributed on the surface. No crack on the erosion surface is found, but some microcracks are observed at the junction of the erosion area and non-erosion area. For the laser alloying sample, the erosion holes are relatively shallow under the same erosion experimental conditions. No crack is observed at the junction of the erosion area and non-erosion area, which is different from the substrate sample. The

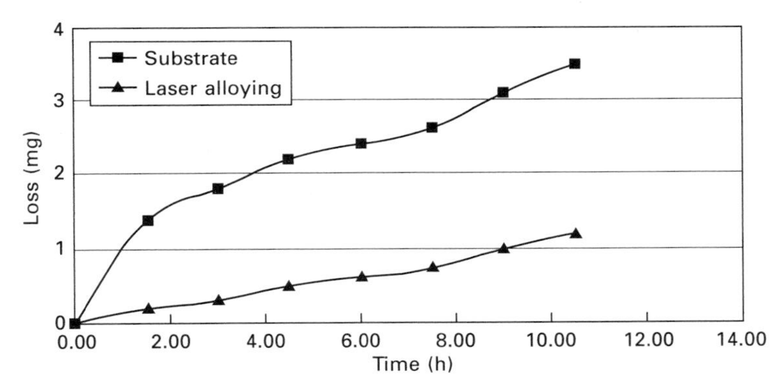

9.8 Cavitation resistance curves of the substrate and laser alloyed sample.

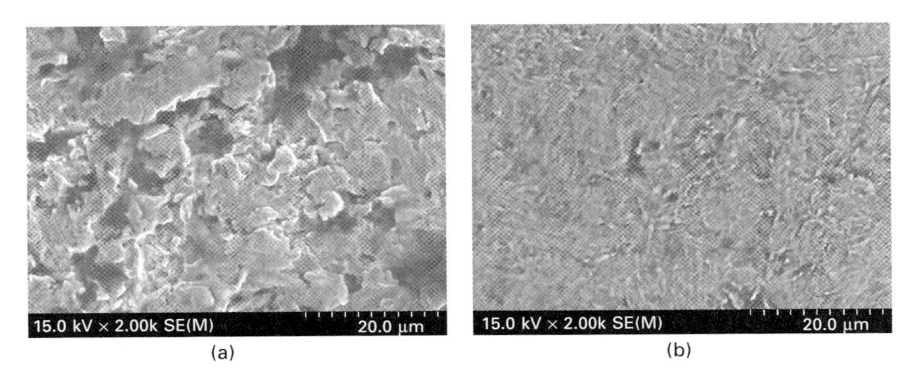

9.9 Surface SEM images after cavitation erosion: (a) the substrate; (b) the laser alloyed sample.

variation is mainly due to the effect of the rapid heating and cooling of the laser beam. In the laser alloyed layer, the grains are refined and a large number of carbide hard phases form. The diffused carbide hardened phases restrict the expansion of cracks and flaking of the surface metal, and then reduce the erosion rate, which results in cavitation resistance of the substrate.

Industrial production using the laser surface alloying technique has been realized instead of induction hardening [14]. The same technique is also applicable to 17-4 PH stainless steel blades [16].

(3) For 17-4 PH stainless steel blades in medium capacity turbine units, the traditional method is to use a solution treatment followed by an aging treatment, which has disadvantages of large distortion and a long aging cycle. This method can now be replaced by laser surface solid solution hardening on the leading edge, with a CO_2 laser power density of 1.7 kW/mm^2, a light spot size of $16 \times 1\,mm^2$, scanning speed of 120 mm/min, and a surface temperature of 1450 °C [17]. The depth of the hardened layer is 2 mm, and the hardness is more than 400 $HV_{0.2}$, which is enough to satisfy technical requirements under normal working conditions.

During the laser heating process, the supersaturated solid solutions exist in martensite and ferrite. After prolonged heat insulation treatment, the solid solutions precipitate in certain forms and disperse in the substrate. Figure 9.10 shows precipitated phases in a ferritic matrix. A large number of precipitated phases in the ferritic matrix can be found, mainly in grain boundaries with the diameter varying from 100 to 1000 nm. According

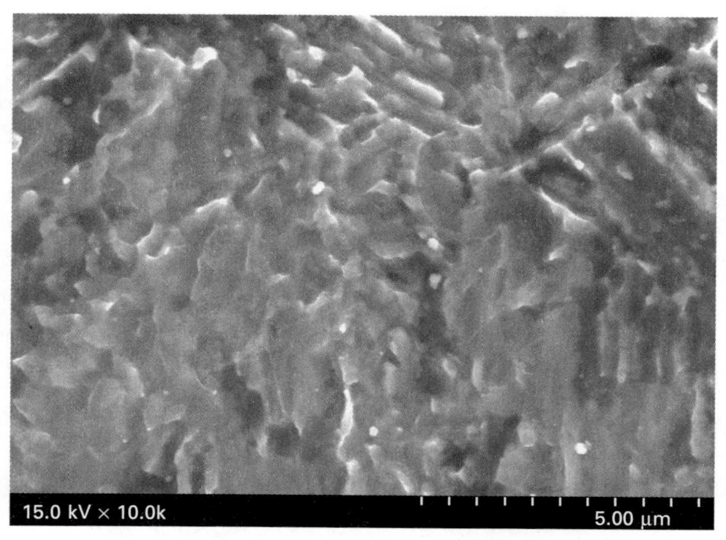

9.10 Precipitated phases in ferritic matrix after laser solid solution hardening.

to a large number of reports and analyses of the EDS of martensite, the precipitation phases can be speculated to be a precipitation phase ε-Cu (face centered cubic, fcc) from the 17-4PH substrate. The hardness gradually reduces from 478 $HV_{0.2}$ on the surface to 301 $HV_{0.2}$ in the substrate, and the distribution is quite uniform. The depth of the hardened layer is about 1.8 mm.

The cavitation resistance of the surface after laser treatment improves to more than 100% of that of the substrate, as shown in Fig. 9.11. The SEM images of cavitation erosion surfaces of the 17-4PH samples before and after laser solution strengthening under the same conditions are shown in Fig. 9.12.

The erosion holes on the 17-4PH substrate are obvious and there are microcracks. However, the erosion holes on the laser hardened layer are reticularly distributed, the depth of the holes is relatively shallow, and no cracks are found on the overall surface. Because of the unbalanced rapid solidification process of laser solution strengthening, the surface is characterized by high hardness and strength and has refined grains. More boundaries between grains can be achieved, which may contribute to buffer the stress. Thus, fatigue cracks cannot extend, and breaking off of grains is prevented so that the toughness and plastic property are improved, which leads to excellent cavitation resistance.

(4) For the 17-4PH stainless steel blades in ultra-supercritical turbine units, the working conditions are severe, so laser surface alloying and laser solid solution strengthening techniques are performed in-step to improve both hardness and depth of the hardened layer [18]. The processing parameters are a laser power of 2.2 kW and a scanning speed of 400 mm/min. The maximum hardness of the alloying layer is 604 $HV_{0.2}$ and the average is 536 $HV_{0.2}$, while the highest hardness of the solutionized layer is 377 $HV_{0.2}$ and the average 361 $HV_{0.2}$. The depth of the hardened layer (the alloyed layer and the solutionized layer) reaches 2.5–3.0 mm.

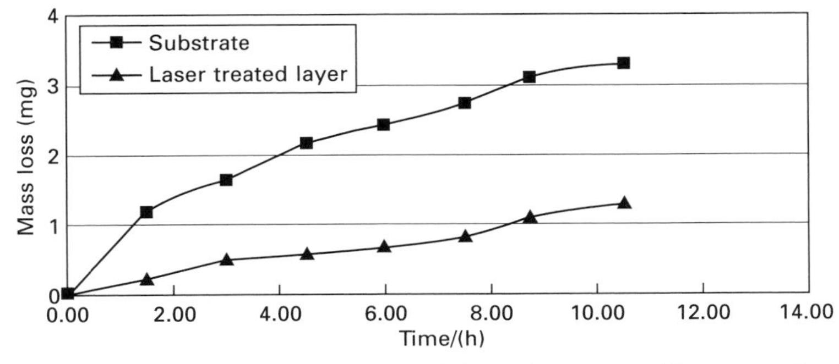

9.11 Cavitation resistance curves of the substrate and laser solution strengthened sample.

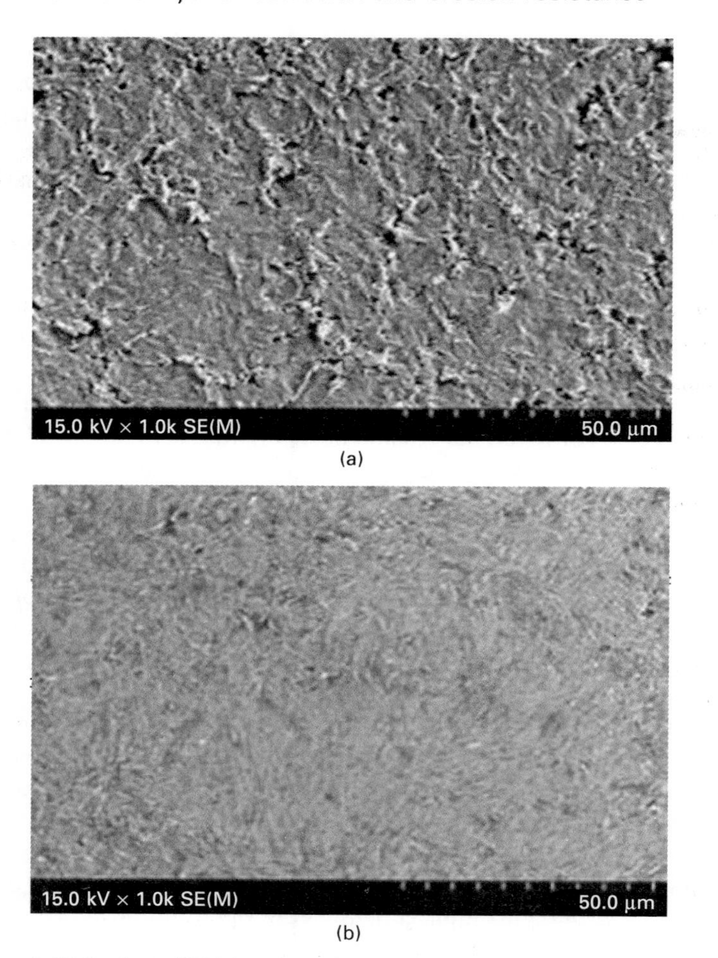

9.12 Surface SEM images after cavitation erosion (a) the substrate; (b) the laser solution strengthened sample.

From Fig. 9.13(a), it can be seen that there are three areas including the alloying area, combining area and solution area. After laser surface alloying and laser surface solution composite strengthening, a smooth hardened layer without cracks, holes and other defects can be achieved. Structure changes in the hardened layer from the surface to the substrate can be found due to the gradual decrease of temperature and cooling velocity, as shown in Fig. 9.13(b). Cellular and cellular dendritic crystals form in the alloying layer. In the solution region, owing to the convection in the upper molten pool and the fast diffusion of the elements, a supersaturated solid solution can form during the rapid cooling process which, after cooling, turns into ferrite and martensite. There are also a few Widmanstätten structures in the local region as a result of overheating.

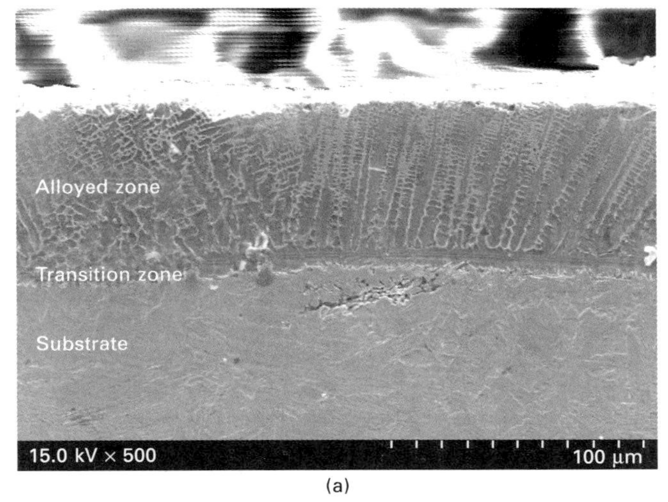

(a)

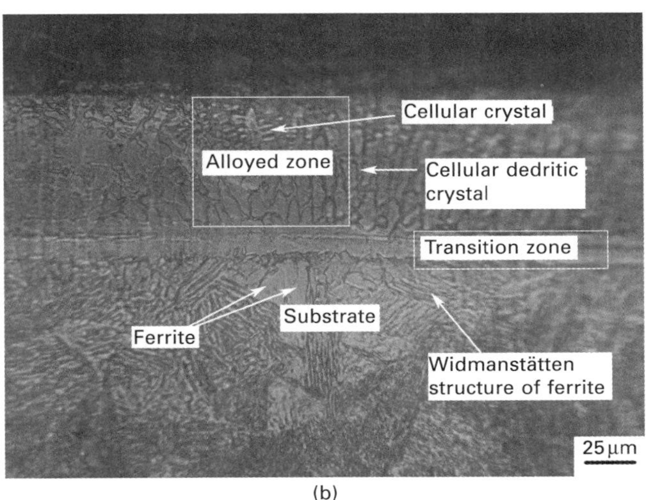

(b)

9.13 Cross-section morphology of laser hardening layer: (a) SEM of laser hardening layer; (b) microstructure of each part in laser hardening layer.

A phase analysis of the laser alloying layer is shown in Fig. 9.14, which reveals that the main phases in the alloying layer comprise Fe–Cr, W_2C and $Al_{0.28}Co_{0.4}Cr_{0.24}W_{0.08}$. The carbothermal reduction reaction between Al and C produces aluminum vapor and aluminum oxide, which react with Co, Cr and W. Finally, hardened phases such as W_2C and $Al_{0.28}Co_{0.4}Cr_{0.24}W_{0.08}$ form, which disperse in the alloying layer and contribute to dispersion strengthening on the surface of the material.

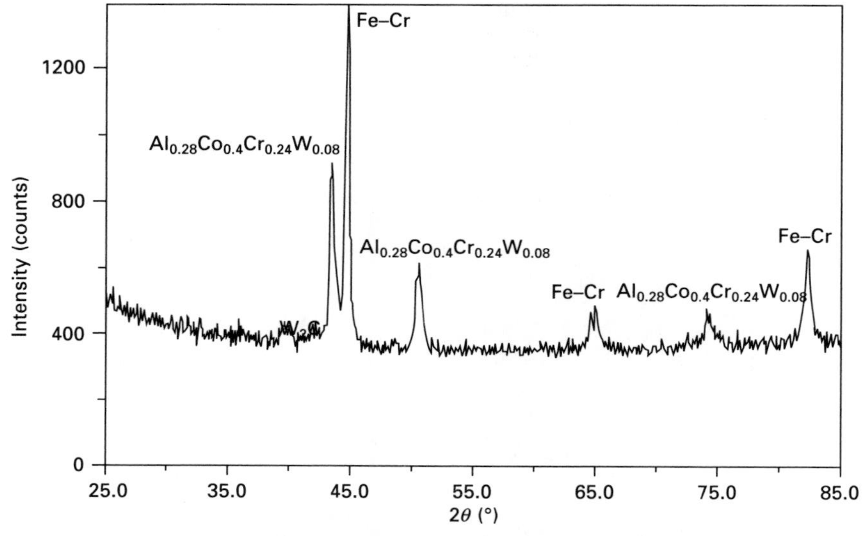

9.14 X-ray diffraction (XRD) patterns of laser alloying layer.

(5) For 17-4PH stainless steel or AISI 420 steel supercritical blades with serious corrosion, laser cladding is adopted to patch the failed surfaces. Special alloy materials for laser remanufacturing should be chosen according to different working conditions and performance requirements. Using a self-made powder feeding device, integrated control system and corresponding parameters, an alloyed layer with high corrosion resistance can be created, as shown in Figs 9.15–9.17 [19,20]. Figure 9.15 shows that the cladding layer has no defects with a depth of 2.3 mm. The average hardness of the cladding layer is about 458 $HV_{0.5}$. The dendrite structure in the cladding layer is much finer and uniform, as shown in Fig. 9.16 and the strong metallurgical bonding between the cladding layer and the substrate can be seen in Fig. 9.17. The failed blade and blade repaired by laser cladding technology are shown in Figs 9.18 and 9.19, respectively.

9.3.4 Application

Laser remanufacturing techniques have been applied in more than 10 turbine companies in China, where hundreds of different kinds of blades have been treated and over two thousand turbine units have been installed. No cracks and no deformations have been found in the laser treated regions and service life increases remarkably after the installations. These techniques have improved the quality and safety coefficient of blades, decreased system faults, and reduced daily maintenance and down-time. These new techniques have been successfully used as substitutes for traditional treatment methods

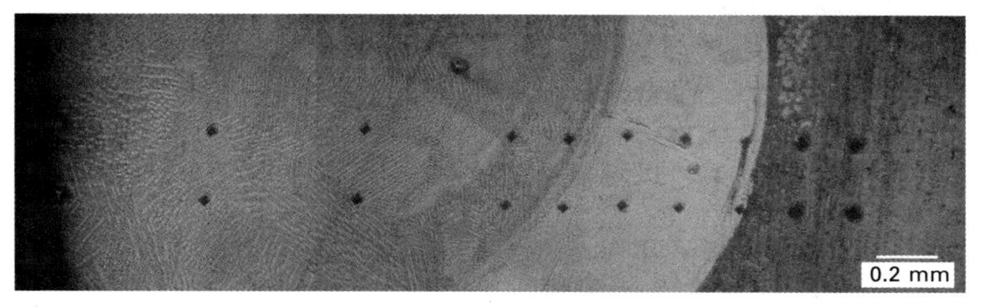

9.15 Cross-section morphology of laser cladding layer with Vickers hardness marks.

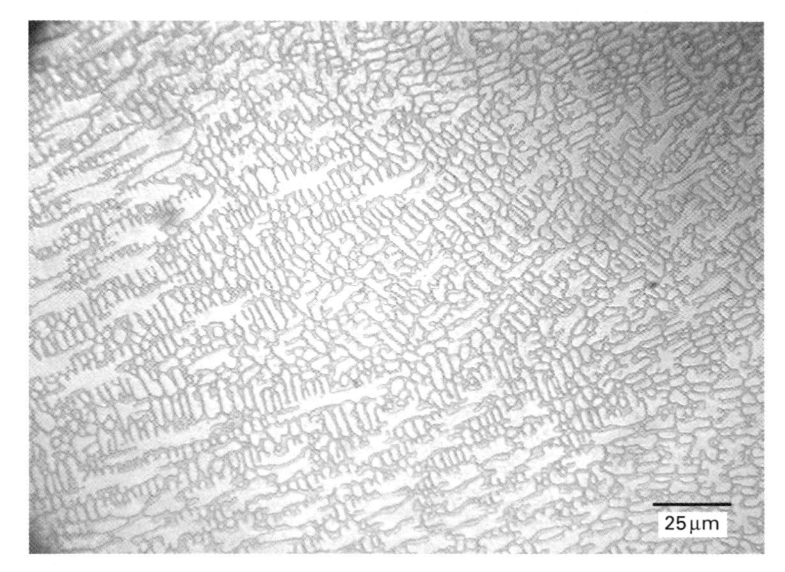

9.16 Microstructure of laser cladding layer.

such as flame hardening, induction hardening, chroming and encasing. Laser remanufacturing technology can also be used to harden key parts of industrial power units, such as rotor shaft necks, stems, impellers and valves.

9.4 Application of laser remanufacturing for corrosion and erosion resistance on injection molding machine screws

9.4.1 Failure of injection molding machine screws

The working conditions of key wearing parts such as screw rods and barrels in plastic (rubber) injection machines are severe, with working temperatures above 400 °C. Barrels bear not only high pressure but also corrosion from

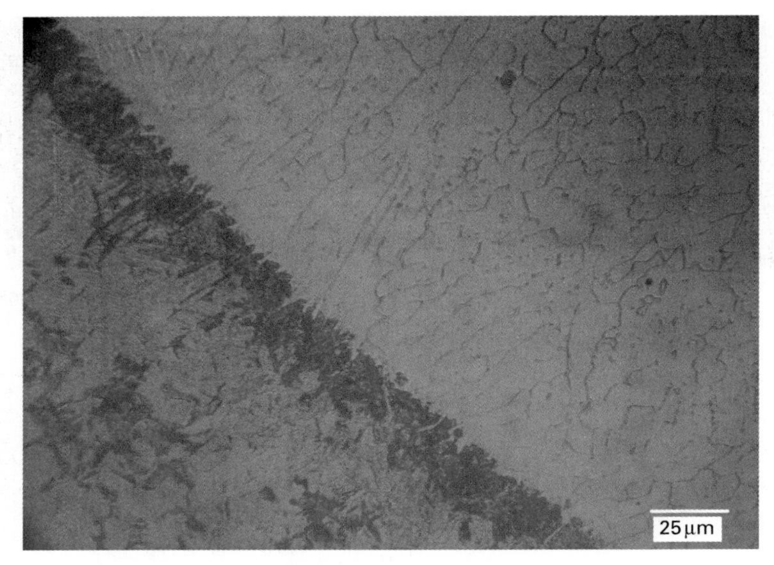

9.17 Microstructure of transition area.

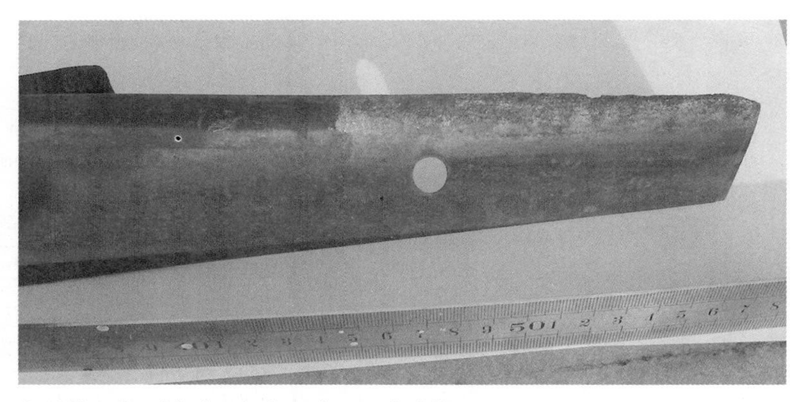

9.18 Erosion blades before laser cladding.

9.19 Repaired blade after laser cladding.

molten materials and frequent starting loads while pre-plasticizing. In addition, screws have to bear pressure and torsion when running. Screw rods and barrels usually fail due to losses from wear. Wear loss can increase the gap between the screw rod and the barrel, which decreases the speed of material melting and the ability to pump out. The temperature of materials therefore becomes unstable and pressure fluctuates, resulting in poor quality, lower productivity and high energy consumption.

High-temperature gas nitriding is the conventional method to strengthen most screw rods and barrels. The thickness of the nitride layer achieved is usually less than 0.3 mm. At the beginning of working, the screw rod is in the cantilever position. The high speed rotating of the screw rod can make a strong scrape between the top of the helical line and the inwall of a barrel, resulting in serious wear of the screw edge. Additionally, being a corrosive medium, the melting plastic (rubber) aggravates the failure of the screw rod, leading to further decrease of service life. Figure 9.20 shows the failure of a screw edge. It can be seen that poor levels of high-temperature adhesion wear and corrosion wear resistance on the surface of the screw rod are the main reasons for failure.

9.4.2 Laser solutions for injection molding machine screws

Carburizing, nitriding and electroplating are usually used to increase the surface hardness and durability of screw rods before installing, but the effects

9.20 Failed screw rod.

are not perfect. Bimetallic screws have been adopted in some countries to enhance service life but they are expensive.

For the failed screw rods themselves, thermal spraying and overlaying processes are usually used to repair and improve their abrasion resistance, but thermal spray coating falls off easily because of its poor bonding with the substrate. During the overlaying process, the screw rod is prone to distortion due to the high heat input, so a post-grinding process is necessary.

There are three laser remanufacturing solutions:

1. Use AISI 4140 to replace ISO 41CrAlMo74 as the screw base material, which reduces the cost by one-third.
2. Perform laser surface alloying on the surface of AISI 4140 screws to enhance their service life.
3. Repair failed screws using laser cladding, making them reach the performances of bimetallic screws.

9.4.3 Laser remanufacturing techniques used on injection molding machine screws

Laser surface alloying

Micron-sized WC based powder (1–6 μm) or Co/W based alloy powder (60–80 μm) with resistance to high temperature, high corrosion and wear are first pre-prepared on the surface of AISI 4140 screws, then the alloying elements diffuse rapidly into the matrix by irradiation by high-power density laser and promptly solidify to form the expected coating [21]. The optimum processing parameters are a laser power of 3 kW (CO_2 laser), beam size of $10 \times 2\,mm^2$ and scanning speed of 0.2 m/min. In order to avoid surface oxidation, argon shielding should be adopted.

The microstructures, hardness profiles and wear loss curves of the samples fabricated by three different treatments (gas nitriding, WC laser surface alloying and Co/W laser surface alloying) are shown in Figs 9.21–9.23 [22], where the processing parameters of conventional gas nitriding are insulation for 12 h at 520 °C, a decomposition rate of NH_4 at 20–30%, and then insulation for 12 h at 560 °C, with a decomposition rate of NH_4 at 30–60%.

Figure 9.21(a) shows the overall morphology of the cross-section of a nitrided sample, where the top thin bright layer is the nitrided layer, with a depth of 0.05 mm. Cr and Al have a high affinity with nitrogen, which can increase the solubility of nitrogen in the ferrite phase (α) and form hard nitride phases such as CrN or $(Cr, Fe)_2N$, for example. Figure 9.21(b) shows the overall microstructure of a Co/W alloyed layer. The brighter contrast is the alloyed layer with a depth of 0.25 mm and the darker contrast is the transition zone and the substrate. It can be seen that the microstructure of the alloyed layer is compact and metallurgically bonded with the substrate. X

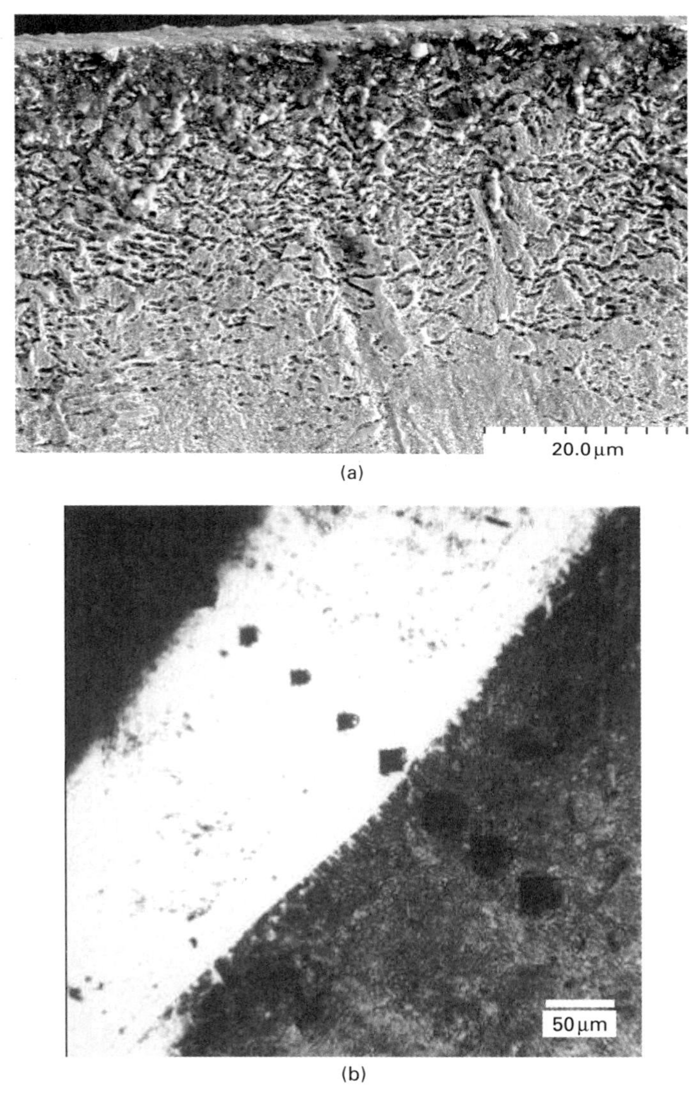

9.21 Microstructure of nitriding and laser alloying: (a) nitriding layer; (b) Co/W laser alloying layer.

ray diffraction (XRD) results illustrate that the existence of the hard phases in the alloyed layer, including W_3C, FeO and Co_3W_3, is the main reason for the increased hardness.

Figure 9.22 shows that the hardness of the outermost nitrided layer (0–0.05 mm) is higher than that of the WC and Co\W laser alloyed layers, while the subsurface hardness (0.05–0.20 mm) of the nitrided layer decreases quickly and is lower than that of the two laser alloyed layers. This is because the

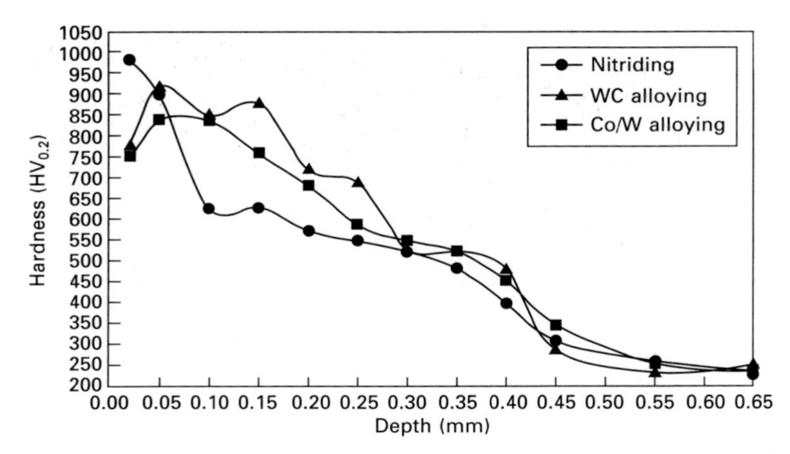

9.22 Hardness curves of hardened layer by different treatment methods.

concentration of nitrogen in the outermost layer is higher than that of the subsurface, while elements in the surface of the laser alloyed layer are easy to overheat. Figure 9.23 shows that wear resistance of WC and Co/W laser alloyed layers are 80% and 40% higher, respectively, than that of AISI 4140, and 60% and 25% higher, respectively, than that of the nitriding layer.

The experimental results above show that the optimal process to promote the corrosion and erosion resistance of screws is laser alloying with WC or Co/W. As for hardness and wear resistance, the former (laser alloying with WC) is much better than the latter (laser alloying with Co/W), while high temperature resistance of the latteris better. The treatment time needed for laser alloying is about 30 mins (including preparation time), while nitriding processing needs 24 hs (excluding preparation time). The thickness of a nitriding layer (≤0.1 mm) is smaller than that achieved by laser alloying.

Laser cladding

Laser cladding can restore the dimensions of failed screws. Depending on the working condition of the screws, Type H powders (as shown in Table 9.1) are applied for laser remanufacturing. Using a powder feeding system and a CO_2 laser power of 1.5–2.5 kW, a scanning speed of 1.0–2.5 m/min, beam size of $10 \times 2\,mm^2$, a powder feeding rate of 8–15 g/min, single or double layers formed by overlaying the melt tracks are prepared by laser cladding. The cross-section morphology of the cladding layer is shown in Fig. 9.24, where the thickness of cladding layer is about 0.85 mm. Figure 9.25 shows the hardness profile of the clad layer to the substrate. It can be seen that the maximum hardness at the surface is 900 $HV_{0.2}$, whilst the average hardness

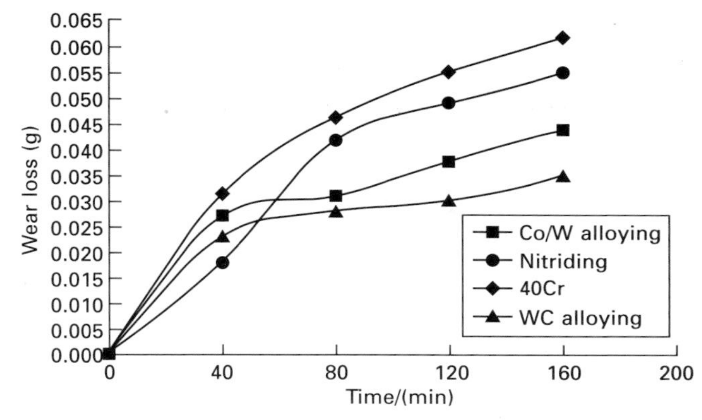

9.23 Wear loss vs. time of hardened layer by different treatment methods.

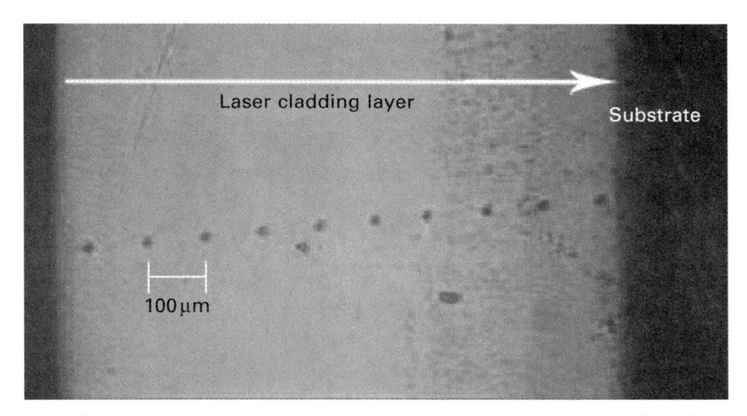

9.24 Cross-section morphology of laser cladding layer with Vickers hardness marks.

is 814 $HV_{0.2}$ under 0.1 mm from the surface, which is uniformly distributed. No cracks or porosities are found in the cladding layer.

The depth of the laser clad layer can be increased by multi-layer cladding. The above experimental results confirm that clad screws exhibit the same performance as bimetallic screws.

9.4.4 Application

Laser cladding has been used to treat Φ150 mm × 160 mm and other types of screws, and barrels of plastic and rubber injection machines. Figure 9.26 shows a screw which, after laser cladding, is smooth and shows no distortion. No cracks or porosity are found after pigmentation detection. After installation, the service life of laser clad screws is 50–60% higher than that of nitrided

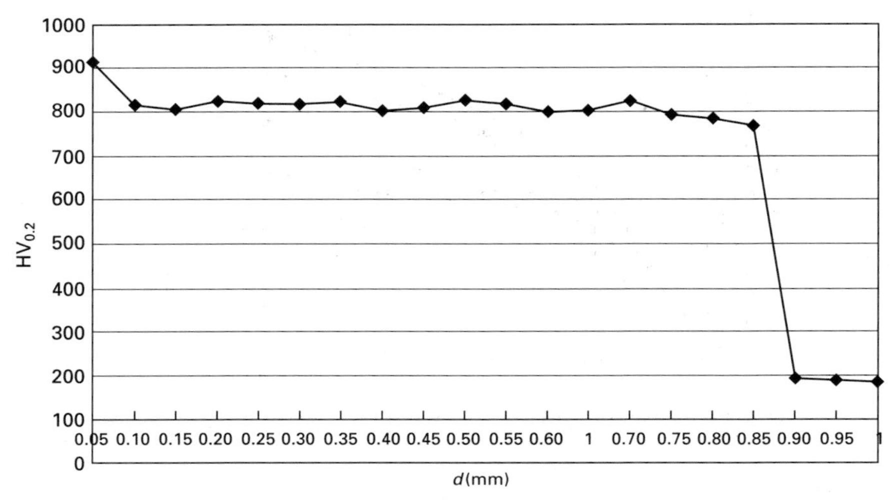

9.25 Hardness distribution of laser cladding layer along the direction of depth.

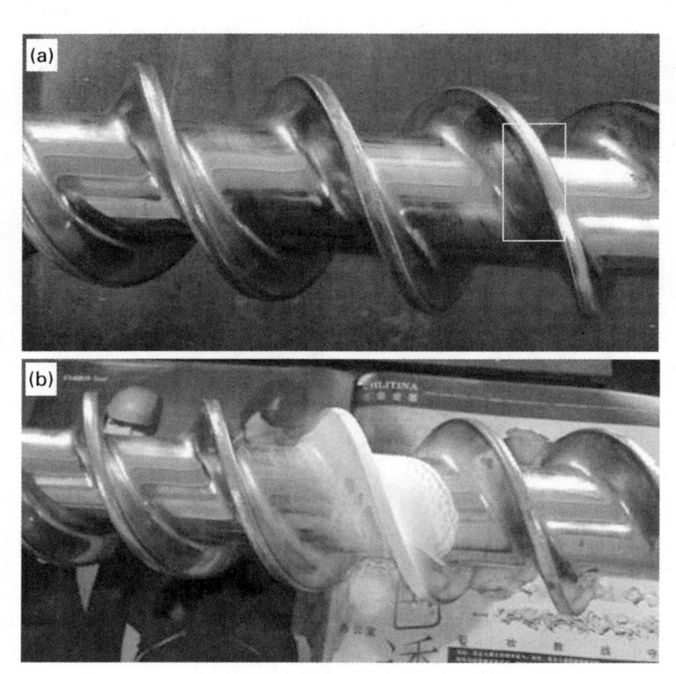

9.26 The screw after laser cladding: (a) screw after laser cladding; (b) cladded screw after pigmentation detection.

high alloy screws under the same conditions. There is no breaking-off of the clad layer during running. The same technique can also be used to repair parts susceptible to corrosion and erosion in other industries.

9.5 Application of laser remanufacturing for corrosion and erosion resistance of petrochemical system alkali filters

9.5.1 Alkali filter failure

Sealed pressurized filters are important pieces of equipment in the production of caustic soda in electrochemical factories, where they are used to filter dissolved salt in electrolytes. The main shaft of the sealed pressurized filter is the most seriously worn and corroded part of the equipment; a failed alkali filter is shown in Fig. 9.27. The length of the main shaft is 3150 mm, and the ends are $\Phi 209 \times 845\,mm^2$ and $\Phi 209 \times 506\,mm^2$, respectively. Wear occurs at the contact surface of the spindle and the bearing, an area of length 350 mm by 250 mm, with a depth of 2–8 mm. The main working conditions are a temperature at the entrance to the alkali cooler of 75°C, filter pressure of ≤0.196 MPa, a liquid level lower than two-thirds of the cavity, a flow ≤3 L/s, and a salt content after filtering ≤1.2%.

There are a number of reasons for failure of the main shaft. First, the mechanical wear of the shaft and bearing shell causes failure; once the mechanical seal is partly worn, the shaft sinks and eccentric wear of the bearing occurs. The second reason is alkali corrosion. Especially in an electrochemical factory, the spindle works in a strong alkali environment. Salt and alkali speed up the corrosion rate of AISI 1045 steel at working temperature and cause serious corrosion and abrasion. Additionally, alkali gas corrosion makes the spindle fail: the sealing surface becomes damaged after the spindle corrodes and causes leaking. High temperature salt and alkali run into the air and evaporate into an alkali gas, which accelerates the wear and corrosion of the spindle, as shown in Figs 9.28 and 9.29 [23].

9.5.2 Laser solutions for alkali filters

Current methods of repairing alkali filter spindles include turning, electroplating, thermal spraying, plasma spraying and overlay welding. The turning method is reliable and has a high surface precision, but turning decreases the diameter of the spindle, takes a long time and has high costs. In addition, the turning

9.27 Image of failed alkali filter spindle.

9.28 Local wear corrosion of alkali filter spindle.

9.29 Local chemical corrosion of alkali filter spindle.

frequency is also limited. Electroplating has poor bonding, insufficient wear resistance and limited thickness (≤0.2 mm). Plasma spraying has the problem of loose coating and non-metallurgical bonding. Overlaying welding causes serious distortion. In the past, lacking any better solutions to repair alkali filter spindles, a failed spindle was repaired by turning one or two times and was then discarded by some companies, whilst other companies replaced the failed spindle with a new one. However, spindles are difficult to produce and are expensive, costing ×100 000 for each piece. Laser cladding and alloying composite strengthening, by comparison, can be successfully used to regenerate and repair failed alkali filter spindles.

9.5.3 Laser remanufacturing techniques for alkali filter ports

If it is only necessary to restore the size of the failed spindle, laser cladding is good enough. If it is also necessary to enhance surface wear and corrosion resistance, laser surface alloying must be adopted after laser cladding. Ni45 alloy powder or powders with similar content are chosen as the cladding material. The laser cladding parameters are: a powder feeding rate of 15 g/min, a CO_2 laser power of 3 kW, scanning speed of 0.2 m/min, beam spot size of Φ7 mm, and argon shielding. Type H alloy powders are used for the laser surface alloying. A CO_2 laser power of 2 kW, scanning speed of 0.5 m/min, beam spot size of Φ5 mm and argon shielding are the optimal laser surface alloying parameters. Figure 9.30 shows a surface topography with smooth and uniform tracks after laser cladding and laser surface alloying. The wear and corrosion resistant performances before and after laser repairing are shown in Figs 9.31 and 9.32. It can be seen that the wear resistance of the hardened layer is 1.2 times higher and the alkali-corrosion resistance is 1.5 times higher than the substrate.

9.30 Surface macro-morphology of restoration layer.

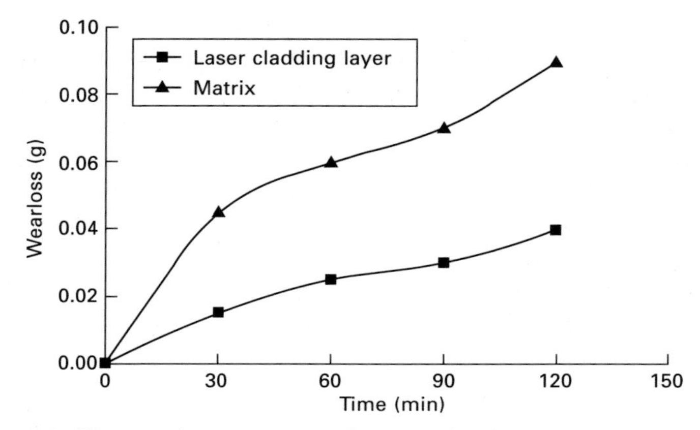

9.31 Wear resistance curve of restoration layer.

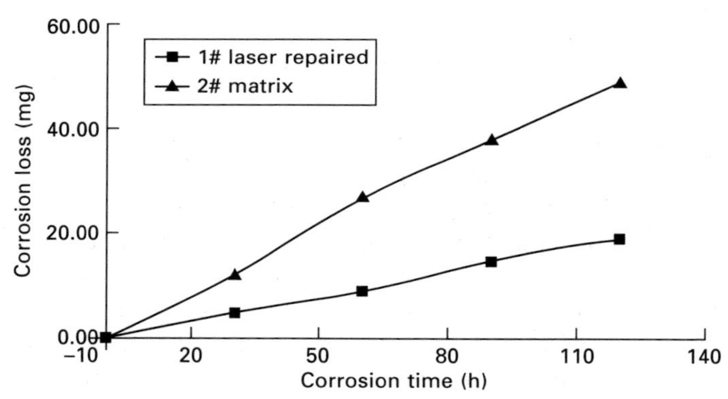

9.32 Alkali-corrosion resistance curve.

9.5.4 Application

After reinstallation, no eccentricity or abrasion is found in a laser repaired filter spindle, and there is no alkali leaking from the two sides of the filter. Filtration properties achieve the requirements of process indicators. The cleanliness of the filtrate in the aggregate bottle showed good properties and the alkali concentration made satisfy technological requirements. This illustrates that laser remanufacturing has successfully solved the regeneration of failed alkali filter spindles. The installation process is shown in Fig. 9.33, and this technology can also be extended to other parts prone to abrasion and corrosion in the petroleum and chemical industries.

9.33 Installation of alkali filter after laser cladding.

9.6 Application of laser remanufacturing for corrosion and erosion resistance of seawater circulating pump sleeves

9.6.1 Failure of seawater circulating pump sleeves

Seawater circulating pump sleeves are prone to wear and are prone to serious corrosion when running in seawater with a high rotation speed. The original hardening method of a seawater circulating pump sleeve is to spray Doloro 6325M or cermet (Cr_2O_3) on to the outer ring of the sleeve. Theoretically, the anti-seawater corrosion resistance and hardness of the coating are pretty good. However, due to inevitable microscopic pinhole problems, loose structure and non-metallurgical bonding between the coating and the substrate, seawater can easily get into the combining gap, causing corrosion, pitting and even punching, finally leading to spalling of the spray coating. Figure 9.34 shows an example of a sleeve with a spalled coating.

9.6.2 Laser solutions for a seawater circulating pump sleeve

The original thermal spraying can be replaced by laser cladding bimetallic coating manufacturing technology. Through the design of the cladding material and the processing parameters, the failed region can be repaired and surface properties can be better than those of a new pump. The coating

9.34 Scene of the sleeve with spalling coating.

can be metallurgically bonded with the substrate and the microstructures of the coating refined and the pinholes reduced, or even disappear, by laser cladding.

9.6.3 Laser remanufacturing techniques for seawater circulating pump sleeves

316L (00Cr17Ni14Mo2) stainless steel is used as a pump sleeve substrate. A special cladding material was designed after consideration of the actual operating conditions, including the working medium (seawater with sediment) and sloshing from long axis movement. High hardness, good toughness and excellent resistance to seawater corrosion are all required in the laser cladding layer. The powder chosen is H_1-3, which is listed in Table 9.1. Under suitable technical parameters, the thickness of the remanufacturing layer can be over 1.5 mm. The effective thickness of the remanufacturing layer can be up to 1.2 mm after surface finishing. The low magnification panorama structure of the laser remanufacturing layer (Fig. 9.35) shows that no cracks, pores, inclusions or other defects can be found. The remanufacturing layer has a good metallurgical combination with the substrate and the average hardness is higher than 450 $HV_{0.2}$. The overheated zone is narrow and the grains are fine, as shown in Fig. 9.36.

9.6.4 Application

According to the working conditions and experimental results achieved, laser remanufacturing of a seawater circulating pump sleeve was undertaken

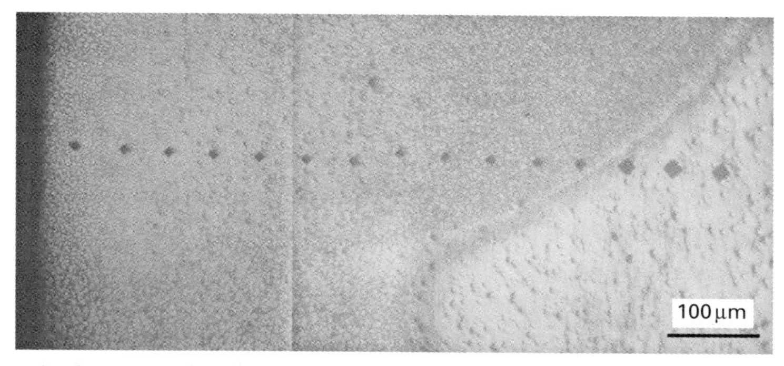

9.35 Cross-section morphology of sleeve after laser cladding layer with Vickers hardness marks.

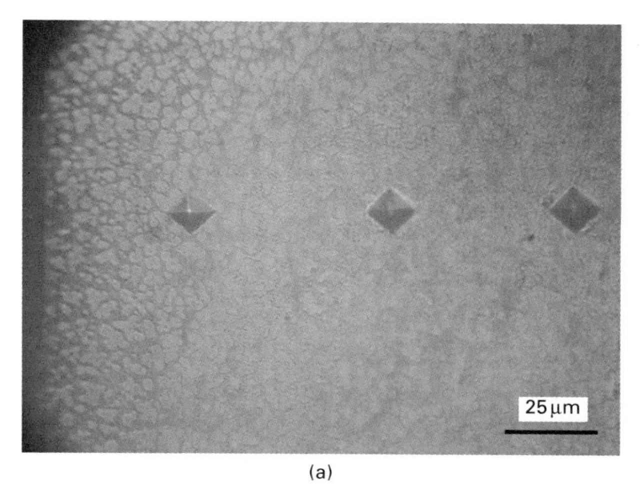

(a)

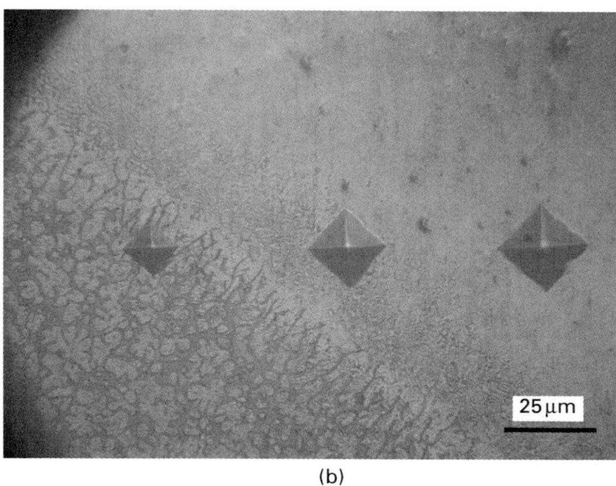

(b)

9.36 Microstructure of laser hardened layer on sleeve: laser cladding area; (b) transition area.

using H_1-3 powder and optimal processing parameters. The results show that the appearance of the axis is smooth and with no visible defects. A uniform remanufacturing layer is formed on the surface as shown in Fig. 9.37. No cracks have been detected through further dye penetrating detection. Three sets of laser treated sleeves have been installed since March 2010, and three sets ordered in 2011. The technique has been comprehensively applied in some important power plants in China.

9.7 Conclusions

- This chapter has introduced laser remanufacturing technology including laser surface alloying, laser solution strengthening and laser cladding used to improve the corrosion and erosion resistance of key mechanical components, and the application of laser remanufacturing technology on typical mechanical parts subjected to corrosion and erosion have been illustrated.
- Laser transformation hardening, laser surface alloying, laser surface solution strengthening, laser surface alloying with laser surface solution strengthening, and laser cladding techniques have each been performed to increase the erosion resistance of different types of turbine blade. The results show that laser remanufacturing technology can be applied successfully to prolong the lives of turbine blades.

9.37 Image of sleeve after laser remanufacturing.

- In the case of injection molding machine screws, high-alloy nitriding steel (ISO 41CrAlMo74) can be replaced by carbon steel (AISI 4140) as the screw substrate, and the traditional nitrogen process can be substituted by laser cladding and laser surface alloying, which lowers the material costs obviously, and improves the anti-wear and corrosion performances of screws, finally producing bi-metal screw rods.
- In the case of alkali filters and seawater circulating pump sleeves, wear and corrosion failure exist at the same time when they are running. Special alloy powders can be used and laser cladding processing performed on the failed regions. The experimental results and tests show that after laser remanufacturing, the failed parts can be repaired effectively and used as new, with even better properties. Laser remanufacturing is, therefore, a promising technology for application in parts subject to corrosion and erosion.

9.8 References

[1] Johan Östlin, Erik Sundin, Mats Björkman. Product life-cycle implications for remanufacturing strategies. *Journal of Cleaner Production*, 2009, **17**(11): 999–1009.

[2] Geraldo Ferrer, Robert U. Ayres. The impact of remanufacturing in the economy. *Ecological Economics*, 2000, **32**(3): 413–429.

[3] Guiping Hu, Shuyan Wang, Binshi Xu. The engineering of green remanufacture and its application prospects in China. *Water Conservancy & Electric Power Machinery*, 2001, **23**(12): 33–35 (in Chinese).

[4] Binshi Xu, Shining Ma, Shican Liu *et al.* Application of surface engineering and remanufacturing engineering. *Materials Protection*, 2000, **33**(1): 1–4 (in Chinese).

[5] Binshi Xu. *Scientific research selections of Armoured Engineering Institute*. Beijing: Mechanical Industry Press, 2001, **3**: 25–27 (in Chinese).

[6] Gholam Reza Gordani, Reza ShojaRazavi, Sayed Hamid Hashemi, Ali Reza Nasr Isfahani. Laser surface alloying of an electroless Ni–P coating with Al-356 substrate. *Optics and Lasers in Engineering*, 2008, **46**(7): 550–557.

[7] Jyotsna Dutta Majumdar. Development of wear resistant composite surface on mild steel by laser surface alloying with silicon and reactive melting. *Materials Letters*, 2008, **62**(27): 4257–4259.

[8] Laser hardening processing of turbine blade inlet side, Invention patent of China, No. 200510060816.4.

[9] Liang Wang. Laser compound strengthening mechanism on precipitation hardening martensite stainless steel steam turbine blades. Dissertation Submitted to Zhejiang Unniversity of Technology for the Degree of Master, April 2009.

[10] L. Sexton, S. Lavin, G. Byrne, A. Kennedy. Laser cladding of aerospace materials. *Journal of Materials Processing Technology*, 2002, **122**(1): 63–68.

[11] Shengzuan Chen, Jianhua Yao. Research of organize and performance on laser cladding Ni coats nano oxidation niobium. *Applied Laser*, 2004, 06, **24**(3) (in Chinese).

[12] The metal surface nano coating processing, Invention patent of China, No. 2005100061928.1.

[13] Jianhua Yao, Haiming Lai. Surface quench-hardening by laser of confined zones of steam turbine blades *Power Equipment*, 2005, (2): 101–103. (in Chinese).

[14] Jianhua Yao, Chunyan Yu, Fanzhi Kong, Chenhua Lou, Dongyue Sun. Laser alloying and quenching of steam turbine blades. *Journal of Power Engineering*, 2007, **27**(4): 652–656 (in Chinese).

[15] Wei Zhang, Jian-hua Yao, Chen-hui Dong, Xiao-dan Tang, Wei Peng. Repairing and strengthening of eroded turbine blades by laser technology. *Journal of Power Engineering*, 2008, **28**(6): 967–971 (in Chinese).

[16] Jianhua Yao, Liang Wang, Qunli Zhang, Fanzhi Kong, Chenghua Lou, Zhijun Chen. Surface laser alloying on 17-4PH stainless steel for steam turbine blades. *Optics and Laser Technology*, 2008, **40**(6): 838–843.

[17] N. Vardar, A. Ekerim. Failure analysis of gas turbine blades in a thermal power plant. *Engineering Failure Analysis*, 2007, **14**(4): 743–749.

[18] A kind of laser surface alloying precipitation hardened stainless steel and its preparation technology and application, Invention patent of China, No. ZL200610052899.7, 17 June 2009.

[19] Jianhua Yao, Wei Zhang, Mingxia Gao, Qunli Zhang. Study of Fe–Ni(Cr) Alloy and nano-Al_2O_3 particles composite coating prepared by laser cladding. *Solid State Phenomena*, 2006, **118**: 585–590.

[20] Wei Zhang, Jianhua Yao, Qunli Zhang, Fanzhi Kong. Microstructure and mechanical characteristic of nano-WC composite coating prepared by laser cladding, *Solid State Phenomena*, 2006, **118**: 579–583.

[21] Metal alloy screw component surface coating processing, Invention patent of China, No.ZL2006100531006, 23 August 2006.

[22] Wei Zhang, Jianhua Yao. Surface laser alloying and its application on 40Cr steel screw. *Heat Treatment of Metals*, 2007, **32**(11): 59–64 (in Chinese).

[23] Weijia Zhou. Research of repairing the abraded shafts of MB40 filter using laser processing technology. Dissertation Submitted to Zhejiang Unniversity of Technology for the Degree of Master, November 2004.

10 Laser surface remelting to improve the erosion–corrosion resistance of nickel–chromium–aluminium–yttrium (NiCrAlY) plasma spray coatings

B. SINGH SIDHU, Punjab Technical University, India

Abstract: The chapter deals with the role of NiCrAlY plasma spray coatings and subsequent laser surface remelting to improve the characteristics of these coatings. NiCrAlY plasma spray coating has been formulated on boiler tube steels, namely low carbon steel ASTM-SA210-Grade A1, 1Cr–0.5Mo steel ASTMSA213-T-11 and 2.25Cr–1Mo steel ASTM-SA213-T-22. The coated steels have also been laser remelted by Nd : YAG (yttrium aluminium garnet) laser. The coatings have been characterised and the degradation behaviour of as-sprayed and laser remelted coatings have been evaluated in actual conditions in a coal-fired boiler for 1000 h at 755 °C. The laser remelting was found to be effective at increasing the degradation resistance of plasma sprayed boiler steels. The decrease in porosity and better surface structure after laser remelting probably contributed to improve the performance of the laser remelted coatings.

Key words: erosion–corrosion, laser remelting, plasma spray coatings, Ni–Cr coatings.

10.1 Introduction

Metallic components in coal gasification pilot plants are exposed to severe corrosive atmospheres and high temperatures. The corrosive nature of the gaseous environments, which contain oxygen, sulphur and carbon, may cause rapid material degradation and result in the premature failure of components [1, 2].

The description of complex industrial atmospheres, with respect to its composition and degradation under oxidation, carburisation and sulphidation, is important in the design and operation of experimentation for materials testing in the laboratory. Coal gasification atmospheres contain species such as oxygen, sulphur and carbon and the reliable performance of various components is strongly dependent on the sulphur contents of the gas phase, duration and temperature of exposure.

The focus is on the development of innovative coatings that are durable, long lasting and user-friendly as a remedial measure of problematic areas. A new hybrid protective coating has been developed to extend the lifetime of power plant components. Metallic coatings sprayed on superheater tubes have been noted to reduce erosion and corrosion in selected cases.

Plasma spray is widely used to prepare coatings including intermetallic, ceramics and composites to protect substrates against wear, corrosion and high-temperature environment [3–5]. High deposition efficiency is very important to reduce the coating cost and the deposition efficiency of plasma spray is considered to be closely related to the melting state of the spraying powders. Plasma spraying is also a simple process from practical point of view.

Laser surface melting is a promising process for improving the performance of plasma sprayed coatings. The use of high-power lasers to remelt the surface coating zone and subsequent surface alloying is an effective way to obtain the required material surface properties. It is one of the surface modification techniques used to improve the surface properties of materials such as corrosion, erosion, wear, etc.

10.2 Need and role of post-coating treatments

10.2.1 Post-coating treatments

Plasma spray techniques are widely used for coating deposition as a result of ease of application and ability to deposit various coating materials on substrate alloys. However, the residual porosity in plasma spray coatings allows corrosive liquids to penetrate through the coating. This led to the debonding or spallation of the coatings associated with the accumulation of corrosion products at the coating/substrate interface. Therefore the coatings may require post-deposition treatments to improve these properties [6–10].

Laser surface remelting is a promising process for improving the performance of plasma sprayed coatings. During the resolidification of the laser melted layer, the cold substrate is always in contact with the melt, thus serving as a virtually infinite heat sink. Large temperature gradients exist across the interface between the melt and the substrate and this produces rapid self-quenching. The rapid quenching from the liquid phase can lead to the formation of the fine microstructures and extended solid solutions. Metallurgical bonding is also established at the coating/substrate interface, thus permitting better adhesion between the coating and the substrate alloy after laser remelting [8].

Although the range of laser processes is extensive for the high-temperature corrosion protection of metallic material, three approaches are possible: laser surface remelting, laser surface alloying and laser cladding.

Laser surface remelting essentially entails modifying the microstructure of a surface layer, although the distribution of elements in the melted zone may also change additionally. Owing to the new surface structure, the nucleation and growth of corrosion products are affected, in some instances leading to improved oxide adherence. Laser remelting has also been applied successfully to plasma-sprayed coated surfaces to reduce coating porosity, long-range chemical inhomogeneities and coating/substrate adherence [11].

10.2.2 Laser remelting of coatings

For metal surface processing, the laser beam is treated as an intense heat source. For efficient coupling, the beam's energy must be easily absorbed by the metals. Because the 1.06 μm radiation is absorbed somewhat better by most metals than 10.6 μm light, it is claimed that only yttrium aluminium garnet (YAG) lasers could be used for metal processing, but this is true only as long as the metal surface is not melted. A dramatic increase in absorption is observed when melting occurs and then both YAG and CO_2 light absorptions become similar [12].

Helium/argon is generally used as an inert gas shielding to avoid oxidation of the materials during the laser treatment. Melting of the whole surface is processed by passes of successive melt stripes, the laser beam rapidly melting a small volume of metal which subsequently solidifies by conducting heat to the bulk specimen. Each pass results in the processing of a ribbon of material, the width and depth of which can be varied by changing the processing conditions such as sweeping speed, laser power or spot size. The complete coverage of a large surface is obtained by successive passes spaced a fraction of pass width from one another.

10.3 Applications of laser remelted coatings to combat erosion and corrosion

The research work published on the effect of laser surface treatment on the high-temperature behaviour of protective coatings or alloys used in these coatings can be classified into two categories depending on the main aspect of modification involved: one deals with the structural modification of coatings and the other with consolidation of coatings by the elimination of porosity.

However, the presence of trapped oxygen in atmospheric plasma sprayed coatings causes unfavourable oxide inclusions during the laser melting process. These oxide inclusions may develop into sites of oxidation attack. Atmospheric plasma sprayed coating seems therefore not to be the best candidate for laser treatment. For these reasons laser surface melting was applied to low-pressure plasma sprayed coatings. These coatings have no residual porosity and the laser treatment affects only the phase composition and distribution within the coating.

10.4 Advantages of laser remelting

Diverse laser surface processing techniques aimed at modifying material surface properties such as corrosion, high temperature oxidation, wear, erosion, etc. have been developed, over the past two decades. The highly localised interaction between lasers and materials makes laser processing an attractive alternative in the realisation of coating strategy for engineering materials. The advantages of laser treatment include the following:

- Because of the formation of only a small heat-affected zone, minimal distortion is induced.
- Owing to rapid quenching, a non-equilibrium and refined microstructure is formed.
- Laser treatment has a high processing speed and ease of automation.
- An excellent metallurgical bond between the alloying material and the substrate is formed.
- A dense surface layer with no porosity is produced.
- There is no fundamental restriction on component shape.
- Surface microstructures which are unobtainable by any of the main coating or surface treatment technologies can be produced. Novel microstructures such as microcrystalline or amorphous surfaces have outstanding relevance for enhancement of corrosion resistance.
- Localised areas of a component surface can be processed without affecting surrounding areas and hence these processes are applicable to the localised area processing of critical regions of a component for corrosion resistance applications.

10.5 Role of nickel–chromium (Ni–Cr) coatings in aggressive environments

Alloys that are developed for heat and oxidation resistance typically form a protective layer of chromia or alumina. The more rapidly this layer is established, the better protection is offered. As this layer grows or as it reforms over areas from which the original layer was removed, it must withdraw chromium or aluminium from the metal in order to provide for further scale growth [13].

Wu *et al.* [14] have studied the oxidation behaviour of laser remelted plasma sprayed coatings of NiCrAlY and NiCrAlY–Al_2O_3 using continuous CO_2 laser. Authors could obtain homogeneous dense remelted layer without voids, cavities, unmelted particles and microcracks. After isothermal oxidation tests at 1000 °C the weight gain of laser remelted NiCrAlY–Al_2O_3 coatings were observed to be lower than plasma sprayed coatings.

Liang and Wong [15] have investigated the microstructure and chemical composition of different parts of the laser melted zone of plasma sprayed

Ni–Cr–B–Si coatings on Al–Si alloy after laser surface remelting by a 5 kW CO_2 laser. Hardness distribution in the laser melted zone was also measured. Experimental results showed that the chemical composition of the sample was not a well-distributed gradient as compositional segregation exists in the laser melted zone. The authors further observed higher aluminium concentration in the surface of the laser melted zone.

10.6 Experimental procedure

The plasma sprayed coatings were obtained on the steel boiler tube and subsequently remelted with the purpose of eliminating the porosity and obtaining improved resistance against hot corrosion. Some of the samples were tested under the actual working conditions of boilers. Some results regarding the behaviour of laser remelted plasma sprayed NiCrAlY coating subjected to the actual working condition of boiler are discussed.

10.6.1 Plasma spraying

Boiler steels, namely low-carbon steel ASTM-SA210-Grade A1 (GrA1), 1Cr–0.5Mo steel ASTM-SA213-T-11 (T11) and 2.25Cr–1Mo steel ASTM-SA213-T-22(T22), were used as substrate steels. Steels were cut to form $20 \times 15 \times 5\,mm^3$ size specimens and were grit blasted by Al_2O_3 (Grit 60) before plasma spraying. Forty kilowatt Miller Thermal Plasma Spray Apparatus available with Anod Plasma Spray Kanpur (India) was used for plasma spraying of Ni–22Cr–10Al–1Y (NiCrAlY) powder up to around 350 μm thick coating.

10.6.2 Laser remelting

The coatings were surface melted using 300 W, Nd : YAG laser at the Atomic Fuel Division (AFD), Bhabha Atomic Research Centre (BARC), Trombay, India. The process parameters are reported in Table 10.1. Complete areas were remelted by making a series of parallel overlapping tracks.

Table 10.1 Process parameters for laser remelting of coatings

Type	Pulsed Nd : YAG laser
Power (W)	300
Pulse energy (J)	6
Pulse width (ms)	12
Repetition rate (Hz)	10
Defocus (mm)	5
Traverse speed (mm/s)	2
Shielding gas	Ar
Tracks overlapping (%)	60

10.6.3 Characterisation of coatings

The coatings were characterised before the studies in the actual environment of a coal-fired boiler. The porosity measurements for as-plasma sprayed and laser remelted coatings have been made with an image analyser using software of Dewinter Material Plus 1.01 based on ASTM B276. Each reported value for porosity is the mean of five measurements performed. The coated as well as coated and laser remelted samples were polished, etched and subsequently typical microstructures of the samples were photographed.

10.6.4 Industrial environment studies

The samples were tested in the middle zone of the platen superheater of the Stage-III Boiler of Guru Nanak Dev Thermal Plant, Bathinda, Punjab (India). The samples were inserted through the soot blower dummy points at 31.5 m height from the base of the boiler with the help of stainless steel wires passed through a predrilled hole of 1 mm to hang of the samples. The average temperature was around 755 °C with variation of ±10 °C. The average volumetric flow of flue gases was 231 m^3/sc. The chemical analysis of the flue gas and ash present inside the boiler is given in Table 10.2. The samples were exposed to combustion gases for 10 cycles each of 100 h duration and followed by 1 h cooling at ambient conditions. After each cycle the samples were visually observed for any change and weight of samples were measured subsequently. At the end of cyclic study the samples were subjected to X-ray diffraction (XRD), scanning electron microscopy (SEM), energy-dispersive X-ray (EDAX) and electron probe microanalysis (EPMA) analyses.

The extent of degradation was monitored by measuring the thickness of unreacted samples, as simultaneous spalling and ash deposition makes the

Table 10.2 Chemical analysis of ash and flue gases inside the boiler

Ash		Flue gases (Volumetric flow, 231 m^3/s)	
Constituent	Wt%age		
Silica	54.70	Constituent	Value relative to flue gases
Fe_2O_3	5.18	SO_x	236 mg/m^3
Al_2O_3–Fe_2O_3/Al_2O_3	29.56	NO_x	1004 μg/m^3
Calcium oxide	1.48	CO_2	12%
Magnesium oxide	1.45	O_2	7%
SO_3	0.23		
Na_2O	0.34	40% excess air was supplied to the boiler for the combustion of coal.	
K_2O	1.35		
Ignition loss	4.31		

weight change data irrelevant. From the thickness loss data the degradation rates in mils/year (mpy) were evaluated.

10.7 Experimental results

10.7.1 Thickness and depth of laser remelted coating

Backscattered electron images (BSEI) for the as-coated as well as coated and laser remelted samples are shown in Fig. 10.1. The micrographs of the coating after laser remelting are very smooth and dense. The thickness of as-sprayed coatings is measured to be 384 μm and melt depth of laser remelted coating is 232 μm from this BSEI micrograph.

10.7.2 Porosity analysis

The porosity of coatings is of prime importance in the hot corrosion studies. The dense coatings are supposed to provide very good corrosion resistance compared with porous coatings. The porosity measurements were made for as-sprayed and laser-remelted coatings. The porosity for as-sprayed coatings was found to be up to 4%. A considerable decrease in the porosity had been observed after laser remelting and it was found to be less than 0.5%.

10.7.3 EDAX analysis

The SEM micrograph presented in Fig. 10.2 indicates the matrix with very fine precipitates for NiCrAlY coating after laser remelting. The matrix mainly contains nickel and iron with some amount of chromium, whereas the precipitates contain 81.12% Ni, 11.83% Cr and 6.32% Al. The matrix is

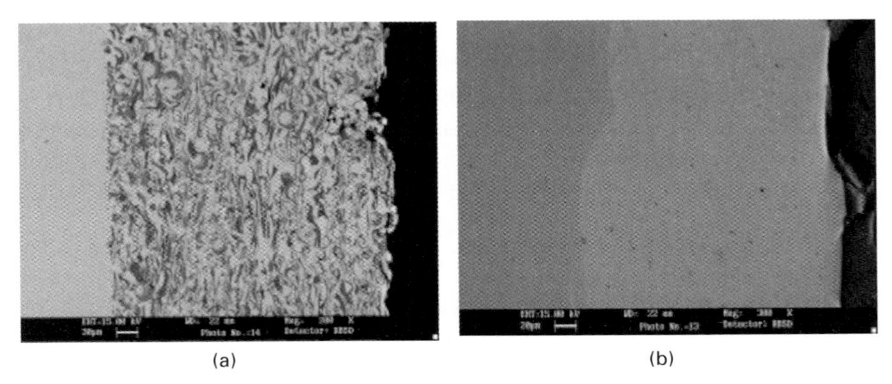

10.1 SEM micrographs showing cross-section morphology of NiCrAlY coating on T22 steel: (a) as-coated, 200 × and (b) laser remelted (300 W, 2 mm/s), 300×.

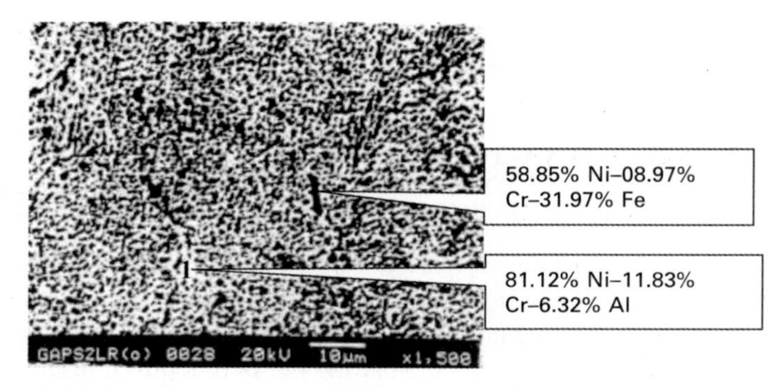

10.2 SEM/EDAX analysis of NiCrAlY coating after laser remelting showing elemental composition (wt%) at various points, 1000×.

rich in iron and this iron may have come through diffusion from the substrate at the time of laser remelting.

10.7.4 Metal thickness loss monitoring

The minimum thickness loss after 1000 h of exposure to the actual condition of a coal-fired boiler was observed in case of NiCrAlY coating and laser-remelted steels. The thickness lost of the plasma sprayed and subsequently laser-remelted steels is a small fraction of the thickness lost of the bare steel and as-sprayed coating on steel (Fig. 10.3). The erosion–corrosion rates after laser remelting of coatings were found to be negligible.

10.7.5 Surface morphology

NiCrAlY plasma sprayed steel subjected to boiler environment exhibits thick scale consisting of mainly NiO with embedded ash particles rich in silica and alumina. There is distinct indication of spalling and the spalled region contains higher percentage of aluminum (51%, Al_2O_3) along with nickel (17%, NiO) and silicon (19%, SiO_2) as shown in Fig. 10.4(a). The scale of the same coated steel when exposed to the given boiler environment after laser remelting consists only of ash deposition and the matrix is rich in Fe_2O_3 (60.6%) having 12% Al_2O_3 and 22.9% SiO_2 as shown in Fig. 10.4(b).

10.7.6 Cross-sectional analysis

Elemental variations across the cross-section of NiCrAlY coated GrA1 steel after 1000 h exposure in a coal-fired boiler are presented in Fig. 10.5. The top scale is rich in nickel and the middle scale has higher amounts of chromium and nickel. Iron has diffused from the substrate steel in to the

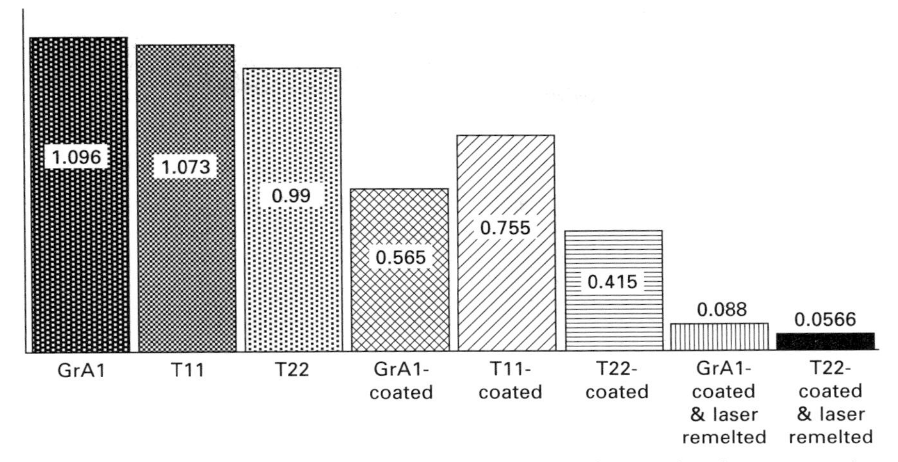

10.3 Bar chart indicating the thickness lost in mm for the uncoated and NiCrAlY coated steels after 1000 h exposure to the coal-fired boiler at 755 °C.

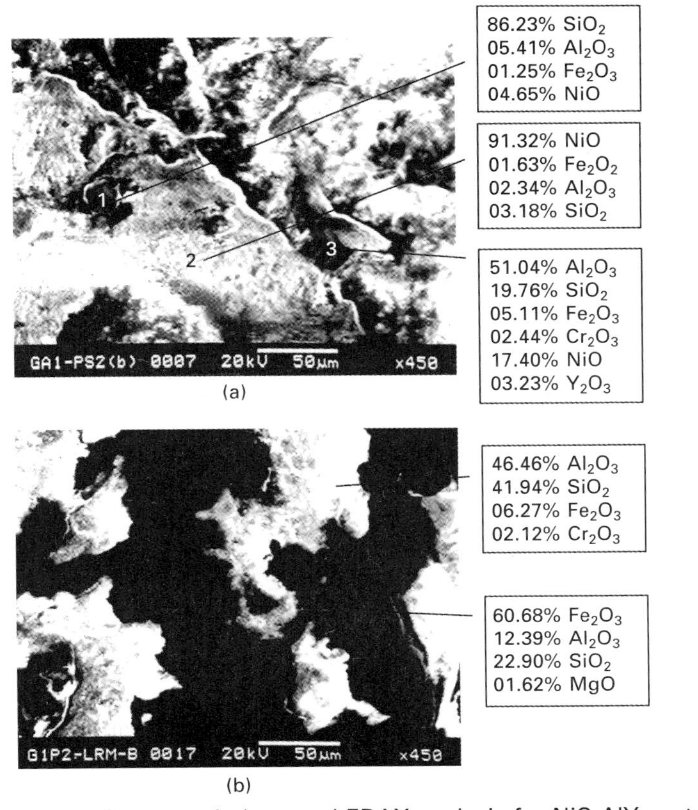

10.4 Surface morphology and EDAX analysis for NiCrAlY coated steel exposed to platen superheater of the coal-fired boiler for 1000 h at 755 °C: (a) as-sprayed and (b) laser remelted.

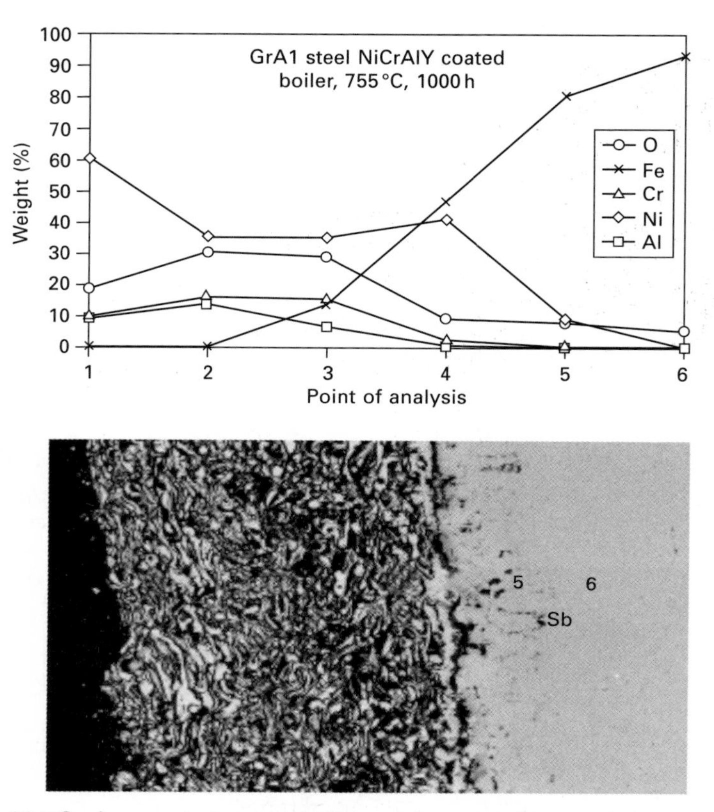

10.5 Scale morphology and elemental composition variation across the cross-section of NiCrAlY coated GrA1 steel exposed to platen superheater of the coal-fired boiler for 1000 h at 755° C, 200×.

scale up to point 3. Some nickel seems to have penetrated into the substrate steel from the bond coat. The probable erosion–corrosion mechanism is shown in Fig. 10.6.

10.8 Conclusions

The measured number of porosities for as-sprayed coating is almost identical to the findings of Chen *et al.* [16], Hidalgo and co-workers [16–18] and Belzunce *et al.* [19]. After laser remelting the number of porosities was observed to be negligible (only 0.5%). Laser remelting has been observed to increase the density of coatings by eliminating microstructural inhomogeneities [20]. The structures for the as-sprayed and laser-remelted coatings are in good agreement with the measured number of porosities. Some porosities present in the as-sprayed coatings after laser remelting might be eliminated and the structure became very dense and homogeneous.

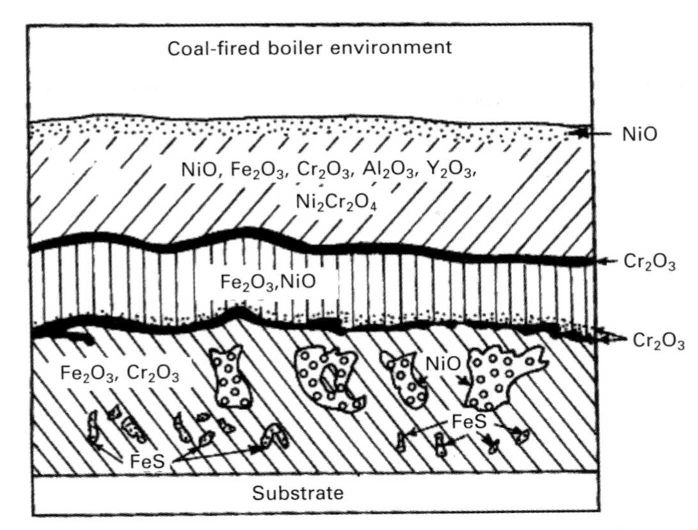

10.6 Schematic diagram showing probable hot corrosion mechanism for NiCrAlY coated T22 steel exposed to the coal-fired boiler environment at 755 °C for 1000 h.

Plasma sprayed NiCrAlY coatings can be used to protect the boiler steels in the coal-fired boiler environment. The erosion–corrosion resistance of the coating is further improved after laser remelting, probably due to lower porosity and compact structure.

Surface alloying at the time of laser remelting may contribute to the formation of spinels, which may be responsible for providing better protection after laser remelting of coatings.

10.9 References

[1] S. Danyluk and J. Y. Park, 'Technical Note: Corrosion and Grain Boundary Penetration in Type 316 Stainless Steel Exposed to a Coal Gasification Environment,' *Corrosion*, Vol. 35, No. 12, (1979), pp. 575–76.

[2] D. Wang, 'Corrosion Behavior of Chromized and/or Aluminized $2^1/_4$Cr–1Mo Steel in Medium-BTU Coal Gasifier Environments,' *Surf. Coat. Technol.*, Vol. 36, (1988), pp. 49–60.

[3] M. Danielewski, S. Sroda, Z. Zurek, R. Gajerski and I. Lalak, *Valtion Teknillinen Tutkimuskeskus Symposium*, Vol. 233, (2004), pp. 223–232.

[4] P. Makkonen and M. Makipaa, *Valtion Teknillinen Tutkimuskeskus Symposium*, Vol. 233, (2004), pp. 197–207.

[5] J. F. Li, H. Liao, B. Normand, C. Cordier, G. Maurin, J. Foct and C. Coddet, 'Uniform design method for optimization of process parameters of plasma sprayed, TiN coatings,' *Surf. Coat. Technol.*, Vol. 176, (2003), pp. 1–13.

[6] W. Aihua, Z. Beidi, T. Zengyi, M. Xianyao, D. Shijun and C. Xudong, 'Thermal-shock Behaviour of Plasma-Sprayed Al_2O_3-13wt.%TiO_2 Coatings on Al-Si

Alloy Influenced by Laser Remelting,' *Surf. Coat. Technol.*, Vol. 57, (1993), pp. 169–172.

[7] C. Wu and M. Okuyama, 'Evaluation of High Temperature Corrosion Resistance of Al Plasma Spray Coatings in Molten Sulfates at 1073 K by Electrochemical Measurments,' *Mater. Trans., JIM*, Vol. 37, No. 5, (1996), pp. 991–997.

[8] S. C. Tjong, J. S. Ku and N. J. Ho, 'Corrosion Behaviour of Laser Melted Plasma Sprayed Coatings on 12% Chromium Dual Phase Steel,' *Mater. Sci. Technol.*, Vol. 13, (1997), pp. 56–60.

[9] B. Wielage, U. Hofmann, S. Steinhauser and G. Zimmermann, 'Improving Wear and Corrosion Resistance of Thermal Sprayed Coatings,' *Surface Engg.*, Vol. 14, No. 2, (1998), pp. 136–38.

[10] L. C. Erickson, R. Westergard, U. Wiklund, N. Axen, H. M. Hawthorne and S. Hogmark, 'Cohesion in Plasma-Sprayed Coatings: A Comparison Between Evaluation Methods,' *Wear*, Vol. 214, (1998), pp. 30–37.

[11] M. F. Stroosnijder, R. Mevrel and M. J. Bennet, 'The Interaction of Surface Engineering and High Temperature Corrosion Protection,' *Mater. High Temp.*, Vol. 12, No. 1, (1994), pp. 53–66.

[12] R. Streiff, 'The Effect of Laser Surface Treatment of High-Temperature Oxidation and Corrosion Resistance of Materials and Coatings,' *Proc. of 10th ICMC*, Madras, India, Vol. II, (1987), pp. 1315–24.

[13] R. J. Link, N. Birks, F. S. Pettit and F. Dethorey, 'The Response of Alloys to Erosion-Corrosion at High Temperatures,' *Oxid. Met.*, Vol. 49, No. 3–4, (1998), pp. 213–236.

[14] Y. N. Wu, G. Zhang, Z. C. Feng, B. C. Zhang, Y. Liang and F. J. Liu, 'Oxidation Behavior of Laser Remelted Plasma Sprayed NiCrAlY and NiCrAlY-Al_2O_3 Coatings,' *Surf. Coat. Technol.*, Vol. 138, (2001), pp. 56–60.

[15] G. Y. Liang and T. T. Wong, 'Microstructure and Character of Laser Remelting of Plasma Sprayed Coating (Ni–Cr–B–Si) on Al–Si Alloy,' *Surf. Coat. Technol.*, Vol. 89, (1997), pp. 121–26.

[16] H. C. Chen, Z. Y. Liu and Y. C. Chuang, 'Degradation of Plasma-sprayed Alumina and Zirconia Coatings on Stainless Steel During Thermal Cycling and Hot Corrosion,' *Thin Solid Films*, Vol. 223, No. 1, (1993), pp. 56–64.

[17] V. H. Hidalgo, F. J. B. Varela and A. C. Menendez, 'Characterization and High Temperature Behaviour of Thermal Sprayed Coatings Used in Boilers,' *Proc. of the 15th Int. Thermal Spray Conf.*, 25–29th May, Nice, France, (1998), pp. 617–621.

[18] V. H. Hidalgo, F. B. J. Varela, S. P. Martinez and S. G. Espana, 'Characterization and High Temperature Behaviour of Cr_3C_2–NiCr Plasma Sprayed Coatings,' *Proc. of the United Thermal Spray Conf.*, Germany, (1999), pp. 683–686.

[19] F. J. Belzunce, V. Higuera and S. Poveda, 'High Temperature Oxidation of HFPD Thermal-sprayed MCrAlY Coatings,' *Mater. Sci. & Engg. A*, Vol. 297, No. 1–2, (2001), pp. 162–167.

[20] Y. N. Wu, G. Zhang, Z. C. Feng, B. C. Zhang, Y. Liang and F. J. Liu, 'Oxidation Behavior of Laser Remelted Plasma Sprayed NiCrAlY and NiCrAlY–Al_2O_3 Coatings,' *Surf. Coat. Technol.*, Vol. **138**, (2001), pp. 56–60.

Index